FUNDAMENTALS OF
OCEANOGRAPHY

SECOND EDITION

FUNDAMENTALS OF
OCEANOGRAPHY

Alison B. Duxbury
Seattle Community College

Alyn C. Duxbury
University of Washington

WCB **Wm. C. Brown Publishers**

Dubuque, IA Bogota Boston Buenos Aires Caracas Chicago
Guilford, CT London Madrid Mexico City Sydney Toronto

Book Team

Developmental Editor *Mary Hill*
Designer *K. Wayne Harms*
Art Editor *Jodi K. Banowetz*
Photo Editor *John C. Leland*
Permissions Coordinator *Karen L. Storlie*

 Wm. C. Brown Publishers

President and Chief Executive Officer *Beverly Kolz*
Vice President, Publisher *Jeffrey L.Hahn*
Vice President, Director of Sales and Marketing *Virginia S. Moffat*
Vice President, Director of Production *Colleen A. Yonda*
National Sales Manager *Douglas J. DiNardo*
Marketing Manager *Amy Halloran*
Advertising Manager *Janelle Keeffer*
Production Editorial Manager *Renée Menne*
Publishing Services Manager *Karen J. Slaght*
Royalty/Permissions Manager *Connie Allendorf*

 A Times Mirror Company

Freelance Production Editor Diane L. Calvert

Copyedited by Catherine S. Di Pasquale

Freelance Permissons Editor Karen Dorman

Cover: Background—Digital Stock. Inset—Darryl Torckler/Tony Stone Images.

The credits section for this book begins on page 303 and
is considered an extension of the copyright page.

Library of Congress Catalog Card Number: 94–73651

ISBN 0–697–26671–0

Printed in the United States of America by Times Mirror Higher Education Group, Inc.,
2460 Kerper Boulevard, Dubuque, IA 52001

10 9 8 7 6 5 4 3 2 1

For those who will come to love
the oceans as we do.

CONTENTS

C H A P T E R   5

Water   89

C H A P T E R   6

The Air and the Oceans   111

C H A P T E R   7

Circulation Patterns and Ocean Currents   135

PREFACE

*F*undamentals of Oceanography is intended for professors and students who need a more basic oceanography text to better serve less intensive college oceanography courses, courses tailored for nonscience majors, and advanced placement oceanography programs for high school students. We have extensively rewritten this second edition to incorporate the ideas and suggestions of instructors using the first edition, to provide the student with up-to-date, contemporary information, and to make each chapter as clear and readable as possible without sacrificing scientific accuracy. As in the previous edition, we emphasize basic principles, processes, and properties of the oceans.

Six *Items of Interest* have been added to this edition. These are designed to cover issues that are not addressed directly in the text and can be read independently to discover more about the oceans. Topics include marine archaeology, robotic devices, biological invaders, clouds, marine birds, and current patterns traced by following bathtub toys and tennis shoes. This edition stresses the process approach: Plate Tectonics (Chapter 3) is placed before The Sea Floor (Chapter 4); driving forces for water motion and circulation patterns are discussed before ocean-current models in Chapter 7. The topic of ocean productivity has been given its own chapter and an expanded presentation. Wetlands and environmental issues in the coastal zones are discussed more fully. Red tides are considered in more detail; the section on El Niño has been updated, and new information is provided in all sections on fishing and harvesting. A new section on quantitative problems is included in the Instructor's Manual.

This edition continues to present students with numerous aids to facilitate their study of oceanography. Each chapter opens with learning objectives, and review questions are presented as self-checks for the student at the end of each section. Chapters end with a concise summary to aid in review. Technical terms are greatly reduced, and only an elementary math background is necessary. Selected additional readings are taken from current, nontechnical journals. Throughout this text, information is presented in chart and table form to help the student organize, summarize, and compare. Three appendices are included: methods of deriving latitude and longitude, taxonomic classifications of plankton, nekton, and benthos, and scientific notation and units. All quantities in the text are given in both metric and traditional units. We recommend that instructors supplement the text with films and videos about the earth sciences according to their individual resources.

Because oceanography embraces immense amounts of geological, physical, chemical, biological, and engineering information related to the marine environment, and because of the interdependency between these subject areas, the choice of topics to be included in a fundamentals text presents a complex challenge. We have endeavored to choose those topics that best illustrate basic processes and at the same time answer students' questions about the oceans while encouraging their interest. We encourage instructors to alter the sequence of material to best fit their own presentations and to elaborate on subjects as desired.

There are three ancillary items available free of charge to the instructor: (a) Instructor's Manual with a test item file and examples of quantitative problems by subject area; (b) MicroTest III, a complete testing system available on diskette in DOS, Windows, or Macintosh versions provided with instructions for a call-in, mail-in, or fax service to generate tests; and (c) an acetate transparency set. Also available is the CD-ROM *Interactive Plate Tectonics*—contact your local WCB representative for more details.

Again we acknowledge that this book is a product of many experiences, in the field and in the classroom. We extend our thanks to our many friends and colleagues who have graciously answered our questions, helped us with current information, and provided access to their photo files. We express our sincere gratitude to Wm. C. Brown Publishers and the staff members working with us for their support and encouragement. We also wish to extend a special thanks to the instructors who used and reviewed the first edition of this text and contributed to the development of this second edition:

David Barnes
Western Michigan University

Richard Fluegeman
Ball State University

Jack C. Hall
University of North Carolina-Wilmington

Robert W. Hinds
Slippery Rock University

Glenn Kroeger
Trinity University

Michael E. Lyle
Tidewater Community College

Richard Mariscal
Florida State University

Kevin Mickus
Southwest Missouri State University

Charles R. Singler
Youngstown State University

Terri Woods
East Carolina University

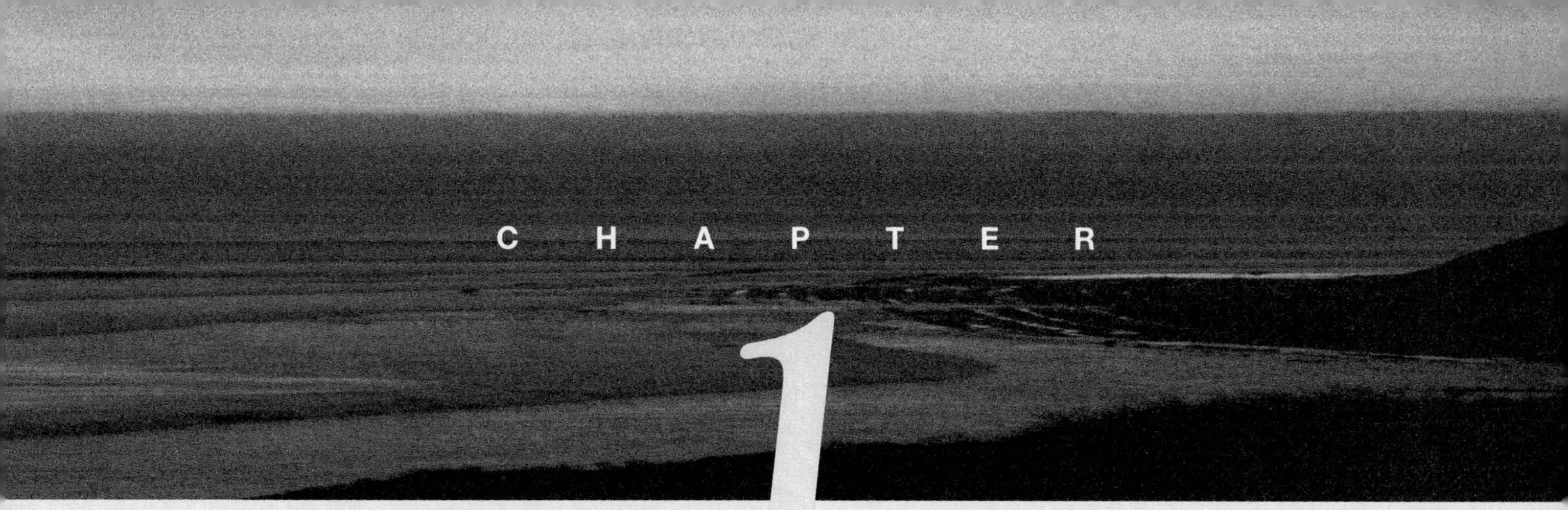

CHAPTER 1

History of Oceanography

Outline

1.1 The Early Times
1.2 The Middle Ages
1.3 Voyages of Discovery
1.4 The Importance of Charts and Navigational Information
1.5 Ocean Science Begins
1.6 The *Challenger* Expedition
1.7 Exploratory Science
1.8 U.S. Oceanography in the Twentieth Century
1.9 Oceanography of the Recent Past
1.10 The Present and the Future

Learning Objectives

After reading this chapter, you should be able to

- Understand the diversity of the sciences collected to form "oceanography."
- Understand the development of oceanography as a science.
- Follow the development of ocean knowledge from early voyages of exploration and discovery.
- Understand the significance of navigation in describing the oceans and making accurate maps of their extent and boundaries.
- Recognize the contributions of early U.S. oceanography to the development of marine commerce.
- Understand the role of early scientific voyages in the investigation of the world's oceans.
- Recognize the role of U.S. agencies in developing oceanography before and after World War II.
- Recognize the relationship between technology, international cooperation, and the development of recent large-scale oceanographic programs.
- Discuss the programs being planned for the future.

◀ Replicas of *La Pinta*, *La Santa Maria*, and *La Nina*, set sail from the southern Spanish port of Huelua, October 13, 1991, in a historical reenactment of Columbus' first voyage.

Oceanography is a broad field in which many sciences focus on the common goal of understanding the oceans. Geology, geography, geophysics, physics, chemistry, geochemistry, mathematics, meteorology, botany, and zoology all play roles in expanding our knowledge of the oceans. Geological oceanography includes the study of the earth at the sea's edge and below its surface and the history of the processes that formed the ocean basins. Physical oceanography investigates how and why the oceans move; marine meteorology, the study of heat transfer, water cycles, and air-sea interactions, is often included in this discipline. Chemical oceanography studies the composition and history of seawater, its processes, and its interactions. Biological oceanography concerns itself with the marine organisms and the relationship between these organisms and the environment of the oceans. Ocean engineering is the discipline of designing and planning equipment and installations for use at sea.

Our progress toward the goal of understanding the oceans has been uneven, and it has frequently changed direction. The interests and needs of nations as well as the intellectual curiosity of scientists have controlled the rate at which we study the oceans, the methods we use to study them, and the priority we give to certain areas of study. To gain some perspective on the current state of knowledge about the oceans, we need to know something of the events and incentives that guided previous investigations of the oceans.

1.1 THE EARLY TIMES

People have been gathering information about the oceans for millennia, passing it on by word of mouth. Curious individuals must have acquired their first ideas of the oceans from wandering the seashore, wading in the shallows, and gathering food from the ocean's edges. As early humans moved slowly away from their inland centers of development, they took advantage of the sea's food sources when they first explored and later settled along the ocean shore. The remains of shells and other refuse found at the sites of ancient shore settlements show that our early ancestors gathered shellfish, and certain fish bones suggest that they also began to use rafts or some type of boat for offshore fishing.

Early information about the oceans was mainly collected by explorers and traders. Around 1500 B.C., the Phoenicians, well known as excellent sailors and navigators, traded from North Africa across the Mediterranean Sea with the inhabitants of Italy, Spain, and Greece. They sailed out of the Mediterranean Sea and north along the Atlantic coast of Europe to trade in the British Isles. They may even have circumnavigated Africa around 600 B.C.

FIGURE 1.1

A navigational chart (*rebillib*) of the Marshall Islands. Sticks represent a series of regular wave patterns (swells). Curved sticks show waves bent by the shorelines of individual islands. Islands are represented by shells.

Between 1500 and 500 B.C., the Arab traders explored the Indian Ocean, and the Polynesians made long voyages of discovery in the Pacific Ocean. These voyages left little in the way of recorded information. Using descriptions passed down from one voyager to another, early sailors piloted their way from one landmark to another, sailing close to shore and often bringing their boats up onto the beach each night. As they began to move away from shore they used birds, waves, cloud formations, and the observation of astronomical bodies in their travels (fig. 1.1). The distinctive smells of land such as flowers and wood smoke alerted them to possible landfalls.

These early sailors did not investigate the oceans; for them the sea was only a dangerous road, a pathway from here to there. This situation continued for hundreds of years. However, the information that they accumulated became a body of lore to which sailors and voyagers added from year to year.

While the Greeks traded and warred throughout the Mediterranean, they observed and also asked themselves questions about the sea. Aristotle (384–322 B.C.) believed that the ocean occupied the deepest parts of the earth's surface; he knew that the sun evaporated water from the sea surface, which condensed and returned as rain. He also began to catalog marine organisms. The brilliant Eratosthenes (c. 265–194 B.C.) of Alexandria, Egypt, mapped his known world and calculated the circumference of the earth to be about 40,250 kilometers or 25,000 miles (today's measurement is 40,067 km or 24,881 mi). Posidonius (c. 135–50 B.C.) reportedly measured an ocean depth to about 1800 meters (6000 ft) near the island of Sardinia, according to the Greek geographer Strabo (c. 63 B.C.– c. A.D. 21). Pliny the Elder (A.D. 23–79) related the phases of

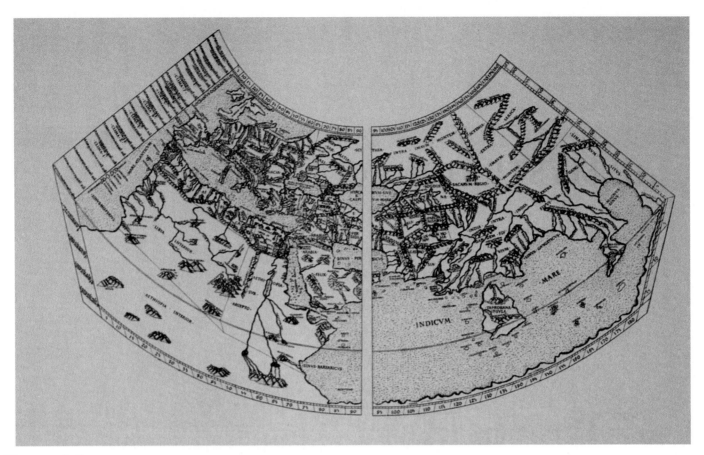

FIGURE 1.2
A chart from an Italian fifteenth-century edition of Ptolemy's *Geographia*.

the moon to the tides and reported on the currents moving through the Strait of Gibraltar. Ptolemy, in A.D. 127–51, produced the first world atlas and established world boundaries: to the north the British Isles, northern Europe, and the unknown lands of Asia; to the south an unknown land, "Terra Australis Incognita," including Ethiopia, Libya, and the Indian Sea; to the east China; and to the west the great Western Ocean reaching around the earth to China on the other side (fig. 1.2). His atlas listed more than 8000 places by latitude and longitude, but his work contained a major error: he had accepted a value of 29,000 kilometers (18,000 mi) for the earth's circumference. This shortened earth distances and allowed Columbus, more than a thousand years later, to believe that he had reached the eastern shore of Asia when he landed in the Americas.

Name the subfields of oceanography.

What did early sailors use for guidance during long ocean voyages?

How long ago was the circumference of the earth first calculated?

How did Ptolemy's atlas contribute to a greater understanding of world geography, and how did it produce confusion?

1.2 THE MIDDLE AGES

After Ptolemy, intellectual activity and scientific thought declined in Europe for about one thousand years. However, shipbuilding improved during this period; vessels became more seaworthy and easier to sail allowing sailors to extend their voyages. The Vikings (A.D. 700–1000) colonized Iceland by 900 and settled Greenland, where they remained until the fourteenth century. Extending their voyages farther west, they reached Vinland, or northeastern North America, in 985. To the south, in the Mediterranean region after the fall of the Roman Empire, the Arabs preserved the knowledge of the Greeks and the Romans, on which they continued to build. The Arabic writer El-Mas'údé (d. 956) gave the first description of the reversal of the ocean currents due to the seasonal monsoon winds. Using this knowledge of winds and currents, the Arabs established regular trade routes across the Indian Ocean. In the 1200s large Chinese junks with crews of 200 to 300 sailed the same routes (between China and the Persian Gulf) as the Arab dhows.

During the Middle Ages, while scholarship about the sea remained static, the knowledge of navigation increased. Harbor-finding charts, or *portolanos*, appeared. These charts carried a distance scale and noted hazards to navigation, but they did not have latitude or longitude. With the introduction of the magnetic compass to Europe from Asia in the thirteenth century,

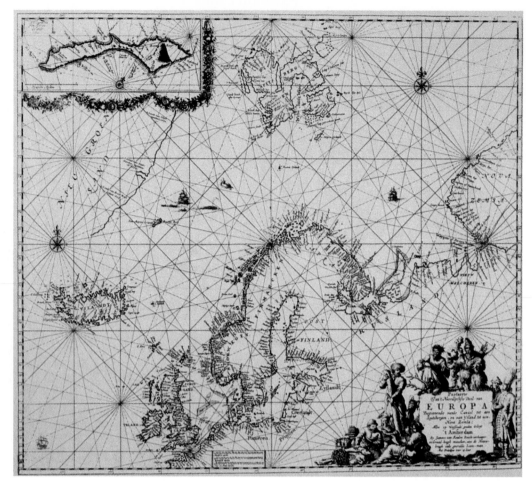

FIGURE 1.3
A navigational chart of northern Europe from Johannes van Keulen's *Sea-Atlas* of 1682–84.

compass directions were added. One example, a Dutch navigational chart from Johannes van Keulen's *Great New and Improved Sea-Atlas or Water-World* of 1682–84, is shown in figure 1.3.

As scholarship was reestablished in Europe, Arabic translations of early Greek studies were translated into Latin, which made them again available to northern European scholars. By the 1300s, Europeans had established successful trade routes, including some partial ocean crossings. An appreciation of the importance of navigational techniques grew as trade routes were extended.

> What advances occurred during the Middle Ages that allowed longer ocean voyages?
>
> During the tenth century, which oceans were explored and by what peoples?

1.3 VOYAGES OF DISCOVERY

Early in the fifteenth century the Chinese organized seven voyages to explore the Pacific and Indian Oceans. More than 300 ships, one more than 122 meters (400 ft) long, participated in these ventures to extend Chinese influence and demonstrate the power of the Ming dynasty. The voyages ended in 1433 when their explorations led the Chinese to believe that other societies had little to offer, and the government of China withdrew within its borders beginning a 400-year period of isolation.

In Europe the desire for riches from new lands persuaded wealthy individuals, often representing their countries, to underwrite the costs of long voyages to all the oceans of the world. The great age of European discovery began when, in 1487 Bartholomeu Dias (1450?–1500) sailed around the Cape of Good Hope into the Indian Ocean, looking for new and faster routes to the spices and silks of the East. Christopher Columbus (1451–1506) made four voyages across the Atlantic Ocean, believing he had found a way to the riches of Cathay or China. Vasco da Gama (1469?–1524) journeyed south and east around the Cape of Good Hope to India, establishing a sea route to the same lands. The Italian navigator Amerigo Vespucci (1454–1512) made several voyages to the New World (1499–1504) for Spain and Portugal; he accepted South America as a new continent, not part of Asia. In 1507, the German cartographer Martin Waldseemüller applied the name "America" to the continent in Vespucci's honor. Vasco Nuñez de Balboa (1475–1519) crossed the Isthmus of Panama and found the

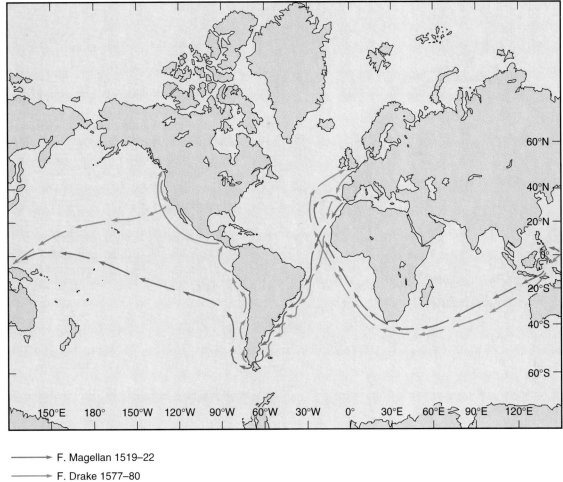

F. Magellan 1519–22
F. Drake 1577–80

FIGURE 1.4

The sixteenth-century circumnavigation voyages by Magellan and Drake.

Pacific Ocean in 1513. All claimed the new lands they found for their home countries. Although they had sailed for riches, not knowledge, they more accurately documented the extent and properties of the oceans, and the news of their travels stimulated others to follow.

Ferdinand Magellan (1480?–1521) left Spain in 1519 with five vessels and discovered and passed through the Straits of Magellan rounding South America in 1520. He crossed the Pacific Ocean with three ships in 1520–21. Magellan was killed in the Philippines, but one of his three ships, *Victoria*, made its way across the Indian Ocean and around the Cape of Good Hope to complete the first circumnavigation of the earth (fig. 1.4). Magellan's skill as a navigator makes his voyage probably the most outstanding single contribution to the early charting of the oceans. In addition, during the voyage he established the length of a degree of latitude and measured the circumference of the earth.

By the latter half of the sixteenth century, adventure, curiosity, and hopes of finding a trading shortcut to China spurred efforts to find a sea passage around the north side of North America. Sir Martin Frobisher (1535?–94) made three voyages in the 1570s, and Henry Hudson (d. 1611) made four voyages between 1607 and 1610, dying with his son when set adrift in

Hudson Bay by his mutinous crew. The Northwest Passage continued to beckon, and William Baffin (1584–1622) made two attempts in 1615 and 1616.

While European countries were setting up colonies and claiming new lands, Francis Drake (1540?–96) set out in 1577 to show the English flag around the world (fig. 1.4). In 1580 he completed his circumnavigation and returned home in the *Golden Hind* with a cargo of Spanish gold, to be knighted and treated as a national hero. Queen Elizabeth I encouraged her sea captains' exploits as explorers and raiders because, when needed, their ships and their knowledge of the sea brought military victories as well as economic gains.

What stimulated the long voyages of the fifteenth and sixteenth centuries?

Who was Amerigo Vespucci, and how was he honored?

Why was Magellan's voyage of such great importance?

What is the Northwest Passage? Why was there an interest in finding it?

History of Oceanography

1.4 THE IMPORTANCE OF CHARTS AND NAVIGATIONAL INFORMATION

As colonies were established far away from their home countries, and as trade and travel expanded, there was renewed interest in developing better charts and more accurate navigation techniques. To obtain the precise location of landfall or ship's position, it is necessary to know the location of the sun or the stars related to time, and because early clocks did not work well on rolling ships, precise navigational measurements were not possible on early voyages. In 1714 the British Parliament offered 20,000 pounds sterling for a clock that could keep time with an error not greater than two minutes on a voyage to the West Indies from England. John Harrison, a clock maker, accepted the challenge and built his first sea-going clock in 1735. It was not until 1761 that his fourth model (fig. 1.5) met the test, losing only 51 seconds, on the 81-day voyage.

Captain James Cook (1728–79) made his three great voyages to chart the Pacific Ocean between 1768 and 1779 (fig. 1.6). During his voyages, he explored and charted much of

FIGURE 1.5

John Harrison's fourth chronometer. A copy of this chronometer was used by Captain James Cook on his 1772 voyage to the southern oceans.

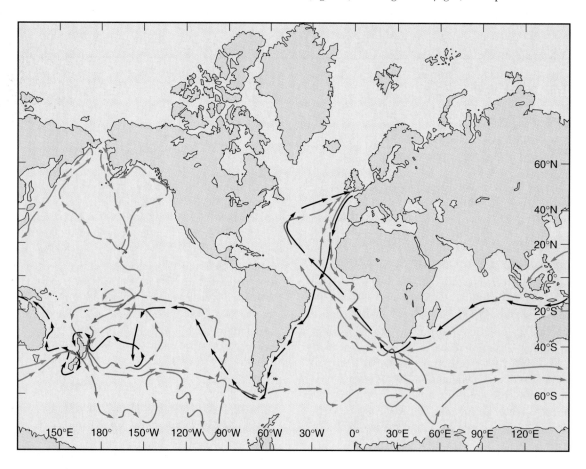

→ Cook's first voyage 1768–71

→ Cook's second voyage 1772–75

→ Cook's third voyage 1776–80

FIGURE 1.6

The three voyages of Captain James Cook.

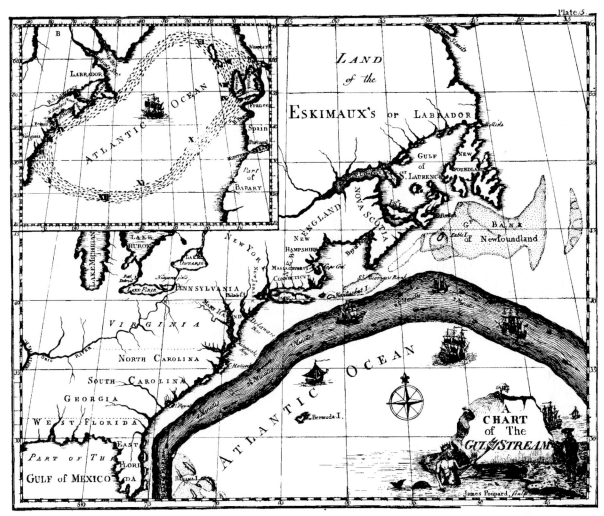

FIGURE 1.7
The Franklin-Folger map of the Gulf Stream, 1769.

the South Pacific and the coasts of New Zealand, Australia, and northwest North America. He took a copy of Harrison's fourth chronometer on his second voyage of discovery to the south seas, and with this timepiece he was able to produce accurate charts of new areas and to correct previously charted positions. He searched for a way to the Atlantic from the Bering Sea and discovered the Hawaiian Islands, where he was killed. He made soundings to depths of 400 meters (1300 ft) and logged accurate observations of winds, currents, and water temperatures. Cook takes his place as one of history's greatest navigators and sailors as well as a fine scientist. His careful and accurate observations produced much valuable information and made him one of the founders of oceanography.

In the United States, Benjamin Franklin (1706–90) became concerned about the amount of time required for news and cargo to travel between England and America. With Captain Timothy Folger, his cousin and a whaling captain from Nantucket, he constructed the 1769 Franklin-Folger chart of the Gulf Stream current (fig. 1.7), which encouraged captains to sail within the Gulf Stream enroute to Europe and to avoid it on the return passage.

The U.S. Naval Hydrographic Office, now the U.S. Naval Oceanographic Office, was set up in 1830. In 1842, Lieutenant Matthew F. Maury (1806–73) was assigned to the Hydrographic Office and founded the Naval Depot of Charts. He began a systematic collection of wind and current data from ships' logs. He produced his first wind and current charts of the North Atlantic in 1847; these became a part of the first published atlases of sea conditions and sailing directions. The British estimated that Maury's sailing directions took thirty days off the passage from the British Isles to California, twenty days off the voyage to Australia, and ten days off the sailing time to Rio de Janeiro. In 1855, he published *The Physical Geography of the Sea*. This work includes chapters on the Gulf Stream, the atmosphere, currents, depths, winds, climates, and storms, and the first contour chart of the North Atlantic sea floor. See figure 1.8 for the Gulf Stream chart from this book and compare the change in detail and style with the Franklin-Folger chart in figure 1.7. Many consider Maury's book the first textbook of what we now call oceanography and consider Maury the first true oceanographer. Again, national and commercial interests were the driving forces behind the study of the oceans.

History of Oceanography

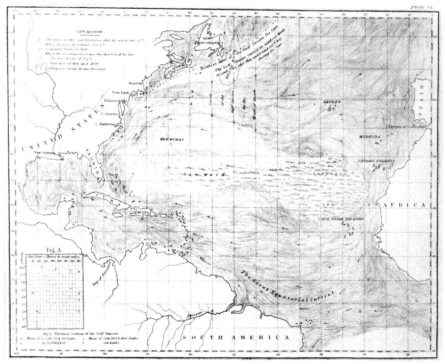

FIGURE 1.8
Maury's chart of the Gulf Stream and North Atlantic Ocean surface currents from *Physical Geography of the Sea*, 1855.

Why was John Harrison's clock important to ocean exploration?

What did Captain James Cook's voyages contribute to the science of oceanography?

Why was Benjamin Franklin interested in the Gulf Stream?

Who was Matthew F. Maury, and what was his contribution to ocean science?

1.5 OCEAN SCIENCE BEGINS

As charts became more accurate and as information about the oceans increased, the oceans captured the interest of naturalists and biologists. Charles Darwin (1809–82) joined the survey ship *Beagle* and served as the ship's naturalist from 1831 to 1836. He described, collected, and classified organisms from the land and sea. His theory of atoll formation (described in Chapter 4) is still the accepted explanation. At approximately the same time, another English naturalist, Edward Forbes (1815–54), began a systematic survey of marine life around the British Isles and the Mediterranean and Aegean Seas. He collected organisms in deep water and, based on his observations, proposed a system of ocean depth zones, each characterized by specific ani-

mal populations. However, he also mistakenly theorized that the environment below 550 meters (1800 ft) was without life. Forbes' systematic attempt to make orderly predictions about the oceans, his enthusiasm, and his influence make him another candidate for a founder of oceanography.

The investigation of the minute drifting plants and animals of the ocean was not seriously undertaken until the German scientist Johannes Müller (1801–58) began to examine these organisms microscopically. Victor Hensen (1835–1924) introduced the quantitative study of these minute drifting organisms and gave them the name *plankton* in 1887. A portion of a plate from an 1899 publication describing these organisms is seen in figure 1.9.

Although science blossomed in the seventeenth and eighteenth centuries, there was little scientific interest in the sea beyond the practical needs for navigation, tide prediction, and safety. By the early nineteenth century, ocean scientists were still few and usually only temporarily attracted to the sea.

How did Edward Forbes advance understanding of the oceans, and in what way was he mistaken?

What kinds of organisms were investigated by Müller and Hensen?

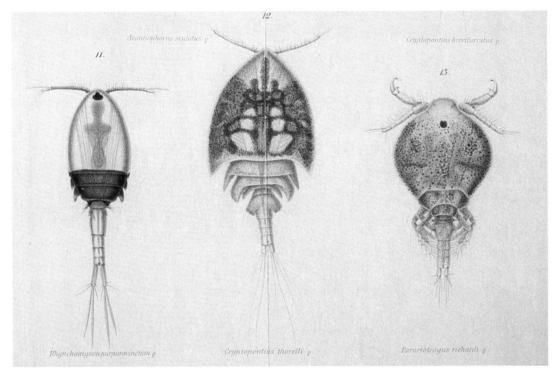

FIGURE 1.9
This illustration of microscopic drifting animals is from volume 25 of *Fauna und Flora des Golfes von Neapel* (*Fauna and Flora of the Gulf of Naples*), published in 1899.

1.6 THE CHALLENGER EXPEDITION

Wyville Thomson (1830–82), a professor of natural history at Edinburgh University in Scotland, participated in a series of deep-sea expeditions during the years 1868–70. The deep-sea organisms recovered by these expeditions caught the public's attention, and with public interest running high, the British Royal Society was able to persuade the British Admiralty to organize the *Challenger* expedition, the most comprehensive single oceanographic expedition ever undertaken. The leadership of this expedition was offered to Wyville Thomson, who named as his assistant a young geologist, John Murray (1841–1914). At this time, Thomson also wrote *The Depths of the Sea*, based on his participation in the previous scientific expeditions. This very popular book was published in 1873 and is regarded by some as the first book on oceanography.

The naval corvette *Challenger* was refitted with laboratories, winches, and equipment; it sailed from England on December 21, 1872, for a voyage that was to last nearly three and a half years (fig. 1.10). During this voyage the vessel logged 110,840 kilometers (68,890 mi), and returned to England on May 24, 1876. The *Challenger* expedition's purpose was scientific research; during the voyage the crew took soundings at 361 ocean stations (the deepest at 8180 m or 26, 850 ft), collected deep-sea water samples, investigated deep-water motion, and made temperature measurements at all depths. Thousands of biological and sea-bottom samples were collected. *Challenger* brought back evidence of an ocean teeming with life at all depths and opened the way for the era of descriptive oceanography that followed.

Although the *Challenger* expedition ended in 1876, the work of organizing and compiling information continued for twenty years, until the last of the fifty-volume *Challenger Reports* was issued. John Murray edited the reports after Thomson's death and wrote many of them himself. He is considered the first geological oceanographer. William Dittmar (1833–92) prepared the information on seawater chemistry for the *Challenger Reports*, confirming the findings of earlier chemists that in seawater around the world, the proportion of the major dissolved elements to each other is constant. Oceanography as a modern science is usually dated from the *Challenger* expedition. The *Challenger Reports* laid the foundation for the science of oceanography.

> Explain the significance of the *Challenger* expedition and the *Challenger Reports*.
>
> The *Challenger* and its expedition are often called unique. Why is this term used?

(a)

(b)

(c)

FIGURE 1.10

The *Challenger* expedition: December 21, 1872–May 24, 1876. Engravings from the *Challenger Reports*, volume 1, 1885, (a) "H.M.S. *Challenger*—Shortening Sail to Sound," decreasing speed to take a deep-sea depth measurement. (b) "Dredging and Sounding Arrangement on board *Challenger*." Rigging is hung from the ship's yards to allow the use of over-the-side sampling equipment. A biological dredge can be seen hanging outboard of the rail. The large cylinders in the rigging are shock absorbers. (c) A biological dredge used for sampling bottom organisms. Note the frame and skids that keep the mouth of the net open and allow it to slide over the sea floor. (d) "H.M.S. *Challenger* at St. Paul's Rocks," in the equatorial mid-Atlantic. (e) The cruise of the *Challenger*, 1872–76, the first major oceanographic research effort.

(d)

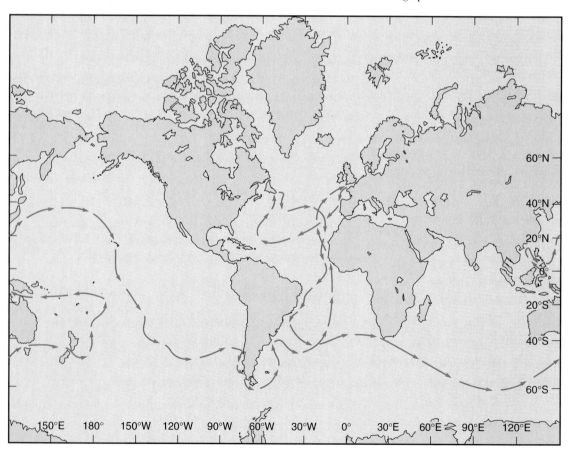

→ Cruise of the *Challenger*, 1872–76

(e)

FIGURE 1.11
The *Fram* frozen in the ice. As the ice pressure increased, it lifted the specially designed and stregthened hull so that the ship escaped being crushed.

1.7 EXPLORATORY SCIENCE

During the late nineteenth century and the early twentieth century, intellectual interest in the oceans increased. Oceanography was changing from a descriptive science to a quantitative one. Oceanographic cruises now had the goal of testing hypotheses by gathering data. Theoretical models of ocean circulation and water movement were developed. The Scandinavian oceanographers were particularly active in the study of water movement. One of them, Fridtjof Nansen (1861–1930), a well-known athlete, explorer, and zoologist, was interested in the current systems of the polar seas. He decided to test his ideas about the direction of ice drift in the Arctic by freezing a vessel into the polar ice pack and drifting with it; he expected in this way to reach the North Pole. To do so he had to design a special vessel that would be able to survive the great pressure from the ice; the 39-meter (128-ft) *Fram* ("to push forward"), shown in figure 1.11, was built with a smoothly rounded, wooden hull and planking over 60 centimeters (2 ft) thick. Nansen with thirteen men departed from Oslo in June 1893.

The ship was frozen into the ice nearly 1100 kilometers (700 mi) from the North Pole and remained in the ice for thirty-five months. During this period, measurements made through holes in the ice showed that the Arctic Ocean was a deep ocean basin and not the shallow sea that had been expected. The crew recorded water and air temperatures, analyzed water chemistry, and observed the great plankton blooms of the area. Nansen became impatient with the slow rate and the direction of drift, and with F. H. Johansen left the *Fram* locked in the ice some 500 kilometers (300 mi) from the pole. They set

off with a dogsled toward the pole, but after four weeks they were still more than 300 kilometers (200 mi) from the pole, with provisions running low and the condition of their dogs deteriorating. The two men turned away from the pole and spent the winter of 1895–96 on the ice living on seals and walrus. They were found by a British polar expedition in June 1896 and returned to Norway. The crew of the *Fram* continued to drift with the ship until they freed the vessel from the ice in 1896 and also returned to Oslo. Nansen's expedition had laid the basis for future Arctic research. His name is familiar today from the Nansen bottle, which he designed to collect and isolate water samples at depth.

The *Fram* paid another visit to polar waters carrying the Norwegian explorer Roald Amundsen (1872–1928) to Antarctica on his successful 1911 expedition to the South Pole. It was also Amundsen who finally made a Northwest Passage in the *Gjoa* (fig. 1.12), leaving Norway in 1903 and arriving in Nome, Alaska, three years later.

Fluctuations in the abundance of commercial fish in the North Atlantic and adjacent seas, and the effect of these changes on national fishing programs, stimulated oceanographic research and international cooperation. As early as 1870, researchers began to realize their need for knowledge of ocean chemistry and physics in order to understand ocean biology. Advances in theoretical oceanography often could not be verified with practical knowledge until new instruments and equipment were developed. Lord Kelvin (1824–1907) invented a tide-predicting machine in 1872 that made it possible to combine tidal theory with astronomical predictions to produce predicted tide tables. Deep-sea circulation could not be

History of Oceanography

FIGURE 1.12

The *Gjoa* preparing for its journey through the Northwest Passage (1903).

systematically explored until approximately 1910, when Nansen's water-sampling bottles were combined with thermometers designed for deep-sea temperature measurements, and an accurate method for determining the water's salt content was devised by the chemist Martin Knudsen (1871–1949). The reliable and accurate measurement of ocean depths had to wait until the development of the echo sounder, which was given its first scientific use on the 1925–27 German cruise of the *Meteor* in the central and southern Atlantic Ocean.

Who was Fridtjof Nansen, and why was his work important?

Give examples of nineteenth- and early twentieth-century advances in technology that helped ocean studies.

1.8 U.S. Oceanography in the Twentieth Century

In the United States, government agencies related to the oceans proliferated during the nineteenth century. These agencies were concerned with gathering information to further

commerce, fisheries, and the navy. After the Civil War, the replacement of sail by steam lessened government interest in studying winds and currents and in surveying the ocean floor. Private institutions became important contributors to oceanography; the Carnegie Institution funded a series of exploratory cruises (fig. 1.13), including investigations of the earth's magnetic field, and maintained a biological laboratory. In 1930, the Rockefeller Foundation allocated funds to stimulate programs and construct laboratories for marine research. At this time, oceanography began to move onto university campuses.

Oceanography mushroomed during World War II, when practical problems of military significance had to be solved quickly. The United States and its allies needed to move men and materials by sea to remote locations, to predict ocean and shore conditions for amphibious landings, to know how explosives behave in seawater, to chart beaches and harbors from aerial reconnaissance, and to find and destroy submarines. Academic studies ceased as oceanographers pooled their knowledge in the national effort.

After the war, oceanographers returned to their classrooms and laboratories with an array of new sophisticated instruments, including radar, improved sonar devices, automated wave detectors, and temperature-depth recorders. They also returned with large-scale government funding for research and education. The earth sciences in general and oceanography in particular blossomed during the 1950s.

The Office of Naval Research funded applied research programs and research vessels, the National Science Foundation underwrote basic research, and the Atomic Energy Commission financed oceanographic work at the South Pacific atoll sites of atomic tests. The National Oceanic and Atmospheric Administration (NOAA) was formed under the Department of Commerce in 1970. NOAA combines the National Ocean Survey and a fleet of research vessels (fig. 1.14), National Weather Service, National Marine Fisheries Service, Environmental Data Service, National Environmental Satellite Service, and Environmental Research Laboratories. NOAA has a broad mandate in many fields relating to the overall objectives of the Department of Commerce—"to aid in the industrial revitalization or economic growth of the nation by increasing exports and the U.S. competitive position in the world market."

How did the U.S. oceanographic research focus change after the Civil War?

How has each of the following affected twentieth-century oceanography? (a) economics, (b) commerce and transportation, (c) military needs.

Compare the U.S. government's institutional support of oceanography before and after World War II.

FIGURE 1.13
The *Carnegie*, built in 1909 by the Carnegie Institute of Washington, was constructed of nonmagnetic materials for use in mapping magnetic forces over the oceans. It was lost in 1929 at Samoa in a gasoline explosion.

FIGURE 1.14
The NOAA research vessel *Discoverer*.

1.9 OCEANOGRAPHY OF THE RECENT PAST

International cooperation brought about the 1957–58 International Geophysical Year (IGY) program, in which sixty-seven nations cooperated to explore the sea floor and made discoveries that completely revolutionized geology and geophysics. In 1963–64 another multinational endeavor, the Indian Ocean Expedition, was in progress. This was followed by the ten-year International Decade of Ocean Exploration in the 1970s, a multinational effort to survey seabed mineral resources, improve environmental forecasting, investigate coastal ecosystems, and modernize and standardize the gathering and use of marine data.

The decade of the 1960s brought giant strides in programs and equipment. As a direct result of the IGY exploration program, special research vessels and submersibles were built to be used by both federal agencies and university research programs. The Deep Sea Drilling Program, a cooperative venture between research institutions and universities, began to sample the earth's crust beneath the sea (fig. 1.15; discussed in Chapter 3). Electronics developed by the space program were applied to ocean research. Computers went aboard research vessels, and for the first time data could be sorted, analyzed, and interpreted at sea; experiments could be adjusted while in progress.

Government funding allowed large-scale ocean experiments; fleets of oceanographic vessels representing many institutions and nations studied ocean chemistry, water motion, and air-sea interaction (fig. 1.16).

Oceanography in the 1970s faced a reduction in funding for ships and basic research; nonetheless, the discovery of deep-sea hot water vents and their associated animal life and mineral deposits renewed the excitement over deep-sea biology, chemistry, geology, and ocean exploration in general. Instrumentation grew ever more sophisticated and expensive as deep-sea moorings, deep-diving submersibles, and the remote sensing of the ocean by satellite became possible. While collection of oceanographic data expanded at sea, data collected by satellites in the 1970s and 1980s increasingly gave researchers the ability to observe worldwide sea surface changes in both space and time. SEASAT, a specialized oceanographic satellite, was launched in June 1978 but remained operational only until October (fig. 1.17a). Its radar could measure the distance between the satellite and the sea surface with an accuracy of about 5 centimeters (2 in), allowing the measurement of wave heights. Sea surface temperatures, wind speeds, sea ice cover, currents, and plant production were also monitored. The orbit of the military satellite GEOSAT was altered to repeat the track of SEASAT; GEOSAT (fig. 1.17b) can detect month-to-month changes in sea level, wave direction, and the movements of surface water masses.

FIGURE 1.15
The *Glomar Challenger*, the Deep Sea Drilling Program drill ship used from 1968 to 1983.

FIGURE 1.16
The research vessel *Atlantis II* operated by the Woods Hole Oceanographic Institute. The research submersible *Alvin* is seen at the stern.

(a)

(b)

FIGURE 1.17
U.S. satellites with ocean sensors. (a) SEASAT, an ocean-observing satellite launched in 1978 by NASA, failed after 104 days of operation. During its active period it monitored sea surface topography, wave heights, and sea surface wind velocities; it measured sea surface temperatures, water vapor, and sea ice distribution. (b) GEOSAT, a U.S. Navy satellite, uses radar altimetry to determine the topography of the ocean.

During this period earth scientists began to recognize the signs of global degradation and the need for management of living and nonliving resources. As more and more nations turned to the sea for food and as technology increased their abilities to harvest the sea, problems of resource ownership, dwindling fish stocks, and the need for fishery management became more and more evident.

How did the emphasis on ocean research change in the 1960s, 1970s, and 1980s?

List the advances in ocean technology during this period.

Why are oceanographers interested in data collected by satellite as well as data from research vessels?

History of Oceanography

1.10 THE PRESENT AND THE FUTURE

The scientific targets for investigation in the 1990s include the effect of ocean circulation on the earth's climate balance, the management of living and nonliving resources, the transport of materials from the land to the deep ocean basins, the chemistry of the interaction of seawater with the earth's crust, the dynamics of the continental margins and the ocean seabed, the energy sources of the sea, the exchange of gases between the oceans and the atmosphere, methods of decreasing the cost of ocean transport, and increasing food availability. At the same time, scientists using refined sensors and techniques will continue to seek answers to the basic questions of how and why ocean processes occur and the relationship of these processes to sea resources and to ourselves.

Scientists increasingly recognize the earth as a complex of systems and subsystems acting as a whole. Driving the unified study of global change are (1) the recognition that individual sciences need to cross disciplines in order to advance, (2) the recognition that more sophisticated satellite sensing has the potential to move the sciences toward an understanding of the total earth, and (3) the worldwide concern over accelerated changes in the earth's environment due to human causes. The success of an integrated approach to earth studies will require that governments, agencies, universities, and national and international programs set priorities and agree on joint programs of global studies. Among the global programs now in progress are a decade-long program to understand the circulation of ocean water and predict changes related to long-term atmospheric changes, a study of the relationship between ocean plant production and solar radiation, an enquiry into the energy transferred to the atmosphere by the tropical oceans, and an analysis of the responses of marine plants and animals to changes in ocean circulation and chemistry. Exploration and investigation of the world's underwater mountain-range systems continue.

Although cooperative ventures have achieved much and will achieve much more, it is important to remember that studies driven by the specific research interests of individual scientists will continue to add knowledge and are essential to point out new directions for oceanography and the other earth sciences.

List some of oceanography's current research targets.

Why are oceanographers interested in a global approach to ocean science?

What role does the individual scientist play in advancing our knowledge?

SUMMARY

Oceanography is a multiscience field in which geology, geophysics, chemistry, physics, meteorology, and biology are all used to understand the oceans. Early information about the oceans was collected by explorers and traders such as the Phoenicians, the Arabs, and the Greeks. Eratosthenes calculated the circumference of the earth, and Ptolemy produced the first world atlas.

During the Middle Ages, the Vikings crossed the North Atlantic, while shipbuilding and chartmaking improved. In the fifteenth and sixteenth centuries the Chinese, Dias, Columbus, da Gama, Vespucci, and Balboa made voyages of discovery. Magellan's expedition became the first to circumnavigate the earth. In the sixteenth and seventeenth centuries, some explorers searched for the Northwest Passage while others set up trading routes to serve developing colonies.

By the eighteenth century, national and commercial interests required better charts and more accurate navigation techniques. Cook's voyages of discovery to the Pacific produced much valuable information, and Franklin sponsored a chart of the Atlantic's Gulf Stream. A hundred years later, the U.S. Navy's Maury collected wind and current data to produce current charts and sailing directions and then wrote the first book on "oceanography."

Ocean science began with the nineteenth-century expeditions and research of Darwin, Forbes, Müller, and others. The three-and-a-half-year *Challenger* expedition laid the foundation for modern oceanography with its voyage, which gathered large quantities of data on all phases of oceanography. Exploration of the Arctic and Antarctic oceans was pursued by Nansen and Amundsen into the beginning of the twentieth century.

In this century, private institutions played an important role in developing U.S. oceanographic research, but the largest single push came from the needs of the military during World War II. After the war, large-scale government funding and international cooperation allowed oceanographic projects that made revolutionary discoveries about the ocean basins. Development of electronic equipment, deep-sea drilling programs, research submersibles, and use of satellites continued to produce new and more detailed information of all kinds. At present, oceanographers are focusing their research on global studies and the management of resources as well as continuing to explore the interrelationships of the chemistry, physics, geology, and biology of the sea.

Bailey, H. S. 1953. The Voyage of the *Challenger*. In *Ocean Science, Readings from Scientific American* 188 (5):8–12.

Brosse, J. 1983. *Great Voyages of Discovery, Circumnavigators and Scientists, 1764–1843*. Facts on File Publications, New York, 232 pp.

Oceanus Fall 1990. 33 (3). Issue devoted to marine education.

Oceanus Winter 1990–91. 33 (4). Issue devoted to naval oceanography.

Oceanus Spring 1991. 34 (1). Issue devoted to ocean engineering and technology.

Oceanus Summer 1991. 34 (2). Issue devoted to Soviet-American cooperation.

Wachsmann, S. 1990. Ships of Tarshish to the Land of Ophir. *Oceanus* 33 (1):70–82. Seafaring in the Mediterranean in biblical times.

ITEM OF INTEREST

Marine Archaeology

More than two thousand years of war and trading have left thousands of wrecks scattered across the ocean floor. These wrecks are great storehouses of information for archaeologists and historians about ships, trade, warfare, and the details of personal lives.

Wrecks that lie in deep water are initially much better preserved than those in shallow water, because deep-water wrecks lie below the depths of the strong currents and waves that break up most shallow-water wrecks. Shallow-water wrecks are often the prey of treasure hunters who destroy the history of the site while they search for adventure and items of market value. Over long periods the cold temperatures and low oxygen content of deep water favor preservation of wooden vessels by slowing decomposition and excluding the marine organisms that bore into wood in shallow areas. Also the rate at which muds and sands falling from above cover objects on the deep sea floor is much slower than the same process in coastal water.

Marine archaeologists are using techniques developed for oceanographic research to find, explore, recover, and preserve wrecks and other artifacts lying under the sea. Sound beams that sweep the sea floor produce images that are viewed on board ship to locate an object of interest, and magnetometers are used to detect iron from sunken vessels. Once the initial contact has been made, the find is verified by divers in shallow water and by research submersibles carrying observers in deeper water. Unmanned, towed camera and instrument sleds or remotely operated vehicles (ROVs) equipped with underwater video cameras explore in deep water or in areas that are difficult or unsafe for divers and submersibles. ROVs and submersibles, as well as divers, collect samples to help identify wrecks.

The Ships

A Bronze Age merchant vessel was discovered in 1983 more than 33 meters (100 ft) down in the Mediterranean Sea off

continued . . .

FIGURE 1

The hull of the *Mary Rose* in it's display hall at Portsmouth Dock-yard, England. To preserve and stabilize the remaining wood, it is constantly sprayed with a preservative solution.

the Turkish coast. Divers have recovered thousands of arti-facts from its cargo, including copper and tin ingots, pottery, ivory, and amber. Using these items archaeologists have been able to learn about the life and culture of the period, trace the ship's trade route, and understand more about Bronze Age people's shipbuilding skills.

In the summer of 1545 the English warship, *Mary Rose*, sank as she sailed out to engage the French fleet. She was first studied in place and then raised in 1982. More than 17,000 objects were salvaged, ranging from a surgeon's walnut medicine chest to archers' longbows and arrows. These objects give archaeologists and naval historians insights into the personal as well as the working lives of the officers and crew of a naval vessel at that time. The portion of her hull that was buried in the mud was preserved and is on display at Portsmouth, England (fig. 1).

The Spanish galleon, *San Diego*, sank December 14, 1600, as she engaged two Dutch vessels off the Philippine coast. The European Institute of Underwater Archaeology and the National Museum of the Philippines began an intensive two-year archaeological excavation in 1992. Relics recovered include 570 stoneware storage jars for water, wine, and oil; 800 pieces of intact Chinese porcelain, including plates, bowls, serving vessels, and

FIGURE 2

The *Vasa*, a 17th century man-of-war, is on permanent display in Stockholm, Sweden. Most of the *Vasa* is original, including two of the three masts and parts of the rigging.

decorative pieces; a bronze astrolabe for determining latitude; and a bronze and glass compass. Most of the hull had been destroyed by shipworms and currents; surviving pieces were measured and then covered with sand for protection.

The Swedish man-of-war, *Vasa*, (fig. 2), sank in 1628 in Stockholm Harbor at the beginning of her maiden voyage. She was located in 1956 and raised to the surface in 1961. Because the water of the Baltic Sea is much less salty than the open ocean, there were no shipworms to destroy her wooden hull, and she is one of the few complete ships ever recovered from the sea floor. The *Vasa* was a great ship over 200 feet long, excluding her bowsprit. Like others of her period, she was fantastically decorated with carvings and statues. Divers searched the seabed every summer from 1963–67 and

(a)

(b)

FIGURE 3

The World War II German battleship *Bismarck* was sunk in 1941 and lies 4750 meters (15,600 ft) below the surface in the North Atlantic. In 1989, the towed camera sled *Argo* photographed sections of the wreck: (a) a 4.1 inch antiaircraft gun, (b) part of the ship's super structure.

recovered 700 sculptures and carved details that had adorned her. Several chests and barrels containing crewmen's possessions, wooden and earthenware utensils used by the seamen, and pewter dishes used by the officers have also been recovered. The *Vasa* is now on exhibit in Stockholm.

Two expeditions have found more modern vessels sunk in the very deep sea. Robert Ballard of the Woods Hole Oceanographic Institute found and surveyed both the *Titanic* in 1985 and the *Bismarck* in 1989. The passenger liner *Titanic* sank in 1912 after striking an iceberg on her maiden voyage; the wreck was located 4000 meters (13,000 ft) down in the North Atlantic. The contact was made by a towed underwater camera sled, computer controlled from the vessel above. A manned submersible was used for direct observation, and further inspection was made using ROVs that could be maneuvered to take pictures inside and outside the vessel. The World War II German battleship *Bismarck* (fig. 3) was located 4750 meters (15,600 ft) below the surface of the North Atlantic. A photographic survey documented both battle damage and destruction due to sinking.

A Harbor

A number of sunken shore towns have also been explored. The remains of a city, Caesarea Maritima, was a major find for marine archaeologists. This city was built by Herod the Great, king of Judaea from 37–4 B.C. It is located on the Mediterranean Sea between present-day Haifa and Tel Aviv. Archaeological studies began in 1978 and have documented the considerable engineering and construction skills required to build the harbor's complex breakwater system and an associated sluice system to flush the harbor of silt.

Readings

Ballard, R. D. 1985. How We Found *Titanic*. *National Geographic* 168 (6)698–722.

———.1986. A Long Last Look at *Titanic*. *National Geographic* 170 (6):698–727.

———.1989. Finding the *Bismarck*. *National Geographic* 176 (5):622–37.

Bass, G. F. 1987. Oldest Known Shipwreck Reveals Splendors of the Bronze Age. *National Geographic* 172 (6):693–734.

Gerard, S. 1992. The Caribbean Treasure Hunt. *Sea Frontiers* 38 (3):48–53.

Goddio, F. 1994. The Tale of the *San Diego*. *National Geographic* 186 (1):35–56.

Hohlfelder, R. L. 1987. Herod's City on the Sea. *National Geographic* 171 (2):261–80.

Marx, R. F. 1990. In Search of the Perfect Wreck. *Sea Frontiers* 36 (5):46–51. 🐚

2

Introduction to Earth

Outline

Learning Objectives

After reading this chapter, you should be able to

■ Describe the development of the earth, its atmosphere, and its oceans.

■ Relate the size, shape, and scale of the earth and its crustal features.

■ Know the age of the earth and understand radiometric dating and the geologic time scale.

■ Describe the location system of latitude and longitude.

■ Differentiate and describe a map projection, a bathymetric chart, and a physiographic map.

■ Understand natural time cycles including day, year, and seasons.

■ Explain the hydrologic cycle.

■ Relate the distribution of oceans and land with elevation.

■ Compare the dimensions of the three major oceans.

■ Know the kinds of electronic navigation aids used in oceanography.

◄ Lava flowing into the sea. Mount Kilauea Hawaii.

Billions of shimmering masses of stars, known as galaxies, move through the space we call the universe. About one-third of the way toward the center of one whirling mass of some 100 billion stars called the Milky Way galaxy, there is a fairly ordinary star, the sun. Around the sun move nine planets in their predictable and nearly constant orbits. The sun and its nine planets are called the solar system, and the third planet from the sun is called Earth (fig. 2.1).

Although we call our planet Earth, it is more accurate to consider it as the water planet. Water did not exist on the earth in the beginning, but its formation on a planet that was not too far from the sun nor too close, not too hot nor too cold, changed the earth and allowed the development of life. In this chapter, we begin to investigate this earth planet on which the largest bodies of water are known as the oceans.

2.1 THE BEGINNINGS

Present theories assign the beginning of our solar system to the collapse of a rotating interstellar cloud of gas and dust about 4.6 billion years ago. As the collapsing cloud's rotation speed increased, its temperature rose, and the gas and dust became disk shaped. At the center of this disk a star, our sun, was formed. The outer region of the disk began to cool and the gases began to interact chemically, producing particles of matter, which grew from collisions with other particles. The earth and the other planets of our solar system had begun to form.

During the early years of the earth's existence, it was bombarded by particles of all sizes; a portion of their energy of motion was converted into heat on impact. Each new layer of accumulated material buried the material below it, trapping this heat. At the same time, the growing weight of the accumulating layers compressed the interior, raising the earth's internal temperature to over 1000°C. Atoms of radioactive elements disintegrated by emitting subatomic particles that were absorbed into the surrounding matter, further raising its temperature.

Soon after the earth formed, its interior reached the melting point of iron and nickel. As the molten iron and nickel migrated toward the earth's center, more heat was generated and lighter substances were displaced. The lighter materials from the partially molten interior repeatedly melted, moved upward, and spread over the surface, cooling and solidifying. In this way the earth became completely reorganized and differentiated into a layered system, the subject of Chapter 3.

The earth's oceans are probably by-products of this heating and differentiation. Water locked in the mineral composition of the planet was released and carried to the surface as water vapor mixed with other gases. As the earth's surface cooled, the water condensed to form the oceans.

It is also believed that during the process of differentiation, gases released from the earth's interior formed the first atmosphere, which was primarily made up of water vapor, hydrogen gas, hydrogen chloride, carbon monoxide, carbon dioxide, and nitrogen. Any free oxygen present would have quickly combined with the metals of the crust. Oxygen gas could not accumulate in the atmosphere until it was produced in amounts large enough to exceed its loss by chemical reactions with the crust. This did not occur until life evolved to a level of complexity in which green plants could convert carbon dioxide and water with the energy of sunlight into organic matter and free oxygen. This process and its significance to life are discussed in Chapter 10.

How and when is our solar system thought to have been formed?

What processes added heat to the early earth, and how was the earth changed by this heat?

What was the source of the early earth's water and atmosphere?

2.2 THE AGE OF THE EARTH

The earth is an active planet, and its original surface rocks no longer exist. The oldest materials on the earth's surface have been dated at more than 4 billion years old. Moon samples returned by the Apollo mission are dated at 4.2 billion years old. Meteorites that have struck the earth have been dated between 4.5 billion and 4.6 billion years old. The substances found in many meteorites appear to represent materials that condensed out of the hot gases present at the beginning of the solar system. These ages agree with theoretical calculations made for the age of the sun. This information sets the accepted age of the earth at about 4.6 billion years.

The method of age-dating rock samples is known as **radiometric dating,** which is based on the decay of radioactive **isotopes** of elements such as uranium and potassium found in the rocks. An atom of a radioactive isotope has an unstable nucleus that changes, or decays. The time at which any single nucleus decays is unpredictable, but with large numbers of atoms, it is possible to predict that a certain fraction of the isotope will decay over a certain period of time. The time over which one-half of the atoms of an isotope decay is known as its **half-life.** The half-life of each isotope is characteristic and constant; for example, uranium-235 has a half-life of 704 million years. If a substance starts out containing only atoms of uranium-235, in 704 million years, the substance will be one-half uranium-235; the other half will be one of several decay products, primarily lead-207. Because each isotope system behaves differently in

FIGURE 2.1
The earth, as seen from space, is the water planet.

nature, data must be carefully tested, compared, and evaluated. The best analyses are those in which different isotope systems give the same date.

> What is the generally accepted age of the earth?
>
> How are rocks dated?
>
> How much of a radioactive isotope would be left after two half-lives had passed?

2.3 GEOLOGIC TIME

To refer to events in the history and formation of the earth, scientists use geologic time (fig. 2.2 and table 2.1). The principal divisions of geologic time are the four eons: the Hadean (4.6 to 3.9 billion years ago), the Archean (3.9 to 2.5 billion years ago), the Proterozoic (2.5 billion to 570 million years ago), and the Phanerozoic (since 570 million years). Fossils are known from other eons, but they are common only in the Phanerozoic.

This eon is divided into three eras: the Paleozoic era of ancient life, the Mesozoic era of middle life (popularly called the

Introduction to Earth

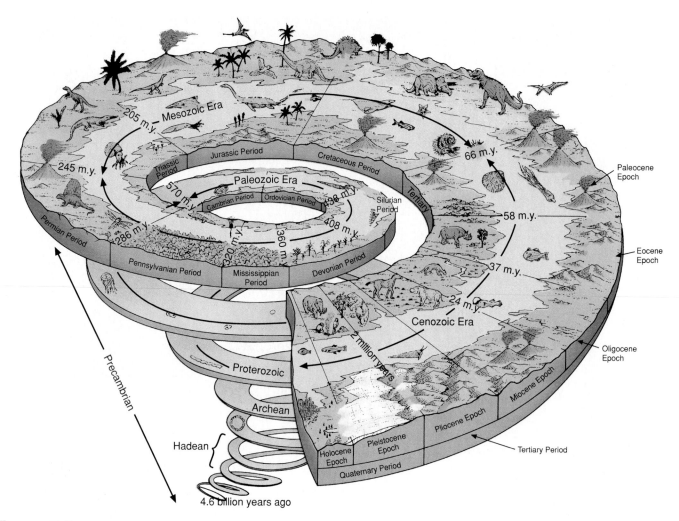

FIGURE 2.2

The history of the earth showing evolution of life-forms.

Source: After *Geologic Time,* U.S. Geological Survey publication.

Age of Reptiles), and the Cenozoic era of recent life (the Age of Mammals). Each of these eras is subdivided into periods and epochs; today, for instance, we live in the Holocene epoch of the Quaternary period of the Cenozoic era. The appearance or disappearance of fossil types was first used to set the boundaries of the time units. Radiometric dating has allowed scientists to assign numbers to these time-scale boundaries.

Very long periods of time are incomprehensible to most of us. We often have difficulty coping with time spans of more than ten years—what were you doing exactly ten years ago today? We have nothing with which to compare the 4.6-billion-year age of the earth. In order to place geologic time in a framework we can understand, let us divide the earth's age by 100 million. Then we can consider that Mother Earth is forty-six years old, a middle-aged lady in whose past we have a strong and intimate interest. What has happened to her over that forty-six years?

The first seven years of her life leave no record; as she entered her eighth year a few fragments of her history can be read in some rocks of Canada, Africa, and Greenland. By the time she was twelve years old, the first living cells of bacteria-type organisms had appeared. Approximately ten or eleven years later, oxygen production by living cells began, but it took eight or more years, until about her thirty-first year, to change the atmosphere sufficiently to support the first complex oxygen-requiring cells. The first hard-shelled fossils were laid down only six years ago, in her fortieth year, and by the time she was forty-one the first vertebrates (animals with backbones) had developed. Eight and a half months later, the first land plants appeared, quickly followed by an age in which fish were the characteristic life-form. By age forty-three the first reptiles had arrived, and dinosaurs became abundant at the beginning of her forty-fourth year, disappearing shortly after her forty-fifth birthday. A little over a year ago, plants with flowers began to develop; seven months ago, the mammals, birds, and insects became the dominant land animals. Our first human ancestors appeared twenty-five days ago, followed two weeks later by the first identifiable member of the genus *Homo.* About half an hour ago, modern humans began the long process we know as recorded civilization, and only one minute ago, the Industrial Revolution began changing the earth and our relationship to her for all time.

Table 2.1 The Geologic Time Scale

Eon	Era	Period	Epoch	Began Millions of Years Ago	Life-Forms/Events
Phanerozoic	Cenozoic	Quaternary	Holocene	0.01	Modern humans
			Pleistocene	1.6	Stone-age humans
					First humans
		Tertiary	Pliocene	5.3	
			Miocene	23.7	
			Oligocene	36.6	Flowering plants
			Eocene	57.8	Mammals, birds, and insects dominant
			Paleocene	66.4	
	Mesozoic	Cretaceous		144	Last of dinosaurs; flowering plants begin
		Jurassic		208	Dinosaurs abundant; first birds
		Triassic		245	First mammals
					First dinosaurs
	Paleozoic	Permian		286	Age of reptiles
		Carboniferous		360	Age of amphibians; first reptiles
		Devonian		408	First seed plants
					Age of fish
		Silurian		438	First land plants
		Ordovician		505	Marine algae; vertebrate fish
		Cambrian		570	Primitive marine algae and invertebrates
Proterozoic	Precambrian			2500	Earliest bacteria and algae
Archean				3900	Oldest surface rocks
Hadean				4600	Oldest meteorites
					Formation of the earth (assumed)

How were the divisions of geologic time established before age-dating of rocks was possible?

What is the significance of the Industrial Revolution occurring "one minute ago" in the history of the earth?

2.4 THE SHAPE OF THE EARTH

As the earth cooled and turned on its axis, gravity and the forces of rotation produced its nearly spherical shape. The earth sphere has a **mean,** or average, radius of 6371 kilometers (3959 mi). It has a shorter polar radius (6356.9 km 3950 mi) and a longer equatorial radius (6378.4 km; 3963 mi), a difference of 21.5 kilometers or 13 miles (fig. 2.3). As the earth spins it tends to bulge at the equator, and because the landmasses are presently concentrated in the middle region of the Northern Hemisphere and centered on the South Pole, the earth's surface is depressed slightly in these areas while it is elevated at the North Pole and

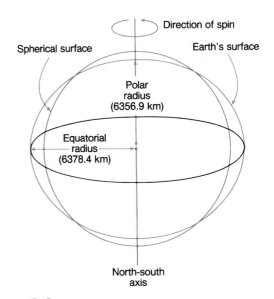

FIGURE 2.3

The rotation of the earth on its axis stretches the equator. Notice that the equatorial radius is larger than the polar radius.

Introduction to Earth

in the middle region of the Southern Hemisphere. This land distribution gives the earth a very slight pear shape, about 15 meters (50 ft) between elevations and depressions.

The top of Mount Everest, the earth's highest mountain, is about 8840 meters (29,000 ft) above sea level, while the deepest ocean depth, the bottom of the Challenger Deep in the Mariana Trench of the Pacific Ocean, is about 11,000 meters (36,000 ft) deep. On a scale model of the earth with a radius of 50 centimeters (19.7 in), Mount Everest would be about 0.07 centimeters (0.027 in) high and the Challenger Deep would be about 0.086 centimeters (0.034 in) in depth. The surface of such a model would feel quite smooth, like the surface of a basketball. The topography of the surface of the earth, its high mountains and its deep oceans, are minor compared to the size of the entire planet.

What is the shape of the planet Earth?

Relate the earth's surface heights and depths to its size.

2.5 LATITUDE AND LONGITUDE

Whether one is an oceanographer, a geographer, an airplane pilot, or a ship's captain, it is necessary to fix one's location on the earth's surface with precision and accuracy. To do so, a grid of reference lines that cross at right angles over the earth's surface is used. These grid lines are determined by internal earth angle measurements and are called lines of **latitude** and **longitude** (see Appendix A). Lines of latitude, also called parallels, begin at the equator, 0° latitude. Other latitude lines encircle the earth parallel to the equator, north to 90°N or the North Pole, and south to 90°S, the South Pole (fig. 2.4). All parallels of latitude must be designated as either north or south of the equator. Since the distance expressed in whole degrees is large (1 degree of latitude equals about 111 kilometers or 60 nautical miles at the earth's surface), each degree is divided into 60 minutes of arc and each minute into 60 seconds of arc.

Lines of longitude, also called meridians, are formed at right angles to the latitude grid and begin at an arbitrarily chosen line; 0° longitude, the **prime meridian,** is a line extending from the North Pole to the South Pole and passing directly through the Royal Naval Observatory in Greenwich, England (fig. 2.5). Meridians of longitude extend east and west of 0° and meet on the other side of the earth at 180°E or W longitude (fig. 2.6). The 180° longitude line approximates the **international date line.** Because longitude lines rotate with the turning earth, it is necessary to know the position of the sun or the stars with time to determine longitude. The reference time used is clock time, set to 12 noon when the sun is directly above 0° longitude. This is **Greenwich mean time (GMT),** now called **Universal Time** or ZULU time. Since sun time changes by one hour for each 15° of longitude, the earth has been divided into 24 time zones that are 15°

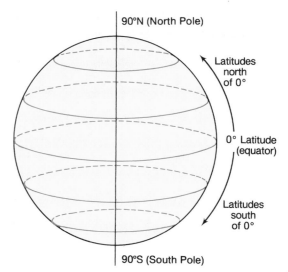

FIGURE 2.4

Latitude lines are drawn parallel to the equatorial plane.

FIGURE 2.5

The Royal Naval Observatory at Greenwich, England. The brass strip set into the courtyard marks the prime meridian, the division between east and west longitudes.

of longitude wide (fig. 2.7). Longitude is determined by comparing local sun time and Universal Time. If the sun appears directly overhead at 1000 UT, the observation location is 30°E of the prime meridian.

2.6 CHART PROJECTIONS

Maps and charts show the earth's three-dimensional surface on a flat or two-dimensional surface. Maps usually show the earth's land features, while charts depict the sea and sky. Making flat maps or charts from a curved earth surface produces distortion, and the user must select the most accurate, most convenient, and least distorted type for his or her purpose.

In order to understand this problem of distortion, imagine a transparent globe with the continents and the latitude and longitude lines painted on its surface. Place a light in the center of the globe and let the light rays shine through it. Hold a piece of paper up to the outside of the globe. The light will project the shadows of the continents and the latitude and longitude lines onto the paper forming a map or chart **projection.** Different projections are obtained by varying the position of the light and the shape of the surface on which the projection is made. Most projections are modifications of three basic types: cylindrical, conic, and tangent plane, all shown in figure 2.8.

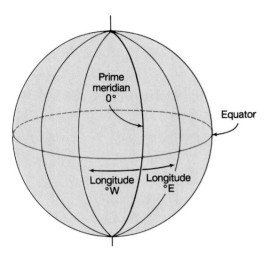

FIGURE 2.6
Longitude lines are drawn east and west of the prime meridian.

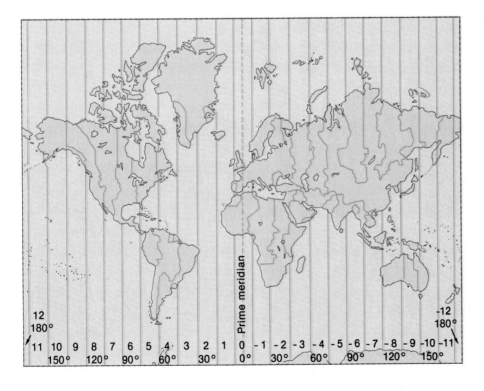

FIGURE 2.7
Distribution of the world's time zones. Degrees of longitude are marked along the bottom of the diagram. Time zones are positive (west zones) or negative (east zones). Time at Greenwich is determined by adding the zone number to the local time.

FIGURE 2.8

The three basic types of map projections: (a) An equatorial cylindrical projection. (b) A simple polar conic projection. (c) A polar tangent plane projection.

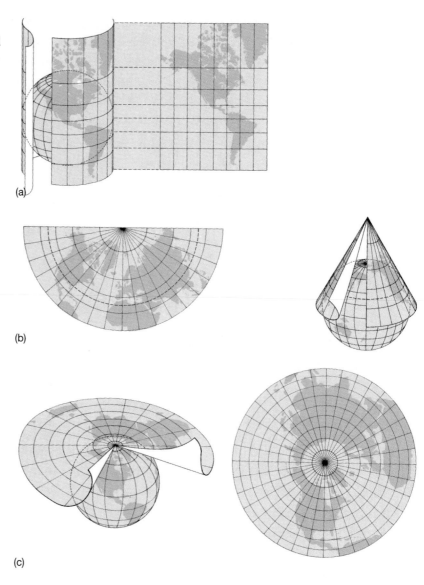

FIGURE 2.9

A bathymetric, or contour, chart of the sea floor along a section of generalized coast. Changes in the pattern and the spacing between contour lines indicate changes in depth.

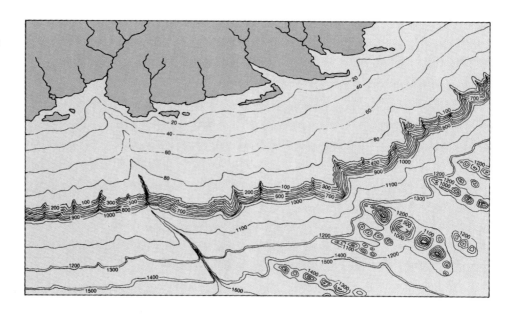

FIGURE 2.10
A physiographic map of the same area shown in figure 2.9.

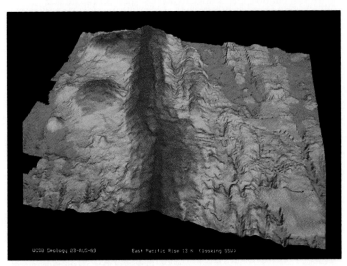

FIGURE 2.11
Three-dimensional computer processed image of a section of the East Pacific Rise. Data for this image were obtained by using the Sea Beam echo sounder.

Compare the three parts of figure 2.8 and notice that distortion on a map or chart increases as the distance from its place of contact to the globe increases. Consider Greenland. In figure 2.8c its size and shape are very close to its true form on earth. In figure 2.8b the island has grown larger, and in figure 2.8a both its size and shape are greatly distorted. Notice that each type of chart or map has its own characteristics.

Charts of the ocean that show lines connecting points of similar depth below the sea surface are **bathymetric** charts (fig. 2.9). The connecting lines are known as **contours** of depth or elevation. Color, shading, and perspective drawing may be added to indicate elevation changes and produce a **physiographic** map. Compare the physiographic map in figure 2.10 with the bathymetric chart in figure 2.9. Today computers use electronic ocean depth measurements to form detailed bathymetric charts that are converted into three-dimensional, color images of the sea floor from any angle (fig. 2.11).

Which of the projections shown in figure 2.9 would provide the most accurate map of a. Alaska? b. Central America? c. Northern Europe? d. Cape Horn?

How are bathymetric charts and physiographic maps similar? How do they differ?

2.7 NATURAL CYCLES

Natural cycles affect the earth's atmosphere, its water, and its land. An understanding of these cycles helps us understand processes that occur at the ocean surface and are discussed in later chapters: climate zones, winds, currents, vertical water motion, plant life, and animal migration. The average time for the earth to make one rotation relative to the sun is one day or twenty-four hours. The earth completes one orbit about the sun in about 365 and one-fourth days or one year. For convenience, the year is set at 365 days with an extra day added every four years, except years ending in hundreds and not divisible by 400.

As the earth follows its nearly circular orbit around the sun, those who live in temperate zones and polar zones are very conscious of the seasons and of the differences in the lengths of the periods of daylight and darkness. The reason for these seasonal changes is seen in figure 2.12; the earth moves with its axis tilted 23½° from the vertical to its orbit. Therefore, during the year the earth's North Pole is sometimes tilting toward the sun and sometimes tilting away from it. The

Introduction to Earth

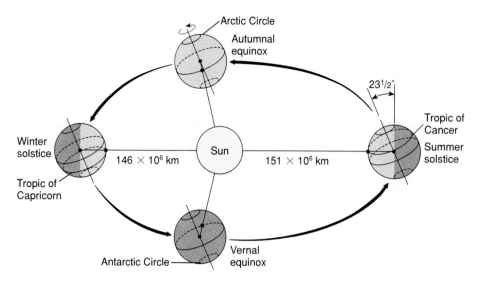

FIGURE 2.12

The earth's yearly seasons. The fixed orientation of the earth's axis during its orbit of the sun causes different portions of the earth to remain in shadow at different seasons. The mean distance between earth and sun is 149×10^6 km, but note the change in this distance during the year.

Northern Hemisphere receives its maximum hours of sunlight when the North Pole is tilted toward the sun; this is the Northern Hemisphere's summer. During the same period the South Pole is tilted away from the sun, so that the Southern Hemisphere receives the least sunlight; this period is winter in the Southern Hemisphere. At the other side of the earth's orbit, the North Pole is tilted away from the sun, creating the Northern Hemisphere's winter, and the South Pole is inclined toward the sun, creating the Southern Hemisphere's summer. Note also that during summer in the Northern Hemisphere it is always light around the North Pole and always dark around the South Pole; the opposite is true during the Northern Hemisphere's winter.

The periods of daylight in the Northern Hemisphere increase as the sun moves north to stand above 23½°N latitude, the **Tropic of Cancer.** This occurs at the **summer solstice** on or about June 22, the day of the year with the longest period of daylight in the Northern Hemisphere; on this day the sun does not sink below the horizon above 66½°N latitude, the **Arctic Circle,** nor does it rise above 66½°S latitude, the **Antarctic Circle.** Then the sun appears to move southward until on or about September 23, the **autumnal equinox,** it stands directly above the equator. On this day the periods of daylight and darkness are equal all over the world. The sun continues its southward movement until about December 21, when it stands over 23½°S latitude, the **Tropic of Capricorn,** at the **winter solstice.** On this day the daylight period is the shortest in the Northern Hemisphere, and above the Arctic Circle the sun does not rise. In the Southern Hemisphere the daylight is longest, and above the Antarctic Circle the sun does not set. The sun then begins to move northward, and about March 21, the **vernal equinox,** it stands again above the equator, and the periods of daylight and darkness are once more equal around the world.

The greatest annual variation in intensity of direct solar illumination occurs in the **temperate** zones. In the polar regions, the seasons are dominated by the long periods of light and dark, but the intensity of solar heating is small, as the sun is always low on the horizon. Between the Tropics of Cancer and Capricorn, there is little seasonal change in solar radiation levels, as the sun never moves beyond these boundaries.

Notice on figure 2.12 that the sun is not in the center of the path traced by the earth in its orbit. During the Northern Hemisphere's summer, the earth is farther away from the sun (151×10^6 km or 91×10^6 mi) than it is in winter (146×10^6 km or 88×10^6 mi).

Diagram the earth's orbit of the sun, and explain why the seasons change during the orbit.

What is the latitude of the Tropic of Cancer, Tropic of Capricorn, Arctic Circle, and Antarctic Circle?

Why are the Arctic and Antarctic Circles displaced from the poles by 23½°?

How will the seasons change over a calendar year at each of these latitudes?

2.8   THE HYDROLOGIC CYCLE

The earth's water is found as a liquid in the oceans, rivers, lakes, and below the ground surface; it occurs as a solid in glaciers, snow packs, and sea ice; it takes the form of droplets and gaseous water vapor in the atmosphere. The places in which water resides are called **reservoirs,** and each type of reservoir, when averaged over the entire earth, contains a fixed amount

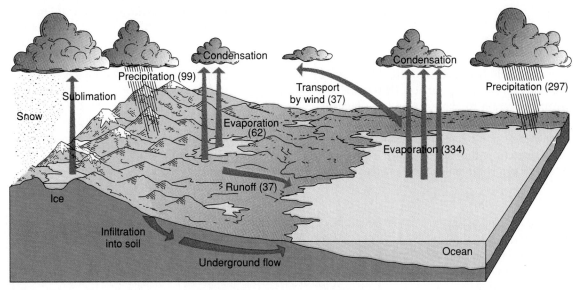

FIGURE 2.13

The hydrologic cycle and annual transfer rates for the whole earth. Snow and rain together equal precipitation. Sublimation, the direct transfer of ice to water vapor, is included under evaporation. Surface flow and underground flow are considered as land runoff. Annual transfer rates are in thousands of cubic kilometers (10^3 km^3).

of water at any one instant. But water is constantly moving from one reservoir to another, as liquid water evaporates from the oceans, rains fall on the land, and rivers flow back to the sea. This movement of water through the reservoirs, diagrammed in figure 2.13, is called the **hydrologic cycle.**

Evaporation takes water from the surface of the ocean into the atmosphere. Most of this water returns directly to the sea, but air currents carry some water vapor over the land. Precipitation transfers this water to the land surface, where it percolates into the soil, fills rivers, streams, and lakes, or remains for longer periods as snow and ice in some areas. Melting snow and ice, rivers, groundwater, and land runoff move the water back to the oceans to complete the cycle and maintain the oceans' volume. For a comparison of the water volume stored in the earth's reservoirs, see table 2.2.

The properties of climate zones are principally determined by their surface temperature (mean earth surface temperature is 16°C) and their evaporation-precipitation patterns: the moist, hot equatorial regions, the dry, hot subtropic deserts, the cool, moist temperate areas, and the cold, dry polar zones. Differences in these properties, coupled with the movement of air between the climate zones, moves water through the hydrologic cycle from one reservoir to another at different rates. The transfer of water between the atmosphere and the oceans alters the salt content of the oceans' surface water and, with the seasonal latitude changes in surface temperature, determines many of the characteristics of the world's oceans, which will be explored in Chapters 6 and 7.

Table 2.2 The Earth's Water Supply

Reservoir	Volume (km³)	Percent of Total Volume
Atmospheric moisture expressed as water	15×10^3	0.001
Rivers and lakes	510×10^3	0.036
Groundwater	5100×10^3	0.365
Glacial and other land ice	$22,950 \times 10^3$	1.641
Oceanic water and sea ice	$1,370,323 \times 10^3$	97.957
Total	$1,398,898 \times 10^3$	100

Use the annual transfer amounts in figure 2.13 to show that ocean volume does not change during the year.

Why does the whole earth cycle shown in figure 2.13 not apply to a specific region of the earth?

What is the relationship between the hydrologic cycle and climate zones?

2.9 DISTRIBUTION OF LAND AND WATER

The oceans cover 361 million square kilometers (361×10^6 km^2 or 139×10^6 mi^2) of the earth's surface. (If you are unfamiliar with scientific notation to express very large numbers, see Appendix C.) Because these numbers are so large, they do not convey a clear idea of size; therefore, an easier concept to remember is that 71 percent of the earth's surface is covered by the oceans, and only 29 percent is land above sea level.

The volume of water in the oceans is enormous: 1.37 billion cubic kilometers (1.37×10^9 km^3). A cubic kilometer of seawater is very large indeed. Consider this: the largest building in the world in cubic capacity is the main assembly plant of the Boeing Company in Everett, Washington. This building is used for the manufacture of Boeing airplanes and has a cubic capacity of 0.0847 km^3; in other words, 11.8 of these buildings would fit into one cubic kilometer. Another way to express the oceanic volume is to think of a smooth sphere with exactly the same surface area as the earth uniformly covered with the water from the oceans. The ocean water volume divided by earth's

FIGURE 2.14

Continents and oceans are not distributed uniformly over the earth. (a) The Northern Hemisphere contains most of the land; (b) the Southern Hemisphere is mainly water.

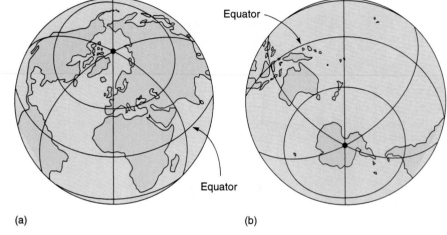

(a)   (b)

FIGURE 2.15

Distribution of land and ocean by latitude. In the Northern Hemisphere, middle latitude areas of land and ocean are nearly equal. Land is almost absent at the same latitudes in the Southern Hemisphere. The areas are calculated on the basis of 5° latitude intervals.

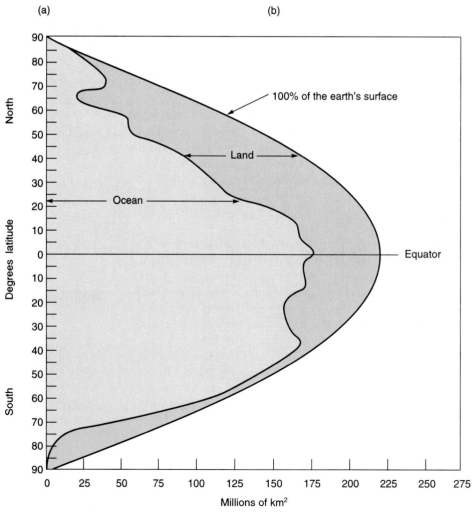

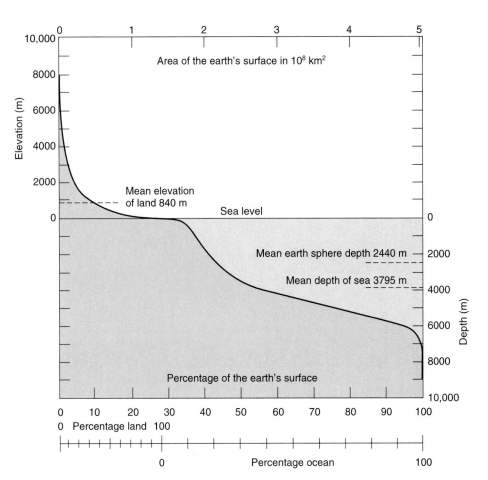

area, would be 2680 meters (8800 ft) deep. If the water from all other sources in the world were added, the depth would rise to 2743 meters (9000 ft).

To understand the distribution of land and water over the earth consider the earth when it is viewed from the north (fig. 2.14a) and from the south (fig. 2.14b). About 70 percent of the earth's landmasses are in the Northern Hemisphere, and most of the land lies in the middle latitudes. The Southern Hemisphere is the water hemisphere, with its land located mostly in the tropic latitudes and in the polar region. The details of this distribution are presented in figure 2.15.

Another method used by oceanographers to depict land-water relationships is shown in figure 2.16. This graph of depth or elevation versus area is called a **hypsographic curve.** Since volume is the product of height and area, the hypsographic curve can also be used to show both the volume of land above sea level and the volume of the oceans below sea level.

> How much of the earth's surface is land, and how much is covered by the oceans?
>
> How do the land-water distributions of the Northern and Southern Hemispheres differ?
>
> What does the hypsographic curve show us about the relationship of the earth's land and water?

2.10   THE OCEANS

Oceanographers view the world ocean as three fingers—the Atlantic, Pacific, and Indian Oceans—stretching up from a common source around Antarctica. The Arctic Ocean is considered an extension of the North Atlantic, and the Antarctic Ocean may be considered as the circumpolar Southern Ocean at latitudes greater than 50°S, or it may be divided along chosen lines of longitude to become the southern portions of the Atlantic, Pacific, and Indian Oceans (fig. 2.17). Each of these three oceans has its own characteristic surface area, volume, and mean depth. The Pacific Ocean has more surface area, a larger volume, and a greater mean depth than either the Atlantic or the Indian Oceans. The Atlantic Ocean is the shallowest ocean and has the greatest number of shallow adjacent seas, such as the Arctic Ocean, Gulf of Mexico, Caribbean Sea, and Mediterranean Sea. The Indian Ocean is a Southern Hemisphere ocean; it is the smallest in terms of area, but it is quite deep. The volume of the Pacific Ocean is over twice that of either the Atlantic Ocean or the Indian Ocean. Use table 2.3 to help compare the three oceans.

> Name the oceans of the world. Which has the greatest surface area? Which has the greatest volume of water below 5000 meters? Which has the least total volume of water? Which has the greatest volume of water between the surface and 1000 meters?

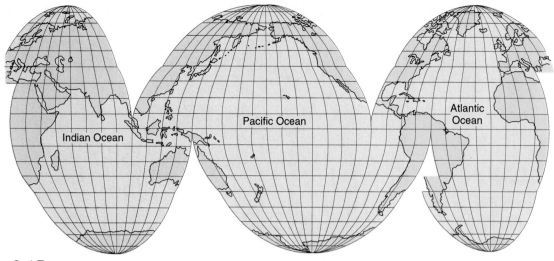

FIGURE 2.17
The world's three major oceans on an equal-area map projection.

Table 2.3 Ocean Areas and Volume versus Depth

Depth (m)	Atlantic (mean depth 3332 m)		Pacific (mean depth 4028 m)		Indian (mean depth 3897 m)		All Oceans (mean depth 3795 m)	
	Area[1]	Volume [2]	Area	Volume	Area	Volume	Area	Volume
0	106.4	354.7	179.7	723.7	74.9	291.9	361.0	1370.3
1000	84.7	259.1	164.0	550.5	69.4	219.7	318.1	1029.3
2000	79.1	177.2	156.9	388.8	66.9	151.6	302.9	717.6
3000	69.7	102.8	147.5	236.6	61.3	87.5	278.5	426.9
4000	50.0	43.0	114.3	105.7	43.4	35.1	207.7	183.8
5000	22.6	6.7	51.0	23.1	14.8	6.0	88.4	35.8
6000	0.64	0	3.2	0.2	0.3	0.1	4.14	0.3
7000	0	0	0.36	0	0	0	0.36	0

[1]All areas are given in 10^6 km^2.

[2]All volumes are given in 10^6 km^3.

2.11 MODERN NAVIGATION

To determine positions at sea, today's navigators may still wait for clear skies to "shoot" the sun or stars with a sextant, but such measurements are primarily used to check their modern electronic navigational equipment. The three main systems in use by any vessel, oceanographic, commercial, or military, are **radar, loran,** and the **satellite navigation systems.** When vessels are near land, radar (radio detection and ranging) bounces radio pulses off a target (the shoreline or another vessel). These reflected pulses indicate the position of the target relative to the sending vessel. The travel time of the pulse is converted electronically to create an image on a display screen. Loran (long-range navigation) measures the difference in arrival time of radio signals from pairs of stations. The position of the re-

ceiving ship is plotted on a chart that shows the time lines for these stations. Recent improvements in loran include receivers with computers that can be programmed with the latitude and longitude of the desired destination. The loran receiver monitors the signals from the stations and directly reads out the course to be sailed and the distance to the destination. The computer can also monitor the signals and continuously calculate the latitude and longitude, allowing the navigator to know the ship's position at all times. The most accurate and sophisticated navigational aids available today use satellite navigation. Satellites orbiting the earth emit signals which are picked up by a receiver on the ship. The ship's receiver monitors the signal and determines the position of the vessel relative to the satellites. The computer determines the ship's position to within 30 meters (100 ft).

The U.S. Navstar Global Positioning System (GPS) consists of satellites that emit continuous radio signals and pinpoint the position of GPS receivers within a few feet, including latitude, longitude, and altitude. Using the system to determine the distance between receiving stations produces data that can determine earth positions within a few centimeters. The system is able to map the land surface, storms, and sea surface features across the oceans. When GPS is tied into radio telescope networks receiving signals from space, very accurate positioning of the satellites will be possible. Once these systems are combined, the GPS satellites will be able to sense variations in the earth's surface elevations associated with storms, tides, currents, earth movements, and changes in glacial ice volume and sea level.

Shipboard computers can now store an electronic atlas that includes surface charts and seafloor bathymetry. Oceanographic vessels use devices that draw charts showing the vessel's changing position and, at the same time, conduct a seafloor bathymetric survey and keep track of water measurements (temperature and salt content) automatically as the ship moves along. Today's ocean scientists are able to return more and more accurately to the same places in the sea to repeat measurements and follow changes in ocean processes.

> Why is electronic navigation more important to oceanographers than celestial navigation?
>
> How do radar, loran, and satellite navigation systems differ from each other?

SUMMARY

The solar system began as a rotating cloud of gas. Over millions of years, the earth heated, cooled, changed, and collected a gaseous atmosphere and an accumulation of liquid water.

Because it rotates, the earth's shape is not perfectly symmetrical. Its surface is relatively smooth when compared to its size.

Reliable age dates for rocks are obtained by radiometric dating. The accepted age of the earth is about 4.6 billion years. The geologic time scale is used to compile the earth's history.

Latitude and longitude are used to locate positions on the earth's surface. To determine longitudinal position requires accurate time measurement. Different types of map and chart projections have been developed. Bathymetric charts and physiographic maps use depth contours and elevation to depict the earth's surface.

Natural cycles are based on the motions of the earth and the sun. Because of the tilt of the earth's axis as it orbits the sun, the sun moves annually between 23½°N and 23½°S, producing the seasons.

There is a fixed amount of water on earth. Evaporation and precipitation move the water through the reservoirs of the hydrologic cycle. Seventy-one percent of the earth's surface is covered by oceans. The Northern Hemisphere is the land hemisphere, and the Southern Hemisphere is the water hemisphere. The hypsographic curve is used to show land-water relationships of depth, elevation, area, and volume. The earth has three large oceans extending north from the Southern Ocean. Each has a characteristic surface area, volume, and mean depth.

Modern navigational techniques make use of radar, radio signals, computers, and satellite systems. The GPS allows more accurate position readings and maps land and ocean surface features.

KEY TERMS

radiometric dating	Greenwich mean time	summer solstice	reservoir
isotope	(GMT)	Arctic Circle	hydrologic cycle
half-life	Universal Time	Antarctic Circle	hypsographic curve
mean	projection	autumnal equinox	radar
latitude	bathymetric	Tropic of Capricorn	loran
longitude	contours	winter solstice	satellite navigation systems
prime meridian	physiographic	vernal equinox	
international date line	Tropic of Cancer	temperate	

SUGGESTED READINGS

Bowditch, N. 1984. *American Practical Navigator*, vol. 1. U.S. Defense Mapping Agency Hydrographic Center, Washington D.C., 1414 pp. History of navigation, chart projections, and navigation aids are covered in chapters 1–5 and 41–46.

Gore, R. 1985. The Planets, Between Fire and Ice. *National Geographic* 167(1):4–51.

Livermore, B. 1993. Bottoms Up. *Sea Frontiers* 39(3):40–45. Satellite images and seafloor mapping.

Maranto, G. 1991. Way Above Sea Level. *Sea Frontiers* 37(4):16–23.

CHAPTER

3

Plate Tectonics

Outline

Learning Objectives

After reading this chapter, you should be able to

- Describe the interior of the earth, and explain how this information has been obtained.

- Explain the relationship between the lithosphere and the asthenosphere.

- Trace the development of the theory of drifting continents.

- Relate mantle convection cells to seafloor spreading.

- List the major discoveries of the 1960s that provided the evidence for seafloor spreading.

- Explain the relationship between seafloor spreading and magnetic reversals.

- Define plate tectonics, and outline the main lithospheric plates.

- Describe the processes that occur at plate boundaries.

- Relate the direction and rate of motion of lithospheric plates.

- Describe a hot spot and its results on moving plates.

- Trace the breakup of Pangaea.

- Explain how deep-sea drilling and the use of submersibles have changed our ideas of the deep sea floor.

◀ The San Andreas fault crossing the Carrizo Plain in California.

Many features of our planet have presented scientists with contradictions and puzzles. The remains of warm-water coral reefs are found off the coast of the British Isles; marine fossils occur high in the Alps and the Himalayas; and coal deposits that were formed in warm tropical climates are found in northern Europe, Siberia, and northeast North America. Great mountain ranges divide the oceans, volcanoes border the coasts of the Pacific Ocean, and deep-ocean trenches are found adjacent to long island arcs. No single coherent theory explained all these features until the new technology and new scientific discoveries of the 1950s and early 1960s combined to trigger a complete reexamination of the earth's history.

3.1 THE INTERIOR OF THE EARTH

Although we cannot observe the interior of the planet, scientists have been able to learn a great deal about the structure and composition of the earth's interior using indirect methods. Because the earth wobbles only very slightly as it rotates, and because at its surface gravity is nearly constant, the earth must have its mass distributed uniformly about its center, as a series of concentric layers. It is possible to calculate the internal pressures and temperatures that can be reached under these pressures, and because there is a magnetic field around the earth, the earth's core must include materials that produce magnetic fields. Radiometric dating of meteorites gives a maximum age of 4.6 billion years, the same as the age of the solar system. These fragments of unknown planets allow us to directly analyze the density, chemistry, and mineralogy of the nickel-iron cores and stony shells of bodies that we believe to have a similar composition to that of the earth.

If the earth were cut and a piece removed, the layers would appear as in figure 3.1. At the center there is a solid **inner core,** surrounded by a thick layer of liquid material forming the **outer core.** The next layer, the **mantle,** contains the largest mass of material of any of the layers, and about 70 percent of the earth's volume. A thin layer of the outermost mantle is thought to be rigid; it is underlain by a mantle layer that is deformable and flows slowly over the deeper, more rigid inner mantle. The earth's outermost layer is the cold, rigid, rocky surface, the **crust.** See table 3.1 for a comparison of these features and their properties. This table introduces **density,** a measure of mass per unit volume usually given in grams per cubic centimeter, written g/cm³, to further compare the layers.

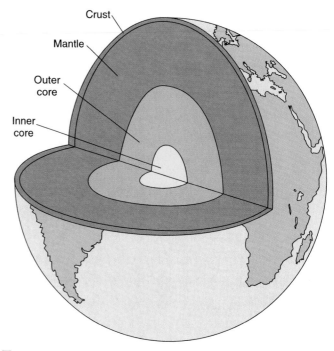

FIGURE 3.1
The layered structure of the earth.

Table 3.1	Layers of the Earth					
Layer	Depth (km)	Thickness Average (km)	State	Composition	Density Average (g/cm³)	Temperature (°C)
Crust						
continental	0–65	45	Solid	Silicates rich in magnesium and aluminum	2.67	−40–1000
oceanic	0–10	7	Solid	Silicates rich in magnesium and iron	3.0	0–1100
Mantle	45–2900	2855	Solid and mobile	Silicates rich in magnesium and iron	4.5	1100–5000
Outer Core	2900–5300	2400	Liquid	Iron, nickel	11.5	5000–6000
Inner Core	5300–6370	1070	Solid	Iron, nickel	13.0	6000–6600

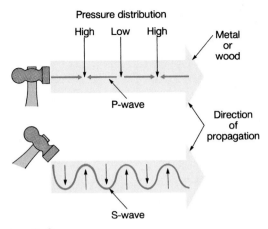

FIGURE 3.2

Waves passing through a solid are of two types: pressure waves, or P-waves (top), and shear waves, or S-waves (bottom).

Seismic waves or underground vibrations produced by earthquakes, volcanic eruptions, or underground explosions are used to sense the structure of the earth. All over the surface of the earth, geologists and geophysicists monitor recording stations that measure the type, strength, and arrival time of seismic wave vibrations. These data are used to precisely locate the origin of each disturbance and to aid scientists in understanding the interior of the earth.

Two types of seismic waves travel through the earth's interior. These are pressure waves, or **P-waves,** which oscillate in the same direction as they move, and shear waves, or **S-waves,** which oscillate at right angles to their direction of movement. These waves are shown in figure 3.2. P-waves are able to move through solids and liquids. S-waves cannot pass through a liquid, so they cannot pass through the outer core.

As the seismic waves move through one earth layer and into another, their speed of travel changes and the waves bend, as shown in figure 3.3. Earth surface measurements of travel time, wave direction, and shadow zones where waves are not detected confirm that there are abrupt changes of speed and direction at certain depths, indicating a layered interior. The dimensions, structure, and physical properties of each of the internal layers are determined from the seismic evidence. If the interior of the earth had uniform properties, the waves would follow straight lines and their speed would not change, as shown in figure 3.3c.

The data from ever more densely spaced and sophisticated seismic recording stations and the computer capacity to analyze the travel times of thousands of seismic waves from earthquakes have been used to produce three-dimensional maps of the boundary between the core and the mantle. The abundance of new data indicates that the outer core is not smooth, but has major peaks and valleys over its surface. At this time, it is thought that the regions above a peak are areas where the mantle has excess heat and mantle material rises toward the crust, drawing the liquid core upward. Cooler, more viscous mantle material sinks to cause depressions in the core's surface. In addition, three-dimensional images indicate that there may be slabs of crust and rigid upper mantle material that have sunk, unmelted, deep into the mantle.

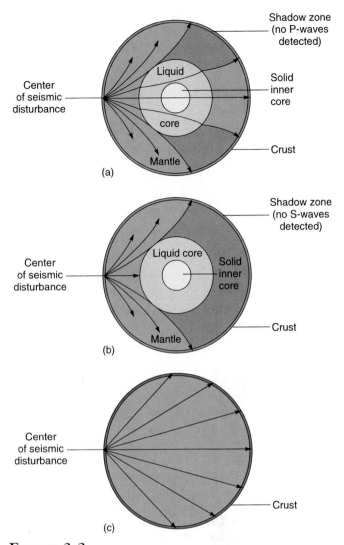

FIGURE 3.3

Movement of seismic waves through the earth. (a) Refraction of P-waves and shadow zones produced by the earth's interior structure. (b) Refraction of S-waves and shadow zones produced by the earth's interior structure. (c) No refraction and no shadow zones occur in an earth with a uniform structure.

Describe the internal structure of the earth.

How are S-waves and P-waves the same; how are they different?

What does the refraction of S-waves and P-waves tell us about the interior of the earth?

How have new data modified our concept of the core and mantle?

3.2 THE LITHOSPHERE

The large continental landmasses are formed primarily from **granite**-type rock, which has a high content of aluminum, magnesium, potassium, and silica. The crustal rock layer lying under the ocean is a denser **basalt**-type rock, which is low in

Plate Tectonics

FIGURE 3.4

The lithosphere is formed from the fusion of crust and upper mantle. Notice that the Moho is relatively close to the earth's surface under the ocean's basaltic crust but is depressed with the mantle under the granitic continents.

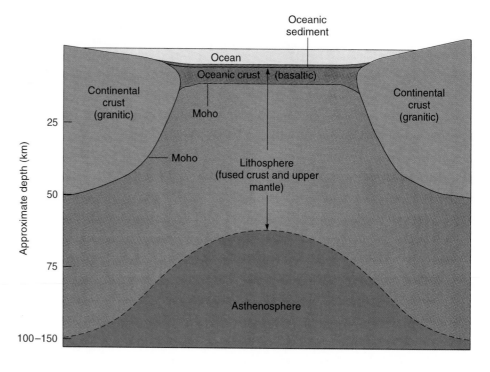

silica and high in iron, magnesium, and calcium. The rock of the continental crust is highly variable in composition and distribution. On the average, this continental rock is less dense than the more uniform crustal material of the sea floor.

Use figure 3.4 as you read the following. The boundary between the crust (both continental and oceanic) and the mantle is the **Moho,** at which there is a sudden change in the speed of seismic waves. Once some thought that the crust moved relative to the mantle at the Moho. Current research places the zone of movement within the upper mantle, from about 70 kilometers (40 mi) below the ocean crust to about 150 kilometers (90 mi) below the continental crust. The mantle just below the crust is considered to be fused to the crust at the Moho. This rigid layer of crust and upper mantle is the **lithosphere.** Seismic waves pass through it at high speeds, implying that the lithosphere is both strong and rigid. The subregion of the mantle extending about 250 kilometers (150 mi) below the lithosphere is the **asthenosphere;** this region of the mantle passes seismic waves more slowly and is considered to be partly melted, flowing slowly when stressed. The lithosphere, made up of rigid crust and upper mantle, rides on the asthenosphere but has a density close to it. As oceanic lithosphere cools, it thickens, and its density increases as it ages. This denser oceanic lithosphere can fracture and sink as rigid slabs into the asthenosphere. The continental lithosphere remains less dense than the asthenosphere and continues at the surface.

How does continental rock differ from seafloor rock?

Distinguish between the lithosphere and the crust; between the asthenosphere and the mantle.

What and where is the Moho?

3.3 HISTORY OF A THEORY

The shapes of the continents on either side of the Atlantic Ocean have intrigued observers for a long time. The possible "fit" of the bulge of South America into the bight of Africa was noted by the English scholar Francis Bacon (1561–1626), the French naturalist Georges de Buffon (1707–88), the German scientist and explorer Alexander von Humboldt (1769–1859), and others in later years. As scientists studied the earth's crust, patterns of rock formation, fossil distribution, and mountain range placement, they recognized even greater similarities between continents. In a series of volumes published between 1885 and 1909, an Austrian geologist, Eduard Suess, proposed that the southern continents had once been joined into a single continent he called **Gondwanaland.** He assumed that portions of the continents sank and created the oceans between the continents. At the beginning of this century, Alfred L. Wegener and Frank B. Taylor independently proposed that the continents were slowly moving about the earth's surface. Taylor soon lost interest, but Wegener, a German meteorologist, astronomer, and Arctic explorer, continued to pursue this concept until his death in 1930.

Wegener's theory, often called **continental drift,** proposed the existence of a single supercontinent he called **Pangaea** (fig. 3.5). He thought that forces arising from the rotation of the earth began Pangaea's breakup. First, the northern portion composed of North America and Eurasia, which he called **Laurasia,** separated from the southern portion formed from Africa, South America, India, Australia, and Antarctica, for which he retained the earlier name Gondwanaland. The continents as we know them today then gradually separated and moved to their present positions. Wegener based his ideas on the geographic fit of the continents, the scouring of landforms,

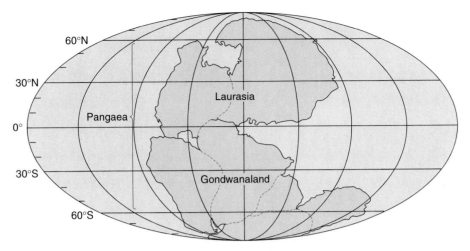

FIGURE 3.5

Pangaea 200 million years before the present. Wegener's supercontinent is composed of the two subcontinents, Laurasia and Gondwanaland.

and the way in which some of the older mountain ranges and rock formations appeared to relate to each other when the landmasses were assembled as Pangaea. The presence of ancient coral reefs at high latitudes and traces of climate changes in the fossil record also supported his idea. He noted that fossils more than 150 million years old collected on different continents were remarkably similar, implying the ability of land organisms to move freely from one landmass to another. Fossils from different places dated after this period showed quite different forms, suggesting that the continents and their evolving populations had separated from one another.

Wegener's theory provoked considerable debate in the decade of the 1920s, but most geologists agreed that it was not possible to move the continental rock masses through the rigid basaltic crust of the ocean basins. The theory was not regarded very seriously in the scientific community and became a footnote in geology textbooks.

> Explain the evidence used by Wegener to propose the theory of drifting continents.
>
> What modern landmasses make up Pangaea, Laurasia, and Gondwanaland?

3.4 EVIDENCE FOR A NEW THEORY

Armed with new sophisticated instruments and technologies developed during World War II, earth scientists returned to a study of the earth's crust in the 1950s. For the first time, scientists were able to examine the deep ocean floor in detail and explore the mid-ocean ridge system. For the first time, detailed seafloor maps, like that shown in figure 3.6, could be made and studied.

In the early 1960s, Harry H. Hess (1906–69) of Princeton University promoted the concept that deep within the earth's mantle there are currents of molten material heated by the earth's natural radioactivity. When these upward-moving mantle currents reach the lithosphere, they move along under it, cooling as they do so until they become cool enough and dense enough to sink down toward the core again. These patterns of moving material are called **convection cells** (fig. 3.7). The convection cells are generally believed to occur in the asthenosphere, although there is no definite evidence that the convection cells are confined to this layer.

Underwater vulcanism occurs when the upward-moving molten **magma** breaks through the seafloor lithosphere, and undersea mountain ranges or **ridges** form along cracks in the crust. As the magma oozes out, it becomes lava, cools, hardens, and is added to the earth's surface as new oceanic-type basaltic crust. If new crust is being produced in this manner, a mechanism is needed to remove old crust since there is no measurable change in the size of the earth. The deep, narrow, steep-sided **trenches** of the Pacific were proposed as areas where the older, cooler, and denser oceanic lithosphere depresses and fractures the sea floor as it sinks into the earth's interior. Figure 3.7 shows this process.

These mid-ocean ridges and deep trenches may be seen in figure 3.6. Follow the north-south ridge in the center of the Atlantic Ocean; it connects around Africa to the ridge in the central Indian Ocean. There is a ridge in the eastern South Pacific that can be followed south of Australia into the Indian Ocean and around South America's Cape Horn into the Atlantic. Conspicuous trenches may be seen along the west coast of South America, seaward of the Aleutian Islands and along the eastern coast of Asia into the western South Pacific.

Although the ascending magma breaks through the crust and solidifies, forcing the crust apart, most of the rising material is turned aside under the rigid lithosphere. This allows the lithosphere to slide away from the spreading center toward the descending sides of the convection cells. The lateral movement

FIGURE 3.6

A physiographic chart of the world's oceans.

World Ocean Floor map by Bruce C. Heezen and Marie Tharp, 1977. © 1977 by Marie Tharp. Reproduced by permission of Marie Tharp, 1 Washington Ave., South Nyack, NY 10960.

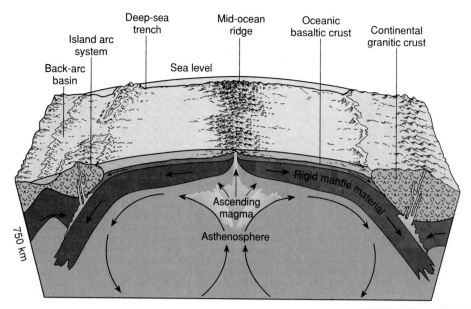

FIGURE 3.7
Seafloor spreading creates new crust at mid-ocean ridges and loses old crust in deep-sea trenches. This process is shown being driven by convection cells in the asthenosphere.

of the crust produces **seafloor spreading** (fig. 3.7). Areas in which new crust is formed above rising magma are **spreading centers;** areas of descending older crustal material are **subduction zones.** The seafloor spreading mechanism provides the forces causing continental drift, with the continents carried as passengers on the lithosphere, like boxes on a conveyor belt.

What causes convection cells in the earth's mantle?

How do convection cells cause seafloor spreading?

Contrast the crust's movement at spreading centers with its movement at subduction zones.

3.5 EVIDENCE FOR CRUSTAL MOTION

As oceanographers, geologists, and geophysicists explored the earth's crust on land and under the oceans, additional evidence to support the idea of seafloor spreading began to accumulate. Earthquakes were known to be distributed around the earth in narrow and distinct zones; see figure 3.8. These zones were found to correspond to the areas along the ridges (or spreading centers) and the trenches (or subduction zones). Analysis of these seismic records shows earthquakes at shallow depths along ridges and at trenches where the plates begin to slide below the crust. Earthquakes deeper than 100 kilometers (60 mi) are produced in subduction zones beneath the continents and island arcs.

Researchers sank probes into the sea floor to measure the heat from the interior of the earth moving through the crust. The measured heat flow shows a regular pattern, highest in the vicinity of the mid-ocean ridges where the crust is thinner and more recent, and decreasing as the distance from the ridge center increases (fig. 3.9). This heat-flow pattern indicates that mantle material is hotter under the ridge system, over the ascending portion of the mantle convection cell.

Cores were drilled down through the loose, particulate materials or **sediments** that cover the ocean bottom and into the ocean floor. (Sediments are discussed in Chapter 4.) This process required the development of a new technology and a new type of ship. The drilling ship, *Glomar Challenger*, was built for a series of studies beginning in 1968. The specialized bow and stern thrusters and propulsion system of a drilling ship respond automatically to computer-controlled navigation, using acoustic beacons on the sea floor. This allows the ship to remain for long periods of time in a nearly fixed position over a drill site in water too deep to anchor. A sonar guidance system enables it to replace drill bits and reenter the same bore holes in water about 6000 meters (20,000 ft) deep. This method is illustrated in figure 3.10.

The cores taken by the *Glomar Challenger* provided much of the data needed to establish the existence of seafloor spreading. No ocean crust older than 180 million years was found, and sediment age and thickness were shown to increase with distance from the ocean ridge system. Figure 3.11 shows that the sediments closest to the ridge system are thin over the new crust while older crust farther away from the ridge system is more deeply buried.

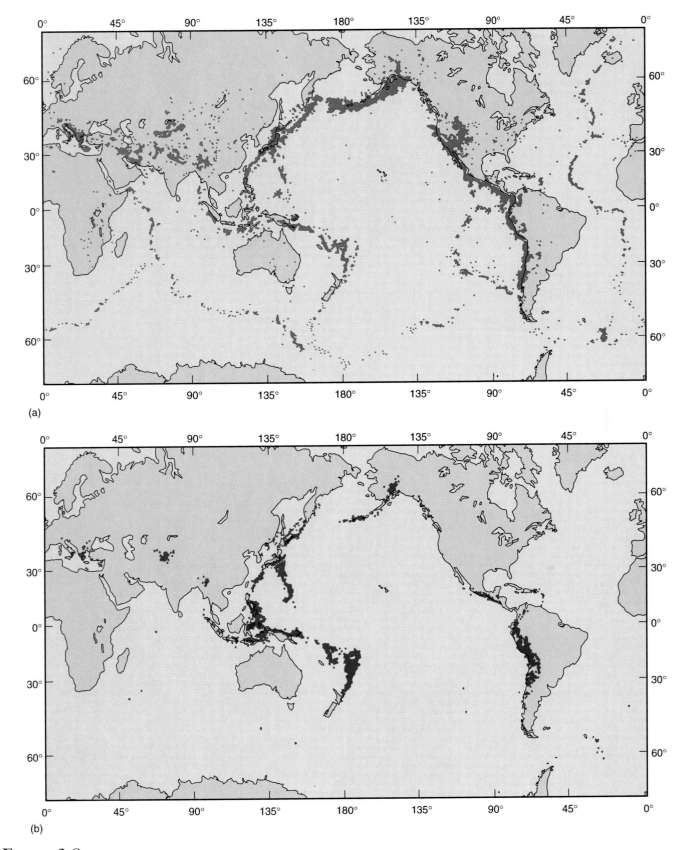

(a)

(b)

FIGURE 3.8

Earthquake epicenters, 1961–67. (a) Epicenters at depths less than 100 kilometers outline plate boundaries. (b) Epicenters at depths greater than 100 kilometers are related to subduction.

Plate Tectonics

FIGURE 3.9

Heat flow through the Pacific Ocean floor. Values are shown against age of crust and distance from the ridge crest.

Source: Data from J. G. Sclater and J. Crowe, "On the Variability of Oceanic Heat Flow Average" in *Journal of Geophysical Research* 81(17): 3004, June 10, 1970.

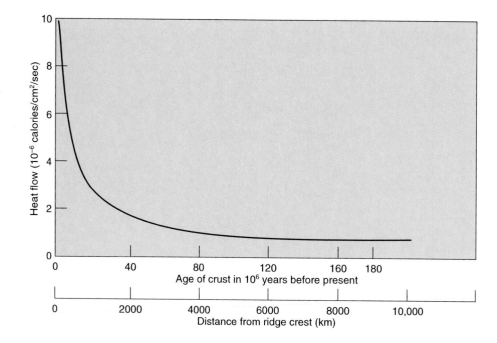

FIGURE 3.10

Deep-ocean drilling technique. Acoustical guidance systems are used to maneuver the drilling ship over the bore hole and to guide the drill string back into the bore hole.

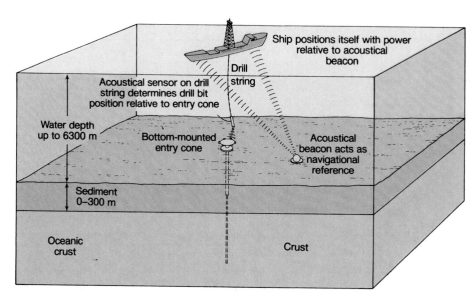

FIGURE 3.11

Age and thickness of seafloor sediments.

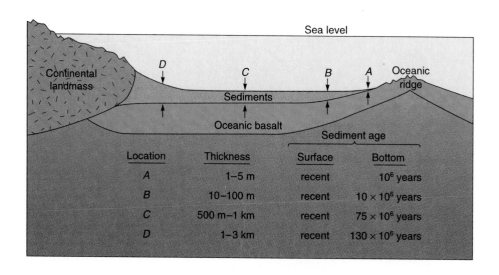

Location	Thickness	Sediment age	
		Surface	Bottom
A	1–5 m	recent	10^6 years
B	10–100 m	recent	10×10^6 years
C	500 m–1 km	recent	75×10^6 years
D	1–3 km	recent	130×10^6 years

Although each of these pieces of evidence fits the theory that the earth's crust produced at the ridge system is new and young, the most elegant proof for seafloor spreading came from a study of the magnetic evidence locked into the oceans' floors.

The earth's familiar north and south geographic poles at 90°N and 90°S latitude mark the axis about which it rotates. Like a great bar magnet, the earth also has a North Magnetic Pole located in the Hudson Bay area and a South Magnetic Pole opposite it in the South Pacific Ocean. Like any magnet, the earth is surrounded by a magnetic field. Its lines of force converge and dip toward the earth at the magnetic poles, as shown in figure 3.12.

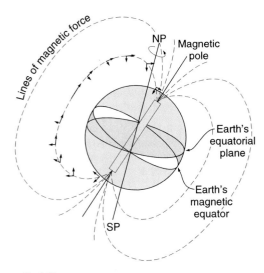

FIGURE 3.12

Lines of magnetic force surround the earth and converge at the magnetic poles. The lines of force have components that are both parallel and perpendicular to the earth's surface. A compass needle orients with the horizontal component to indicate the direction to the magnetic pole. NP and SP indicate the north and south geographic poles, respectively.

When materials that can be magnetized are heated in a magnetic field they become magnetized as they cool, with their north and south magnetic poles lined up with the poles of the magnetic field. In the same way, as molten volcanic material cools and solidifies, its iron-bearing minerals become magnetized and permanently aligned with both the horizontal and vertical components of the earth's magnetic field.

Research done on age-dated layers of volcanic rock found on land shows that the polarity or north-south orientation of the earth's magnetic field reverses for varying periods of geologic time. This means that at different times in the earth's history the present north and south magnetic poles have switched places. Each time a layer of volcanic material solidified, it trapped the magnetic orientation and polarity of the time period in which it cooled. Nearly 170 magnetic reversals have been identified during the last 76 million years; our present magnetic orientation has existed for 710,000 years. The cause of reversals is unknown, but is thought to be associated with changes in the motion of the magnetic material of the earth's liquid outer core. We do not know how long our present magnetic polarity will last.

When magnetometers were towed over the sea floor during the early 1960s, the resulting maps revealed a pattern of stripes that ran parallel with the mid-ocean ridge (fig. 3.13). Their significance was not understood until later in the decade, when Fred J. Vine (1940–) and Drummond H. Matthews (1931–) of Cambridge University proposed that these stripes represented a recording of the **polar reversals** of the earth's magnetic field, frozen into the sea floor. As the molten basalt rose along the crack of the ridge system and solidified, it locked in the direction of the prevailing magnetic field. Seafloor spreading moved this material off on either side of the ridge, to be replaced by more molten materials. Each time the earth's magnetic field reversed, the direction of the magnetic field was recorded in the new crust. Vine and Matthews proposed that if such were the case, there should be a symmetrical pattern of magnetic stripes centered at

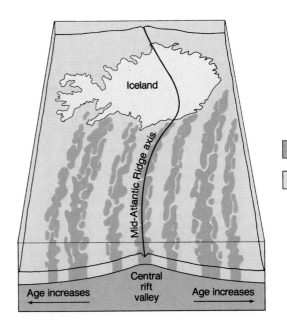

Stronger than regional average magnetic orientation

Weaker than average magnetic orientation (reversal)

FIGURE 3.13

Reversals in the earth's magnetic polarity cause the symmetrically striped pattern centered on the Mid-Atlantic Ridge. The age of the sea floor increases with the distance from the ridge. The spreading rate along the Mid-Atlantic Ridge is about 1 centimeter per year.

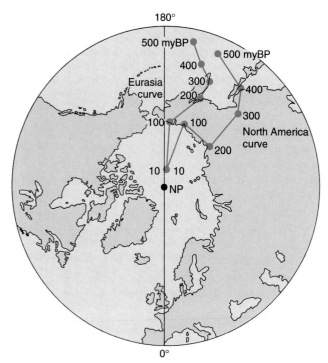

FIGURE 3.14

The positions of the North Pole millions of years before present (myBP), judged by the magnetic orientation and the age of the rocks of both North America and Eurasia. The divergence of the two paths indicates the landmasses of North America and Eurasia have been displaced from each other as the Atlantic Ocean opened.

the ridges and becoming older away from the ridges (fig. 3.13). The polarity and age of these stripes should correspond to the same magnetic field changes found in dated layers on land. Deep-sea core drilling confirmed their ideas and thus dramatically demonstrated seafloor spreading. The present ocean basins are not old but new, created by seafloor spreading during the past 200 million years, or the last five percent of the earth's history.

Records of the relationship between the magnetic poles and the orientation of magnetic particles within crustal rock at the time of the rock's formation are locked within it. If the rock was formed at the magnetic equator (fig. 3.12) the magnetic particles are oriented north-south, parallel to the earth's surface; if the rock was formed at the magnetic pole (fig. 3.12) the orientation is 90° to the earth's surface. Crustal rock showing a magnetic alignment that does not agree with the direction of the magnetic field at its present location is evidence that the rock has moved from its latitude of formation as the lithospheric plate changed its position on the earth's surface.

When magnetic crustal rocks of the same age are found at different locations on a crustal plate, another clue to plate motion has been preserved. The magnetic orientation at the time of formation of each rock sample points to the magnetic pole position at that time. Samples of rock formed at different times show orientation toward a different magnetic pole location. If these magnetic pole positions are mapped and dated, and if it is assumed that the continental plates remained in a fixed position through geologic time, it appears that the earth's magnetic

poles have wandered. The paths constructed by mapping in this way are known as **polar wandering curves** (fig. 3.14), and because all plates do not move along the same path, each plate produces its own polar wandering curve. However, the magnetic poles appear not to have wandered but to have remained close to the geographic poles, even though the magnetic polarity has reversed through time. If this is so, the polar wandering curves are caused by the movement of the lithospheric plates.

This can be demonstrated by considering the relative motion between North America and Europe; see figure 3.14. Think of the polar wandering curves as the open blades of a pair of scissors with one handle attached to North America and the other to Europe. If the scissor blades are closed to make the polar wandering curves coincide, the handles move Europe and North America toward each other, joining the continents, as the polar wandering curves approach each other. This demonstrates the case for the time between 500 and 200 million years before the present, when North America and Europe moved as one landmass across the earth's surface. Two hundred to one hundred million years ago the polar wandering curves diverged, indicating that the continents began to move independently, drifting apart and forming the North Atlantic Ocean. This plan is supported by geological evidence including the joining of ancient mountain ridges and fault systems.

How do earthquake patterns, seafloor heat flow, age of the earth's crust, and thickness of seafloor sediments help to explain seafloor spreading?

What is a magnetic reversal and how is it recorded in the rock of the sea floor?

How do the seafloor patterns of magnetic reversals verify seafloor spreading?

How and why is a polar wandering curve constructed?

3.6 PLATE TECTONICS

When the concept of continental drift was joined with the idea of seafloor spreading, it led to the formation of a single concept of crustal movement: **plate tectonics.** The earth's lithosphere is seen as a series of rigid plates outlined by the major earthquake belts of the world (refer to fig. 3.8). Seven major lithospheric plates are recognized—the Pacific, Eurasian, African, Australian, North American, South American, and Antarctic—as well as numerous smaller plates off South and Central America, in the Mediterranean area, in the East Indies, and along the northwest United States (fig. 3.15).

Define the term "plate tectonics."

Locate the boundaries of the seven major plates in figure 3.6.

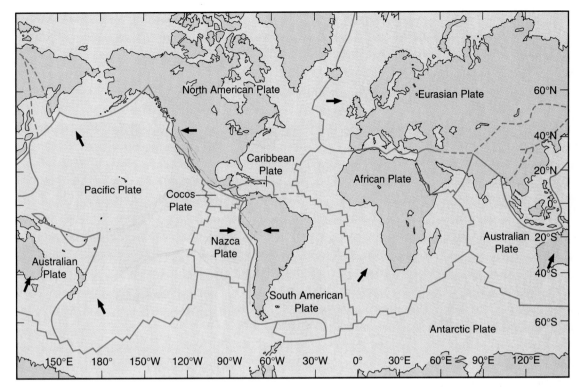

FIGURE 3.15

Major lithospheric plates of the world. Arrows indicate direction of plate motion.

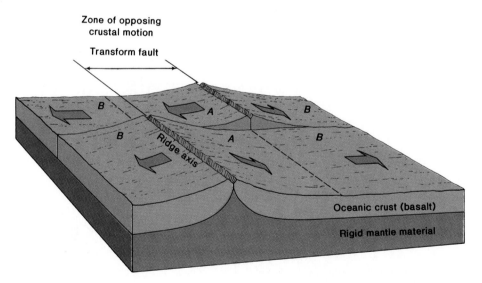

FIGURE 3.16

Relative plate motion and displacement of the crust along a fault. Plates at the transform fault, A, move in opposite directions. Plate motion at B is in the same direction.

3.7   PLATE BOUNDARIES

Each plate is a lithosphere segment that may contain continental or oceanic crust or some of each. The plate boundaries are indicated by trenches, ridges, and **faults.** The direction of plate motion at each boundary depends on whether the plates move apart (ridges), move together and subduct (trenches), or slide past each other either horizontally or vertically (faults). A central **rift valley** runs along portions of the crests of the mid-ocean ridges; these rift valleys are visible in figure 3.6. Sections of these rift valleys are displaced laterally from each other along

a special kind of fault called a **transform fault.** The opposite sides of a transform fault are two different plates that are moving in opposite directions (see area A in fig. 3.16). This movement creates a fault zone that is active and may be the seat of frequent but shallow earthquakes. Where the motion of the plates is in the same direction (see area B in fig. 3.16), the fault area is relatively quiet.

The San Andreas fault system, an excellent example of one of the larger transform faults, extends from the northern end of the East Pacific Rise up through the Gulf of California, to the San Francisco Bay region, and out to sea to the north

FIGURE 3.17

(a) (inset) The San Andreas fault runs north-south along the San Francisco peninsula. (b) The sunken portion of the fault south of San Francisco is used to store water for the city's use.

(fig. 3.17). It terminates at another small ridge system pointed to the northeast, the Gorda and Juan de Fuca Ridges. The land on the west or Pacific plate side, including the coastal area from San Francisco to the tip of Baja California, is actively moving northward past the North American plate on the east side of the fault. The motion is not uniform: when stress accumulates, there is sudden movement and displacement along the fault, causing severe earthquakes. Movement along this fault system caused the famous 1906 and October 1989 earthquakes in the San Francisco Bay area and the 1994 Los Angeles earthquake.

The reason for the many short transform faults that are found associated with the ridge system is not completely understood. These faults probably develop because the crust is subject to different degrees of rotational motion depending on latitude. Changes in the direction of crustal motion, due to variations in the strength or location of convection cells or due to collisions between sections of crust, could also result in transform faults.

What are the boundaries of a lithospheric plate?

What process occurs at each plate boundary, and in what direction does it occur?

Explain why earthquakes of different magnitudes occur along the San Andreas fault.

3.8 RIFTING

Rift zones, or spreading zones, are regions where the lithosphere splits, separates, and is forced apart as new crustal material moves into the crack or rift. The major rift zones are the mid-ocean ridge systems, but rifting is not limited to oceanic areas. Rifting also occurs on land, breaking up landmasses.

The lithosphere is thin under the oceans' newer rift zones (10–100 km or 6–60 mi) and thicker (100–150 km or 60–90 mi) where it is older and farther away from the rift. To crack the lithosphere under the continents, some mechanism must thin and apply tension to the crust. How this mechanism operates is not fully known. Sinking of thicker, cooler, more dense litho-sphere into the mantle at a subduction margin may stretch and thin the plate to allow magma to rise, weakening the plate so that it cracks under tension and produces a fault zone into which magma eventually flows (fig. 3.18).

Another explanation for rifting requires a more active role for the mantle processes. Rising magma in a mantle convection cell may heat the overlying crust and bow it upward. This weakens the crust and results in localized faulting that allows a central block of crust to drop downward, as seen in the Great Rift Valley of Africa, an actively forming continental rift stretching from Mozambique to Ethiopia. Continuing convection cell motion forces the plates apart, thinning the crust and deepening the fault. Gradually the magma penetrates into the crack and widens the rift. If the rift deepens sufficiently, seawater may

enter. The Red Sea appears to be at this more intermediate stage; both oceanic-type basalt and blocks of continental-type crust are present on its floor, much as in figure 3.18b.

When a continental landmass moves away from a spreading center, the continental margin closest to the spreading center becomes a midplate **passive margin** or trailing margin. Because the passive margins have moved away from the spreading center, they are not greatly modified by additional tectonic processes. Old passive margins are often broad and shallow with thick sedimentary deposits; the East Coast of the United States is a passive margin. As the passive margin ages and moves away from the rift zone, it becomes cooler, thicker, and denser. These changes cause the passive plate margin to settle deeper into the mantle and subside slowly over time.

What is a rift zone?

How might a rift zone split a continent and form an ocean between the two new continents?

What is a passive margin and how is it formed?

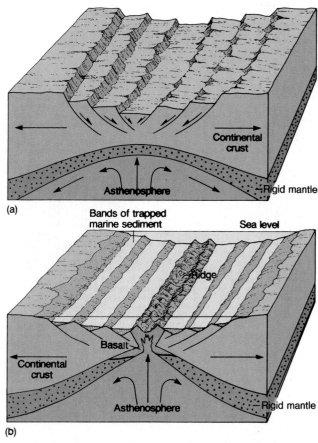

FIGURE 3.18

(a) Rifting is thought to happen when magma rises, causing tension and stretching of the overlying plate. (b) As the spreading continues, the fault deepens and cracks, allowing magma to penetrate and eventually form a ridge. Marine sediments accumulate in the fault, smoothing the sea floor.

FIGURE 3.19
Subduction processes produce (a) island arc
systems with volcanic activity, and
(b) additions to a landmass and explosive
volcanoes.

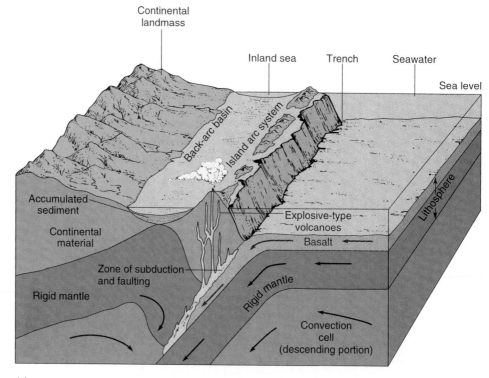

(a)

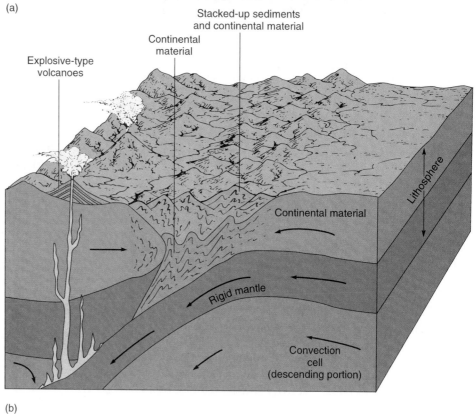

(b)

3.9   SUBDUCTION ZONES

The continental margin at a subduction zone, is the **active margin** or leading margin. Active margins are modified by the processes that occur in these convergence zones. Where oceanic crust descends beneath continental crust, the active continental margin is often narrow and steep due to abrasion along the edge of the landmass. The subducting of the sinking plate depresses the sea floor forming a trench. In some cases the amount of material scraped from the active margin in the convergence zone and the loose materials accumulated on the sea floor fill the trench, masking its appearance. The forcing of one plate margin beneath another often leads to the gradual elevation and compression of the upper plate margin or active margin and, in this case, coastal mountains are formed.

At a subduction zone where crustal plates collide, the denser plate sinks into the mantle. The sinking plate encounters the high temperature of the mantle and melts. The molten material moves upward through crustal fractures in volcanic eruptions. Eruptions of this type are explosive and produce a lot of ash as large amounts of gas and steam are liberated. The volcanic peaks that form the island arc chains of the Aleutians, Japan, the Philippines, and Indonesia are of this type (fig. 3.19a). Other such volcanoes are common in the Andes and the Cascade Range of the Pacific Northwest, including Mount St. Helens (fig. 3.20).

When continents collide at a subduction zone, the landmasses resist sinking because their density is lower than the mantle density (listed in table 3.1). Instead, they are crumpled and deformed at the colliding edges, and continental material from one plate may override the continental material of the other plate, producing a large thickness of continent. The sediments and less dense crust of the intervening sea floor may be compressed together and piled upward. The Himalayas and the Alps in part are examples of such mountain ranges, and this is why old marine sediments and fossils, including the limestone remains of coral reefs, are found on the summits of peaks in these ranges. This situation is illustrated in figure 3.19b.

Studies of the North American crust indicate that the core of the continent was assembled about 1.8 billion years ago by collisions of four or five large pieces of even older continental land. Crustal fragments with properties and a history distinct from adjoining crustal fragments are known as **terranes.** The terranes of Alaska and the West Coast of the United States and

FIGURE 3.20

Mount St. Helens erupted violently on May 18, 1980. The mountain lost nearly 4.1 cubic kilometers (1 cu mi) from its once symmetrical summit, reducing its elevation from 2950 meters to 2550 meters (a 1312-ft change). The force of the lateral blast blew down forests over a 594 square kilometer (229 sq mi) area. Huge mud floes of glacial meltwater and ash flowed down the mountain.

Plate Tectonics

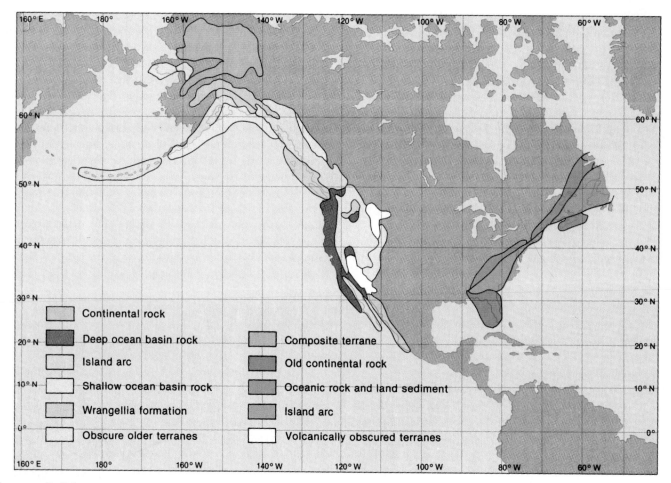

FIGURE 3.21
Terranes of the west and east coasts of North America.

Canada appear to have arrived from the south about 70 million years ago. Fossils collected between Virginia and Georgia indicate that a long section up and down the East Coast was formed somewhere adjacent to an island arc system; the fossils point to a past European connection. It is estimated that 25 percent of North America was formed from terranes; see figure 3.21 for examples. The subcontinent of India is considered by some to be a single, giant terrane; it arrived from Antarctic latitudes and is now firmly attached to the continent of Asia. North of the Himalayas there appear to be terranes that became a part of the continent before the arrival of India.

How does an active margin differ from a passive margin?

Explain what happens when continents collide.

What happens at a subduction zone when a continental plate collides with an oceanic plate?

How does plate tectonics explain sea fossils found high in the Alps and Himalayas?

What does the presence of terranes tell us about the past geological history of an area?

3.10   THE MOTION OF THE PLATES

The sea floor moves away from a ridge system, most commonly at a rate of 1 to 10 centimeters (0.4 to 4 in) per year, and the old crust disappears into the trenches at a comparable rate. Although the rates are small by everyday standards—about as fast as fingernails grow—they produce large changes over geologic time. For example, at the rate of 1.5 centimeters (0.6 in) per year it takes about 100,000 years to move 1.6 kilometers or 1 mile; therefore in the 200 million years since the breakup of Pangaea the crust could move 3200 kilometers, or 2000 miles, which is more than half the distance between Africa and South America. A ridge system with a steep profile (like the Mid-Atlantic Ridge) has a slower spreading rate than a ridge system with less steep sides (like the East Pacific Rise). Spreading rates are estimated at 1.25 centimeters (0.5 in) per year for the Mid-Atlantic Ridge and at 5 centimeters (2 in) per year for the East Pacific Rise. The largest spreading velocity known is 18.3 centimeters (7 in) per year off the west coast of South America. Keep in mind that the process of spreading does not occur smoothly and continuously; it goes on in fits and starts, with varying rates and time periods between occurrences.

Iceland is the only large island lying across a mid-ocean ridge and rift zone; many of the processes of seafloor spreading can here be seen on land. Spreading in Iceland occurs at rates

Chapter 3   56

similar to those found at the crest of the mid-ocean ridge. North-eastern Iceland had been quiet for one hundred years, until volcanic activity began in 1975; in six years this activity widened by 5 meters (17 ft) an 80 kilometer (50 mi) stretch of the ridge.

What is the average rate of seafloor spreading?

Compare spreading rates across ridges in the Atlantic and Pacific.

Relate the spreading rate of Iceland since 1975 to the average spreading rate of the Mid-Atlantic Ridge.

3.11 HOT SPOTS

Scattered around the earth are approximately forty specific fixed areas of isolated volcanic activity known as **hot spots.** They are found under continents and oceans, in the center of plates, and at the mid-ocean ridges. These hot spots periodically channel hot material to the surface from deep within the mantle. The life span of a typical hot spot appears to be about 200 million years. At these sites, mantle material may force its way through the lithosphere and form a seamount directly above. The hot spot may also raise a broad swelling of the ocean floor that will subside as the crust moves off the magma source. Hot spots may also resupply the asthenosphere, which is constantly cooling and becoming attached to the base of the lithosphere.

As the crustal plate moves over a hot spot, successive, and usually nonexplosive, eruptions can produce a linear series of peaks or **seamounts** with the youngest peak above the hot spot and the seamounts increasing in age as the distance from the hot spot increases (fig. 3.22). For example, in the islands and seamounts of the Hawaiian Islands system, the island of Hawaii with its active volcanoes is presently over the hot spot. The newest volcanic seamount in this series is Loihi, found to be erupting in this area in 1981. Loihi lies 45 kilometers (28 mi) east of Hawaii's southernmost tip, rising 2450 meters (8000 ft) above the sea floor but still under water.

The Hawaiian seamount system continues to the west, until just west of Midway Island, the chain changes direction and stretches to the north, indicating that the crust over the hot spot moved in a different direction some 40 million years ago. This is the Emperor Seamount Chain, volcanic peaks that once were above the sea surface as islands but have eroded and subsided over time. **Subsidence** occurs when the plate supporting the seamounts slides down and away from the bulge of the hot spot. This motion, in addition to the contraction and increasing density of the cooling plate and the weight of the seamount, depresses the mantle and carries the seamounts below the surface (fig. 3.22). Refer to figure 3.6 to follow this seamount chain. Recently, sequentially related seamounts from hot spots have been used to reconstruct the opening of the Atlantic and Indian Oceans.

Define "hot spot."

Discuss the formation of the Hawaiian Islands.

What do the Hawaiian Islands and the Emperor Seamount Chain tell us about the motion of the Pacific plate?

3.12 THE BREAKUP OF PANGAEA

Figure 3.23 traces the recent plate movements that led to the configuration of the continents and oceans as we know them today. About 200 million years ago, Pangaea began to break apart. Laurasia moved away from Gondwanaland, and the South Indian Ocean began to form (fig. 3.23b). India drifted away from Antarctica to move northward toward Asia and its eventual collision with the Asian mainland. Water flooded the spreading center between Africa and South America 135 million years ago (fig. 3.23c); North America and Eurasia were still firmly attached, as were Australia and Antarctica. South America separated from Africa 65 million years ago (fig. 3.23d); Africa moved northward, and the Mediterranean Sea was created. India continued to move toward Asia, Australia began to separate from Antarctica, and Madagascar split off from Africa.

During the last 16 million years (fig. 3.23e), North America and Eurasia separated, and Greenland moved away from Europe. North and South America united, and Australia at last became free of Antarctica. India had crumpled up against Asia,

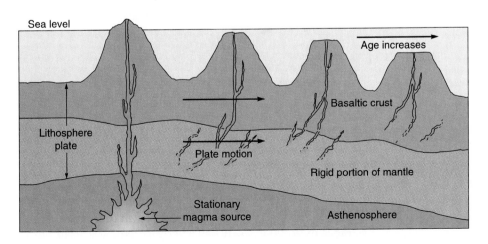

Sea level

Age increases

Basaltic crust

Lithosphere plate

Plate motion

Rigid portion of mantle

Stationary magma source

Asthenosphere

FIGURE 3.22

Chains of islands are produced when a crustal plate moves over a stationary hot spot.

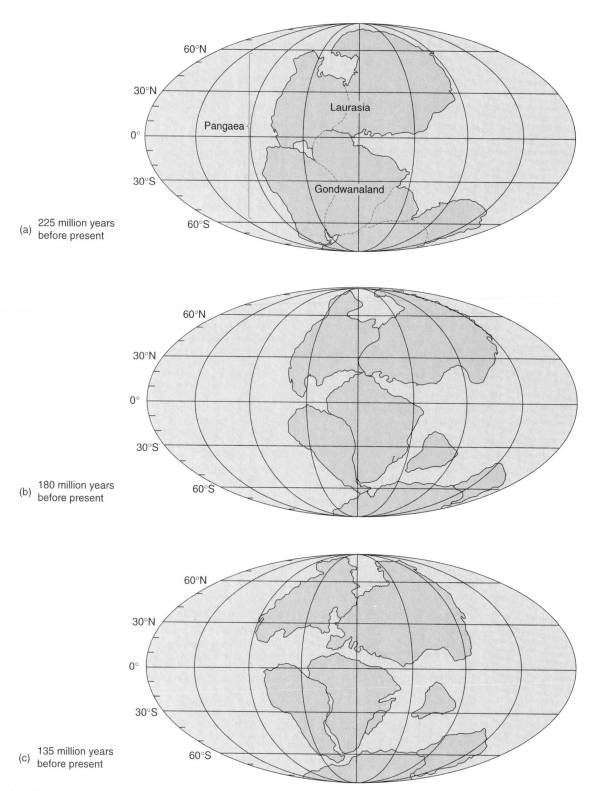

(a) 225 million years before present

(b) 180 million years before present

(c) 135 million years before present

FIGURE 3.23

The movements of the continents from Pangaea to the present.

Source: After R. S. Dietz and J. C. Holden, "Reconstruction of Pangaea: Breakup and Dispersion of Continents, Permian to Present" in *Journal of Geophysical Research* 75(26), September 10, 1970.

resulting in the Himalayas. A similar situation occurred when Italy moved northward across the Mediterranean to collide with Europe and produce the Alps. Twenty million years ago, Arabia moved away from Africa to form the Gulf of Aden and the new and still opening Red Sea.

Since the earth is about 4.6 billion years old, there is no reason to believe that Pangaea's breakup was the first or only time during which the sea floor spread or the continents changed position and recombined. Refer to the section on terranes in section 3.9. It has been suggested that there is an

FIGURE 3.23
(continued)

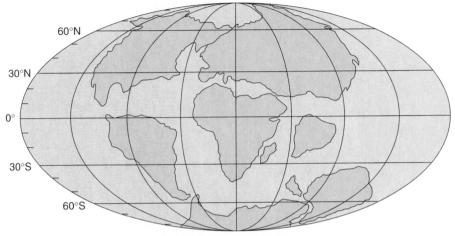

(d) 65 million years
before present

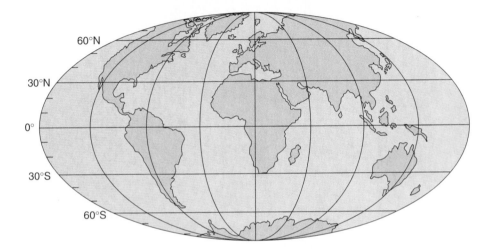

(e) Present

orderly, cyclic, 500-million-year pattern of assembling and disassembling the landmasses powered by the heat from radioactive decay in the earth's interior. Although production of heat is continuous, this model proposes that the heat is lost at varying rates that are related to the movement of the continents. As the continents shift, coalesce, and compress, the sea level drops. When sufficient heat accumulates under the continental mass, the rifting process begins unstitching the continents, and as the continents move apart, cool, and subside, the sea level rises and covers their borderlands.

Trace the breakup of Pangaea over the past 200 million years.

Compare the histories of the Red Sea and the Mediterranean Sea.

Compare possible fates of the North Atlantic and the North Pacific after the next 500 million years.

3.13   RESEARCH PROJECTS

The Ocean Drilling Program, an international effort, using the state-of-the-art drilling ship *JOIDES Resolution* (fig. 3.24)

began in 1985. Since that time the *JOIDES Resolution* has drilled in regions of Baffin Bay and the Labrador Sea, seeking to determine the history of these two basins. In Caribbean waters where the edge of the North American plate moves under the Caribbean plate and along the coast of Peru where the Pacific plate slides underneath South America, samples have been made of the materials forming the plate boundaries. In 1988, drilling in the southern Indian Ocean discovered fragments of fossil wood and soils on a submerged plateau; this indicates that the area was once above sea level. Through 1991 the ship operated in the Pacific Ocean, drilling in the South China Sea, the Japan Sea, the Bering Sea, and the south and eastern tropical waters of the Pacific. During 1992 drilling took place on the East Pacific Rise and on selected South Pacific seamounts and atolls. Later that year a series of holes was drilled across the North Pacific, and in the northeast Pacific samples were obtained from the area between the Juan de Fuca Ridge and the Oregon coast (fig. 3.17). In 1993 the *JOIDES Resolution* returned to the Atlantic to investigate seafloor spreading at the once-connected margins of Spain and Newfoundland. Drilling also took place along the coast of New Jersey to investigate past changes in sea level and at high latitudes along the sea floor connecting the North Atlantic to the Arctic Ocean to look for evidence of changes in sea level and climate in this area.

FIGURE 3.24
The deep-sea drilling vessel, *JOIDES Resolution*.

Drilling in 1994 included taking cores in the delta of the Amazon River to look for information on sea-level changes, studying the transform faults of the ridge system of the equatorial Atlantic, additional drilling in the Mediterranean, and a return to the area between the North Atlantic and Arctic Oceans.

Geological oceanographers and geophysicists have descended into the mid-ocean ridges' rift valleys to study the production of the new crust. In the summer of 1973, French oceanographers aboard the submersible *Archimède* made the first visual study of the rift valley of the Mid-Atlantic Ridge. The following summer, the U.S. submersible *Alvin* (figs. 3.25 and 3.26) joined in the dive program.

In 1977 and 1979, expeditions from the Woods Hole Oceanographic Institution explored the Galápagos Rift between the East Pacific Rise and the South American mainland, west of Ecuador. The most surprising discovery of these expeditions was the unexpected and perplexing presence of large communities of animals concentrated in areas of hot water rising from **hydrothermal vents** along the rift. In 1979, *Alvin* made twenty-four dives to the 2500-meter-deep (8000-ft) rift to study these animal populations, which are discussed in Chapter 12.

Also in 1979, the Mexican-French-American research program RISE (Riviera Submersible Experiment) investigated seafloor hot springs near the tip of Baja California. Two thousand meters (6500 ft) down, *Alvin* found mounds and chimney-shaped vents 20 meters (65 ft) high ejecting hot (350°C) streams of black mud containing particles of lead, cobalt, zinc, silver, and other minerals. Here, too, the vents are surrounded with rich animal life.

Expeditions have continued to search for hydrothermal vents along the spreading centers of the world. In September 1981, researchers first photographed a small section of a rift valley in the ridge system along the small Juan de Fuca plate off the coast of Oregon. In 1988, a multi-institutional program used the submersible *Alvin* to deploy and recover instruments designed to monitor this vent activity. Special time-lapse video cameras (fig. 4.10) were mounted on the sea floor to monitor sulfide chimney growth. Temperature sensors were placed in vents, seismometers were set to record crustal

FIGURE 3.25

The submersible *Alvin* being lowered to the sea surface for a dive along the Juan de Fuca Ridge in the northeastern Pacific. The *Alvin* carries two scientists and a pilot. It is equipped with cameras inside and out, as well as baskets for samples taken with its mechanical claws (fig. 3.26). It dives at a speed of 1.3 meters per second to a depth of 3000 meters (10,000 ft). It requires 3 hours for the round trip and spends 4 to 5 hours on the bottom.

FIGURE 3.26

Alvin's mechanical arm collects a sample.

Plate Tectonics

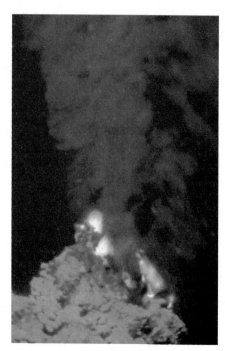

FIGURE 3.27

A color enhanced photo from the Juan de Fuca Ridge showing the glowing vent light produced by a hot vent.

movement, and hot-water flow rates were also measured. An electronic still camera mounted on the *Alvin* was used to investigate the radiant energy from the vents (fig. 3.27). Between 1990 and 1991 the *Alvin* revisited the vent communities that had been discovered since 1979 to look for changes in these areas. Changes in the biological communities indicate changes in the chemistry and amount of hot water supplied to the vent, which in turn point to changes in crustal activity at the plate boundaries.

As the researchers' abilities to monitor the activity of the sea floor increase, additional and unexpected discoveries continue to be made. Oceanographers studying the Juan de Fuca Ridge in 1986 discovered a huge mass of warm water about 1000 meters (3300 ft) above the ridge. Analysis of the water showed that it had recently been formed by a large hydrothermal discharge. Its original volume was estimated at 100 million cubic meters, and it was given the name, **megaplume.** Since that time other plumes of warm water and also fresh lava have been observed in the rift area. Seismic recorders, sensors to monitor temperature and magnetic properties, and acoustical devices to measure seafloor activity have been installed and monitored along this section of the ridge.

In 1993 oceanographers engaged in detailed scanning of the sea floor northwest of Easter Island in the South Pacific discovered an area of intense volcanic activity. Over 1100 seamounts and volcanic cones were discovered, some rising 2300 meters (7000 ft) above the sea floor. The large number of volcanoes in such a small area was surprising, as was the evidence of frequent eruptions often involving multiple cones. The significance of this submarine vulcanism is not fully understood; its large and small-scale effects on the chemistry, marine life and circulation patterns of water in the area are still unknown.

Future research projects are planned for all these areas. Because the availability of manned submersibles limits diving time and study sites, remotely operated submersibles that send information back to surface ships will come into increasing use as seafloor investigations are expanded. See "Undersea Robots."

In future research projects, instruments will be placed on seafloor structures to study the formation of sulfide deposits, the chemistry of vent water, and the energy flow between vents and biological communities.

What kind of information do the *JOIDES Resolution* drilling projects produce?

What kinds of discoveries have been made by deep-sea submersibles, and how do their observations help us to better understand the processes of plate tectonics?

What kinds of areas produce a hydrothermal vent?

What is a megaplume?

SUMMARY

The earth is made up of a series of layers: the crust, the mantle, the liquid outer core, and the solid inner core. Evidence for this structure comes from the ways in which seismic waves change speed and direction as they move through the earth.

Continental crust is formed from granite, which is less dense than oceanic crust formed of basalt. The boundary that separates the crust and the mantle is the Moho. The top of the mantle is fused to the crust to form the rigid lithosphere. The lithosphere floats on the deformable upper mantle or asthenosphere.

Alfred Wegener's theory of drifting continents was based on the geographic fit of the continents and the similarity of fossils collected on different continents. The discovery of the mid-ocean ridge system and the idea of convection cells in the asthenosphere led to seafloor spreading as the mechanism for continental drift. New lithosphere is formed at the ridges, or spreading centers; old lithospheric material descends into trenches at subduction zones. Evidence for lithospheric motion includes the match of the earthquake zones to spreading centers and subduction zones, greater crustal heat flow near ridges, age measurement of seafloor rocks, age and thickness measurements of sediment from deep-sea cores, and the magnetic stripes in the seafloor basalt on either side of the ridge system. Polar wandering curves are also used to trace plate movement. Rates of seafloor spreading range from 1 to 10 centimeters per year. The concept of continental drift and the theory of seafloor spreading combined to produce plate tectonics, in which the plates are made up of continental and oceanic lithosphere bounded by ridges, trenches, and faults.

Rifting has separated landmasses and produced new ocean basins. Passive or trailing continental margins move away from spreading centers to become midplate margins; they are not greatly modified by tectonic processes. Active or leading continental margins move toward subduction zones to be modified by the subduction process. Plate collisions produce earthquakes, island arcs, volcanic activity, and mountain building. Plate movements traced over the last 225 million years show the breakup of Pangaea to form the present continents.

Deep-sea drilling programs investigate tectonic processes, the evolution of oceanic crust, and the sediment records of past ages. Submersibles have been used to explore the rift valley of the Mid-Atlantic Ridge and the hydrothermal vents and animal communities found along the Galápagos Rift, Baja California, and off the Oregon coast.

KEY TERMS

inner core	Moho	trench	transform fault
outer core	lithosphere	seafloor spreading	rift zone
mantle	asthenosphere	spreading center	passive margin
crust	Gondwanaland	subduction zone	active margin
density	continental drift	sediment	terrane
seismic waves	Pangaea	polar reversal	hot spot
P-waves	Laurasia	polar wandering curve	seamount
S-waves	convection cell	plate tectonics	subsidence
granite	magma	fault	hydrothermal vent
basalt	ridge	rift valley	megaplume

SUGGESTED READINGS

Ballard, R. 1975. Dive into the Great Rift. *National Geographic* 147 (5):604–15. Project FAMOUS and photographs from *Alvin*.

Ballard, R. D., and J. F. Grassle. 1979. Return to Oases of the Deep. *National Geographic* 156 (5):689–705.

Bloxham, J., and D. Coubbins. 1989. The Evolution of the Earth's Magnetic Field. *Scientific American* 261 (6):68–75.

Bonatti, E. 1994. The Earth's Mantle Below the Oceans. *Scientific American* 270 (3):44–51.

Continents Adrift and Continents Aground, Readings from Scientific American. 1976. W. H. Freeman, San Francisco, 230 pp.

Golden, F. 1991. Birth of the Caribbean. *Sea Frontiers* 37 (5):20–25.

Hoffman, K. A. 1988. Ancient Magnetic Reversals: Clues to the Geodynamo. *Scientific American* 258 (5):76–83.

Howell, D. G. 1985. Terranes. *Scientific American* 253 (5):116–25.

Jeanloz, R., and T. Lay 1993. The Core-Mantle Boundary. *Scientific American* 268 (5):48–55.

Lonsdale, P., and C. Small. 1991/92. Ridges and Rises: A Global View. *Oceanus* 34 (4):26–35.

Macdonald, K. C., and P. J. Fox. 1990. The Mid-Ocean Ridge. *Scientific American* 262 (6):723–79.

Nance, R. D., T. R. Worsley, and J. B. Moody. 1988. The Supercontinent Cycle. *Scientific American* 259 (1):72–79.

Rabinowitz, P. D., L. E. Garrison, and A. W. Meyer. 1989. Four Years of Scientific Deep Ocean Drilling. *Sea Technology* 30 (6):35–40.

Shaping the Earth—Tectonics of Continents and Oceans, Readings from Scientific American. 1990. W. H. Freeman, New York. 206 pp.

Tivey, M. K. 1991/92. Hydrothermal Vent Systems. *Oceanus* 34 (4):68–74.

Vink, G. E., W. J. Morgan, and P. R. Vogt. 1985. The Earth's Hot Spots. *Scientific American* 252 (4):50–57.

York, D. 1994. The Earliest History of the Earth. *Scientific American* 268 (1):90–96.

Undersea Robots

Historically, deep-sea oceanographic data has been gathered by surface ships that towed instruments and samplers or sent equipment down thousands of meters to collect data from scattered areas. Since the 1970s manned submersibles, such as the *Alvin* (refer to Chapter 3), have allowed researchers to actually experience the deep sea. A single dive in the *Alvin* takes more than eight hours from launch to retrieval, but only four hours are spent at the target depth. The cost of the dive and support costs for the vessel and crew at the surface are about $25,000 per day. In order to explore larger areas faster, decrease the costs, and decrease the risks, dozens of deep-diving robotic devices are presently in use or are being built by the world's centers for oceanographic research. These devices act as eyes, samplers, and manipulators for oceanographers confined to submersibles or surface vessels.

Remotely Operated Vehicles

Remotely Operated Vehicles or ROVs are tethered to a surface vessel, a manned submersible, or a below-the-surface system that isolates the ROV from ship motion at the sea surface. The tether is the umbilical cord between ROV and operator; it supplies power and transmits information over an electrical or fiber-optic cable. ROVs use video, electronic-digital and still cameras accompanied by lights to explore their environment. They are equipped with mechanical hands to manipulate objects and, depending on their mission, other sensors such as sonar and temperature and salinity monitors. The operator "flies" the ROV over the sea floor, and the ROV sends data to the operator and receives back directions to change position, to manipulate or retrieve objects, or to use cameras and other sensors. Full-ocean-depth ROVs, required to perform a variety of tasks over long periods, weigh several tons and must be towed slowly because of the stresses on their long cables. Small, limited-use, shallow-water vehicles that may weigh less than 45 kilograms (100 lbs) can be towed rapidly by small surface vessels. These smaller robotic devices are used to service, inspect, and aid in construction of offshore oil and gas facilities, bridge piers, and dam footings. MkI ROV is shown in figure 1.

ROV *Jason* is a tethered vehicle operated by the Woods Hole Oceanographic Institute. *Jason* weighs about 1350 kilograms (3000 lbs), operates to a depth of 6000 meters (20,000 ft), and provides simultaneous use of up to four color video channels, sonar, electronic cameras, and

FIGURE 1
MkI ROV, a MiniRover, is an example of a small "low-cost" system.

manipulators. A smaller ROV, *Jason Jr.* or *JJ*, operating from the submersible *Alvin,* was used to inspect the wreck of the *Titanic* and photograph its interior spaces (see figure 2). In 1992 ROPOS or Remotely Operated Platform for Ocean Science was used by a team of Canadian and U.S. scientists to map and sample a hydrothermal vent field on the Juan de Fuca Ridge off the Oregon coast. It also successfully retrieved data packages from previously installed samplers.

In March 1994 the Japanese ROV, *Kaiko,* descended 10,911 meters (36,000 ft) into the Mariana Trench. *Kaiko* is the world's deepest diving ROV and the only ROV capable of descending below 6500 meters (20,000 ft). Because of a failure in its video transmission, the mission was aborted, and engineers expect it to take up to a year before *Kaiko* descends again. This ROV is equipped with five television

FIGURE 2

The camera-equipped ROV, *Jason Jr.*, peers into a cabin of the luxury liner *Titanic*, 3800 meters (12,500 ft) below the surface. A tether connecting *Jason Jr.* to the submersible *Alvin* is seen at left.

cameras, still cameras, sonar, manipulators and other sensors. Plans for *Kaiko* include detail mapping of active faults, examining the sediment on top of subducting plates, and observing the ocean's deepest life-forms.

The U.S. Navy has been working with an Underwater Security Vehicle or USV, another ROV, that can be operated from small support craft or land stations. Its mission is to search for, track, and intercept underwater targets that may represent a threat to ships, submarines, and naval shore stations. These ROVs use special acoustical sensors to home in on their targets and then use visual sensors to evaluate and identify them. This ROV may be equipped with response capability to be used if the target is considered a security threat.

ROVs are much less costly than manned submersibles, for there is no need to provide safety for humans. Figure 3 shows the *Ventana* ROV. Supplied with continuous power through their tethers they can operate to deeper depths, work under water for long periods of time, and explore areas of high risk such as vent and volcanic areas. The ROV's two-dimensional video images are a drawback, for they do not measure up to the three-dimensional image seen by a human in a submersible. Therefore, ROVs cannot yet react in the same way as a manned submersible. Also they do require the expensive support of surface vessels. Efforts to further reduce costs and extend deep-sea exploration time has led to the development of Autonomous Underwater Vehicles.

FIGURE 3

The *Ventana* ROV is owned and operated by Monterey Bay Aquarium Research Institute. This ROV is used to study marine life in Monterey Canyon.

Autonomous Underwater Vehicles

AUVs or Autonomous Underwater Vehicles are independent, mobile, instrument platforms with sensors. They are not tethered to another vessel, and therefore, their range of movement is greater than that of ROVs. AUVs may be linked acoustically to an operator in a surface vessel, or they may be preprogrammed to complete surveys and take samples

continued . . .

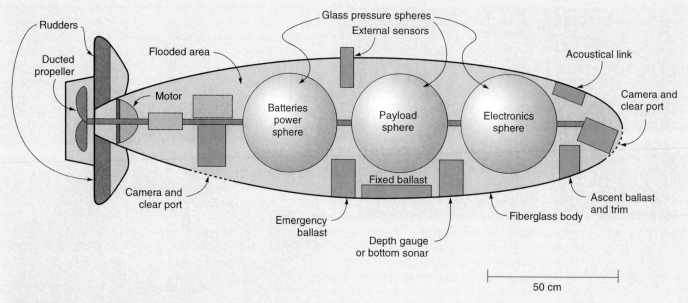

Rudders

Ducted propeller

Flooded area

Motor

Glass pressure spheres

External sensors

Acoustical link

Camera and clear port

Batteries power sphere

Payload sphere

Electronics sphere

Camera and clear port

Fixed ballast

Emergency ballast

Depth gauge or bottom sonar

Fiberglass body

Ascent ballast and trim

50 cm

FIGURE 4

A typical AUV design has a streamlined, external, fiber-glass shell surrounding pressurized internal functional units. Electronics, payload (sensor controls, data recording, etc.), and batteries are housed in glass spheres; each camera and sensor has its own pressure housing. The electrical junctions and drive motor are oil-filled; internal oil pressure changes with depth to equal external pressure. The space between the external shell and the pressure-protected internal units is flooded when the AUV submerges.

with little or no human supervision. Many of these vehicles are small, less than 25 kilograms (50 lbs) and have specialized capabilities and restricted depth ranges. Others work effectively in deep water and in inaccessible regions such as under sea ice. A typical AUV is diagrammed in figure 4.

The Navy's Advanced Unmanned Search System or AUSS locates unidentified objects on the sea floor at depths from 5 meters (17 ft) to 6 kilometers (20,000 ft); it is a small vehicle, 23 meters (77 ft) long, 78 centimeters (31 in) in diameter, and weighs 1270 kilograms (2800 lbs). It follows a predefined search pattern while transmitting sonar images to the surface. Its acoustic link may order suspension of the search pattern, require the AUSS to take a closer look at an object, and then resume its search pattern. The Autonomous Benthic Explorer or ABE is designed to survey deep-sea hydrothermal vent areas for up to one year, making a daily survey of its area. It monitors currents, takes pictures, and makes temperature and other measurements; then it returns to its mooring platform on the sea floor and remains quiet to conserve energy.

An AUV can communicate its data acoustically to a surface vessel, or it can dock with a buoy or beacon, transfer its data to the buoy system, receive new instructions and recharge itself. The data can then be transferred from the buoy system to a satellite link and beamed back to a surface ship or shore station. Some AUVs store their data internally to be kept until the AUV is recovered (fig. 5).

AUVs may be preprogrammed to make measurements while running specific courses, or they may relay their data acoustically to their controller; their course can be changed to improve their sampling. AUV power requirements depend on speed and distance traveled as well as the power required for their sensors and data transmission. Small size, low weight, and low power requirements mean the AUV can spend more time submerged and so can extend its range.

AUVs use internal software and sonar to survey terrain and avoid obstacles. They can use their proximity to the sea floor and its topography as a course guide, or they can navigate using preset acoustical seafloor beacons or acoustical buoys.

FIGURE 5
Retrieving a programmed torpedo (AUV) beneath the Arctic Ocean ice.

Satellites can monitor vast areas, but they track mainly surface features. A fleet of AUVs working in an area can monitor changing conditions and obtain more data in more detail over different depths than a single large research vessel. They are able to operate in any weather, and as the technology for standardizing their components develops, AUVs will cost less, more will be made, and the use of these robotic devices will increase. They are becoming the survey tool of choice for work in the earth's last, largest, and least known frontier.

Readings

Bellingham, J. G., and C. Chryssostomidis. 1993. Economic Ocean Survey Capability with AUVs. *Sea Technology* 34 (4):12–17.

Fletcher, B. 1993. Navy Underwater Security Vehicle. *Sea Technology* 34 (5):39–43.

Uhrich, R. W., and J. M. Walton. 1993. AUSS—Navy Advanced Unmanned Search System. *Sea Technology* 34 (2):29–35.

Yoerger, D. R. 1991. Robotic Undersea Technology. *Oceanus* 34 (1):32–37. 🐚

CHAPTER

4

The Sea Floor

Outline

Learning Objectives

After reading this chapter, you should be able to

- Describe the divisions and features of the continental margin.

- Describe the formation of a submarine canyon.

- Describe the features of the ocean basin floor.

- Understand the use of echo sounding to map the sea floor.

- Compare ocean sediments by source, properties, and distribution.

- Relate sediment particle size and sinking rate.

- Describe the instruments used to sample seafloor sediments.

- Give examples of seafloor mineral resources.

- Understand the political, legal, and economic factors related to harvesting seafloor mineral resources.

◄ Hoskyn Islands, Great Barrier Reef, Australia.

The sea floor was an unknown environment to early mariners and the first curious scientists. They believed that the oceans were large basins or depressions in the earth's crust, but they did not conceive that these basins held features that were as magnificent as the mountain chains, deep valleys, and great canyons of the land. As maps were created in greater detail and as ocean travel and commerce increased, it became essential to map seafloor features in the shallower regions, but it was not until the 1950s that improvements in technology made it possible to sample the deep sea floor routinely and in detail.

The geography and the geology of the sea floor are the products of crustal changes on the geological time scale. The main features of the sea floor are the evidence for plate motion and the results of plate tectonics discussed in the previous chapter.

4.1 THE SEA FLOOR

The mountain ranges under the sea are longer, the valley floors are wider and flatter, and the canyons are often deeper than those found on land. Land features are continually eroded by wind, rain, and ice, and are affected by changes in temperature and rock chemistry. The erosion of undersea mountains and canyons is slow, and occurs only by way of waves and currents along the shore and on steep underwater slopes. Beneath the surface the water dissolves away certain materials, changing the rock chemistry, but the features retain their shape and are only slowly modified as they are covered by a constant rain of particles or **sediment** falling from above. Computer-drawn profiles of elevations across the United States and the Atlantic Ocean are compared in figure 4.1. Notice that the profiles of the mountain ranges on land and ridge systems of the sea floor are similar.

4.2 THE CONTINENTAL MARGIN

The lithospheric plates (converging, diverging, and slipping past each other) have greatly influenced the **continental margins.** The continental margins are the edges of the landmasses at present below the ocean surface and their steep slopes that descend to the deep sea floor. A continental margin is composed of the continental shelf, shelf break, slope, and rise.

The **continental shelf** is a nearly flat land-border of varying width that slopes very gently toward the ocean basins. Shelf widths average about 65 kilometers (40 mi) and vary from only a few tens of meters to more than 100 kilometers (60 mi).

The distribution of the world's continental shelves is shown in figure 4.2. The shelf is narrow along the mountainous coasts of active continental margins and wide along low-lying land at passive continental margins. Refer to sections 3.8 and 3.9. Passive margin continental shelves are often further modified by sea-level changes and storm waves that erode the edges of the continents and, in some cases, natural dams on the shelves trap sediments between the offshore dam and the coast. Such dams include reefs, volcanic barriers, upthrust rock, and salt domes. Notice the narrow shelf along the west coasts of North and South America and the wide shelves off the eastern and northern coasts of North America, Siberia, and Scandinavia.

The continental shelves are geologically part of the continents, and during past ages, they have been covered and uncovered by fluctuations in sea level. When the sea level was low during the ice ages (periods of increased ice on land), erosion deepened valleys, waves eroded previously submerged land, and rivers left their sediments far out on the shelf. When the ice melted, these areas were flooded and sediments built up in areas close to the new shore. Although submerged, these areas still show the scars of old riverbeds and glaciers, features they acquired when exposed as part of the continent. Some continental shelves are covered with thick deposits of silt, sand, and mud derived from the land; for example, the Mississippi and Amazon Rivers deposit large amounts of sediments at their mouths. Other shelves are bare of sediments, such as where the Florida Current sweeps the tip of Florida, carrying the shelf sediments northward to the deeper water of the Atlantic Ocean.

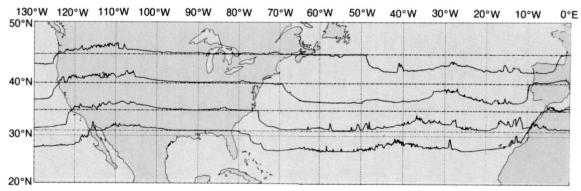

FIGURE 4.1
Computer-drawn topographic profiles from the west coast of Europe and Africa to the Pacific Ocean. The elevations and depths above and below 0 meters are shown along a line of latitude by using the latitude line as zero elevation. For example, the ocean depth at 40°N and 60°W is 5040 meters (16,500 ft). The vertical scale has been exaggerated about one hundred times the horizontal scale. If both the horizontal and vertical scales were kept the same, a vertical elevation change of 5000 meters would measure only 0.05 millimeters (0.002 in).

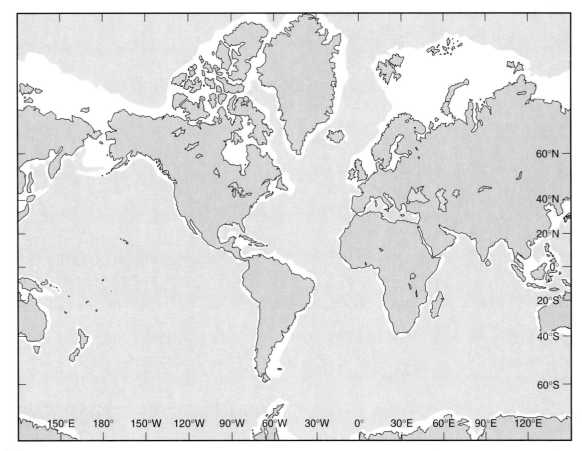

FIGURE 4.2

Distribution of the world's continental shelves (white). The seaward edges of these shelves are at an average depth of approximately 130 meters (430 ft).

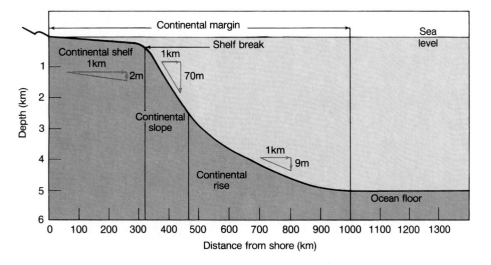

FIGURE 4.3

A typical profile of the continental margin. Notice both the vertical and horizontal extent of each subdivision. The average slope is indicated for the continental shelf, slope, and rise. The vertical exaggeration is one hundred times greater than the horizontal scale.

The boundary of the continental shelf on the ocean side is determined by an abrupt change in slope, leading to a more rapid increase in depth. This change in slope is referred to as the **continental shelf break,** while the steeper slope extending to the ocean basin floor is known as the **continental slope.** These features are shown in figure 4.3.

The angle and extent of the slope vary from place to place. The slope may be short and steep—for example, the depth may increase rapidly from 200 meters (650 ft) to 3000 meters (10,000 ft), as in figure 4.3—or it may drop as far as 8000 meters (26,000 ft) into a great deep seafloor depression or trench, as it does off the west coast of South America. The continental slope may show rocky outcroppings, and it is relatively bare of sediments because of its steepness.

The most outstanding features found on the continental slopes are **submarine canyons.** These canyons sometimes extend

The Sea Floor

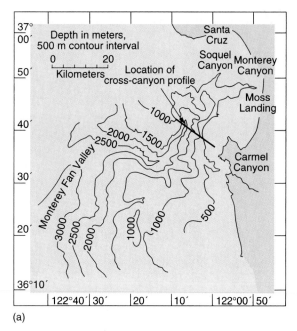

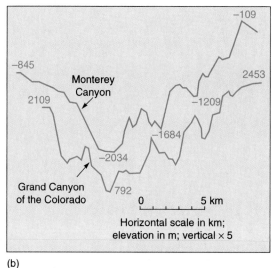

(a)

(b)

FIGURE 4.4

(a) Depth contours depict the Monterey and Carmel canyons off the California coast as they cut across the continental slope and up into the continental shelf. (b) Cross-canyon profile of the Monterey Canyon shown in (a). Compare this profile with that of the Grand Canyon drawn to the same scale.

up, into, and across the continental shelf. A submarine canyon is steep-sided and has a V-shaped cross section, with tributaries similar to those of river-cut canyons on land. Figure 4.4a shows the Monterey and Carmel canyons off the coast of California. Figure 4.4b compares a profile of the Monterey Canyon with a profile of the Grand Canyon of the Colorado River.

Many of these submarine canyons are associated with existing river systems on land and were apparently cut into the shelf during periods of low sea level, when the rivers flowed across the continental shelves. Ripple marks on the floors of canyons and sediments fanning out at the ends of these canyons suggest that they have been formed by moving flows of sediment and water called **turbidity currents.** Turbidity currents are fast-moving avalanches of mud, sand, and water caused by earthquakes or the overloading of sediments on steep slopes. They flow down the slope, eroding and picking up sediment as they gain speed, and over time they erode the slope and excavate the submarine canyon. The abundance of submarine canyons can be seen along the continental slopes in the chart of the sea floor shown earlier in figure 3.6. As the turbidity current flow reaches the sea floor, it slows and spreads, and the sediments settle. During this settling process, deposits of coarse materials overlaid by successive layers of decreasing particle size called **turbidites** are formed; figure 4.5 shows a turbidite preserved in a shore cliff. Turbidity currents have never been directly observed, although similar but smaller and more continuous flows, such as sand falls, have been observed and photographed.

At the base of the steep continental slope, there may be a gentle slope formed by the accumulation of sediment. This portion of the sea floor is the **continental rise** (fig. 4.3); it is a region of sediment deposition by turbidity currents, underwater landslides, and any other processes that carry sand, mud, and silt down the continental slope. The continental rise is a

FIGURE 4.5

This beach cliff shows a series of ancient turbidite flows that have been uplifted and then exposed by wave erosion.

conspicuous feature in the Atlantic and Indian Oceans and around the Antarctic continent. Few continental rises occur in the Pacific Ocean; here great seafloor trenches are often at the base of the continental slope, where material from the active margin is being subducted.

4.3   THE OCEAN BASIN FLOOR

The deep sea floor, between 4000 and 6000 meters (13,000–20,000 ft), covers more of the earth's surface (30 percent) than do the continents (29 percent). In many places, the ocean floor is a flat plain, known as the **abyssal plain.** Abyssal plains form as sediment deposits (from turbidity currents and the waters above) cover the irregular topography of the sea floor. Large wave-like undulations have been discovered in these sediment layers; these "mud waves" are formed by deep-ocean currents flowing across the abyssal plain. **Abyssal hills** and seamounts are scattered across the sea floor in all the oceans. Abyssal hills are less than 1000 meters (3300 ft) high, and seamounts are steep-sided volcanoes that are formed by local vulcanism and over hot spots; they rise abruptly toward the sea surface. Sometimes seamounts pierce the sea surface to become islands. These features are shown in figure 4.6. Abyssal hills are probably the earth's most common topographic feature. They are found over 50 percent of the Atlantic sea floor and about 80 percent of the Pacific floor; they are also abundant in the Indian Ocean.

In the warm waters of the Atlantic, Pacific, and Indian Oceans, coral reefs and coral islands are formed in association with seamounts. Reef-building coral is a warm-water animal that requires a place of attachment and grows in intimate association with a single-celled, plant-type organism. It is therefore confined to sunlit, shallow, warm waters. When a seamount pierces the sea surface to form an island, it provides a base on which the coral can grow. The coral grows to form a **fringing reef** around the island. If the seamount sinks or subsides slowly enough, the coral continues to grow upward at the same rate as the rising water, and a **barrier reef** with a lagoon between the reef and the island is formed. If the process continues, eventually the seamount disappears below the surface and the coral reef is left as a ring, or **atoll.** This process is illustrated in figure 4.7.

These steps required to form an atoll were suggested by Charles Darwin, based on his observations during the voyage of the *Beagle* in 1831–36. Darwin's ideas have been proved substantially correct by more recent expeditions that drilled through the coral debris of lagoon floors and found the basalt peak of the seamount that once protruded above the sea surface.

Submerged flat-topped seamounts, known as **guyots** (fig. 4.6), are found most often in the Pacific Ocean. These guyots are 1000 to 1700 meters (3300 to 5600 ft) below the surface, with many at the 1300 meter (4300 ft) depth. Many guyots show the remains of shallow marine coral reefs and the evidence of wave erosion at their summits. This indicates that at one time they were surface features and that their flat tops are the result of past coral reef growth, wave erosion, or both. Guyots have dropped far below sea level because their weight has helped depress the oceanic crust; refer to section 3.11. They have also been submerged by rising sea level during periods in which the land ice has melted.

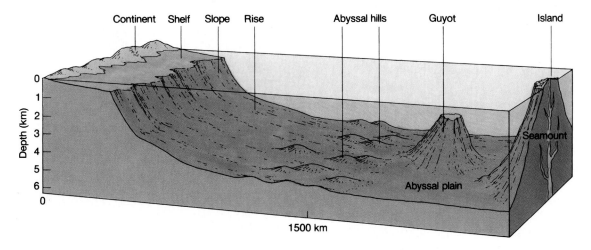

FIGURE 4.6
An idealized portion of ocean basin floor with abyssal hills, a guyot, and a seamount on the abyssal plain. Seamounts and guyots are known to be volcanic in origin (vertical × 100).

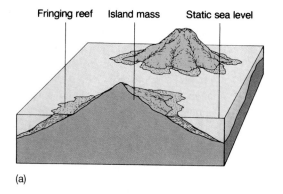

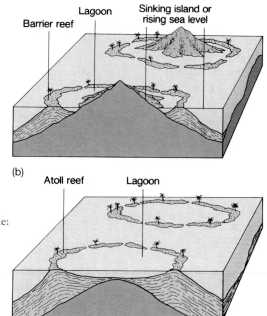

FIGURE 4.7

Types of coral reefs and the steps in the formation of a coral atoll are shown in profile: (a) fringing reef; (b) barrier reef; and (c) atoll reef.

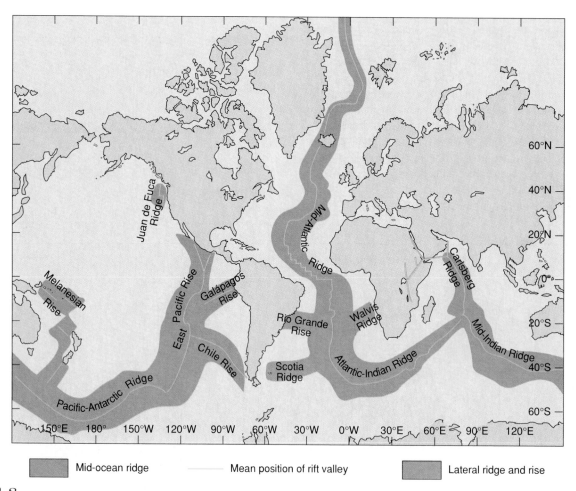

Mid-ocean ridge   Mean position of rift valley   Lateral ridge and rise

FIGURE 4.8

The mid-ocean ridge and rise system with its central rift valley.

4.4 THE RIDGES AND RISES

The most remarkable features of the ocean floor are the mid-ocean **ridge and rise systems** that form at the spreading boundaries of the lithospheric plates. This series of great continuous underwater volcanic mountain ranges stretches for 65,000 kilometers (40,000 mi) around the world and runs through every ocean; see figure 4.8 and refer to figure 3.6. These ridge systems are about 1000 kilometers (600 mi) wide and 1000 to 2000 meters (3500 to 7000 ft) high. If the slopes of these mountain ranges are steep, they are referred to as ridges (such as the Mid-Atlantic Ridge and the Mid-Indian Ridge); if the slopes are more gentle, they are called rises (such as the East Pacific Rise). Along portions of the system's crest is a central **rift valley,** 15 to 50 kilometers (9 to 30 mi) wide and 500 to 1500 meters (1500 to 5000 ft) deep. The rift valley is volcanically very active and bordered by rugged rift mountains. Fracture zones with steep sides run perpendicular to the ridges and rises, connecting offset sections of the mid-ocean ridges to make a stair-step pattern. The faults between the offset sections of rift valley are the transform faults; refer to section 3.7. The mid-ocean ridge and rise systems and their smaller lateral extensions separate the ocean basins into sub-basins, in which bodies of deep water are isolated from each other. Refer to figure 3.6.

Distinguish between a ridge and a rise.

Where would you find examples of each? Use figure 3.6.

Use figures 3.6 and 4.8 to identify South Atlantic ocean basins formed by the Mid-Atlantic Ridge and its lateral extensions.

4.5 THE TRENCHES

The Mid-Atlantic Ridge is the dominant feature of the floor of the Atlantic Ocean, but the narrow, steep-sided, deep-ocean trenches characterize the Pacific. Trenches occur at converging plate boundaries; refer to section 3.9 to review their formation, and see figures 4.9 and 3.6 for their locations. Some of these trenches are located on the seaward side of chains of volcanic islands known as **island arcs.** For example, the Japan-Kuril Trench, the Aleutian Trench, the Philippine Trench, and the deepest of all ocean trenches, the Mariana Trench, are all associated with island arc systems. The Challenger Deep, a portion of the Mariana Trench, has a depth of 11,020 meters (36,150 ft), making it the deepest known spot in all the oceans. The longest of the trenches is the Peru-Chile Trench,

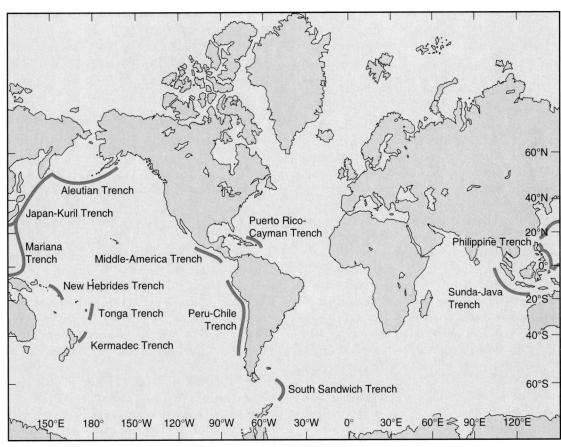

FIGURE 4.9
Major ocean trenches of the world.

The Sea Floor

stretching 5900 kilometers (3700 mi) along the west side of South America. To the north, the Middle-America trench borders Central America. The Peru-Chile and Middle-America Trenches are not associated with volcanic island arcs, but are bordered by land volcanoes. In the Indian Ocean the great Sunda-Java Trench runs for 4500 kilometers (2800 mi) along Indonesia, while in the Atlantic there are two comparatively short trenches, the Puerto Rico-Cayman Trench and the South Sandwich Trench, both associated with chains of volcanic islands.

What land features are associated with deep-sea trenches?

Where are the major deep-sea trenches of the world's oceans? Use figure 3.6.

4.6 MEASURING THE DEPTHS

Early sailors made their soundings, or depth measurements, using a hemp line or rope marked in equal distances (usually in **fathoms,** a 6-foot unit) with a greased lead weight at the end. The change in line tension when the weight touched bottom indicated depth, and the particles from the bottom adhering to the grease confirmed the contact and brought a bottom sample to the surface. This method was quite satisfactory in shallow water, and the experienced captain used bottom samples with depth measurements to aid in navigation, particularly at night or in heavy fog.

Later, piano wire with a cannonball attached was used for deep-water soundings. The heavy weight of the ball made it easier to sense the bottom, but the time (eight to ten hours) and effort to winch the wire out and in for each measurement were so great that by 1895 over all the world's oceans only about 7000 depth measurements had been made in water deeper than 2000 meters and only 550 measurements had been made of depths greater than 9000 meters. It was not until the 1920s, when the **echo sounder** or **depth recorder** was invented, that deep-sea depth measurements became routine. The depth recorder measures the time required for a sound pulse to leave the surface vessel, reflect off the bottom, and return. It allows continuous measurements to be made easily and quickly while the ship is under way. The behavior of sound in seawater and its use as an oceanographic tool are discussed in Chapter 5. Today data from precision depth recorders, underwater photography, and television (fig. 4.10), as well as direct observation from submersibles, are used to produce even more detailed knowledge of the sea floor.

A new method of sensing seafloor features is based on satellite measurements of the oceans' gravity and irregular surface. These irregularities are related to changes in local gravity due to the uneven sea floor. The sea surface may be depressed as much as 60 meters (200 ft) over ocean trenches where the mass of the earth's crust is reduced and may bulge as much as 5 meters (15 ft) over seamounts and ridges where the mass of crust is increased (fig. 4.11).

FIGURE 4.10
This internally recording television camera is placed on the sea floor, where it automatically photographs events until it is retrieved by the researchers.

What kinds of equipment have oceanographers used to develop the detailed seafloor chart shown in figure 3.6?

How was figure 4.11 obtained?

4.7 SEDIMENTS

The margins of the continents and the ocean basin floors receive a continuous supply of particles from many sources. Whether these particles have their origin in living organisms, the land, the atmosphere, or the sea itself, they are called sediment when they accumulate on the sea floor. The thickest deposits of sediment are found near the margins of continents, while the deep sea floor receives a constant but slow buildup of sediment.

Oceanographers study the rate at which sediments accumulate, the distribution of type and abundance of sediments over the sea bottom, their chemistry, and the history they record in layer after layer as they slowly but continuously settle on top of the oceanic crust. In order to describe and catalog the sediments found on the ocean floor, geological oceanographers have classified the sediments by their source, place of deposit, chemistry, particle size, age, and color.

Figure 4.11

Sea surface topography reflects seafloor features. Seventy days' worth of data collected by the short-lived SEASAT satellite were used to produce this computer-generated chart of mean sea surface topography. Marked are (1) the Mid-Atlantic Ridge, (2 and 3) trenches along the west and northwest margins of the Pacific, (4) the Hawaiian Island–Emperor Seamount system, (5) the Louisville Ridge of the South Pacific, (6) fracture zones produced by stresses acting on the earth's crust, and (7) low-amplitude bumps which may mark upwellings of molten rock circulating in the earth's interior.

4.8 SEDIMENT SOURCES AND DEPOSIT PATTERNS

Classification by source relates the sediments to their origins from rocks, seawater, space, and organisms. Sediments derived from rock are **lithogenous sediments.** Rock at or above sea level is weathered by wind, water, and continual freezing and thawing. The resulting particles are transported to the oceans by water, wind, ice, and gravity. It is estimated that 51 percent of the world's total river input and 60 percent of the world's river-borne sediments enter the oceans in the tropics. The tropics lie between approximately 15°N and 15°S, a zone of high precipitation and little seasonal change in the weather. These conditions result in large accumulations of lithogenous sediment in coastal areas. Windblown dusts from arid areas of the continents, ash from active volcanoes, and rocks picked up by glaciers and embedded in icebergs are additional sources of lithogenous materials.

In certain areas sediments are produced in seawater by chemical processes. These are called **hydrogenous sediments** and include **carbonates** (limestone-type deposits), **phosphorites** (phosphorus in phosphate form in crusts and nodules), and **manganese nodules.** The manganese nodules are deposits of manganese, iron, copper, cobalt, and nickel in the form of potato-shaped nodules scattered across the ocean floor and lying on top of the other sediments; see figure 4.12. Nodules develop very slowly (1–10 millimeters per million years), accumulating layer upon layer, often around a hard skeletal piece such as an earbone of a whale or a shark's tooth. The nodules form in areas of scant sediment deposits or rapid bottom currents. If there is a rapid accumulation of other sediments, nodules either do not form or are buried quickly by the constant rain of particles from above.

Particles from space constantly bombard the earth's surface; most of the particles are very small and stay in suspension for a very long time, and most are thought to dissolve as they slowly sink to the sea floor. These iron-rich **cosmogenous sediments** are found in small amounts in all oceans.

Sediments formed from the remains of living organisms are called **biogenous sediments** and may include shell and coral fragments as well as the hard skeletal parts of single-celled plants and animals that live in the surface waters. Deep-sea biogenous sediments are composed almost entirely of the hard outer coverings of single-celled organisms (fig. 4.13). Their chemical composition is either calcareous (calcium carbonate or seashell material) or siliceous (clear and hard). If fine sediments from the deep sea are more than 30 percent biogenous sediment by weight, the sediment is known as an ooze—a **calcareous ooze** or a **siliceous ooze.**

The distribution of calcareous and siliceous oozes in the sediments is related to the supply of organisms, the dissolving abilities of the water through which the organisms fall, water depth, and dilution with other sediment types. In the Atlantic Ocean, the waters from the surface down to a depth of about 4000 meters (12,000 ft) are saturated with calcite, a form of calcium carbonate, and the calcareous materials do not dissolve in these calcite-rich waters. Below 4000 meters, the water is undersaturated with calcite and calcareous materials dissolve read-

(a)

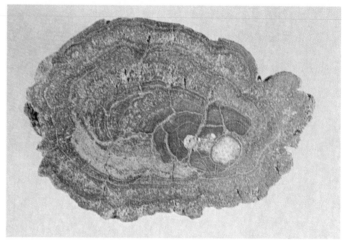

(b)

FIGURE 4.12

(a) Manganese nodules resting on red clay photographed on deck in natural light. (b) A cross section of a manganese nodule showing concentric layers of formation.

ily. In the North Pacific Ocean, water saturated with calcite only occurs from the surface to a few hundred meters in depth. This means that calcareous materials in this ocean change very little in the surface water but begin to dissolve at depths greater than 500 meters (1600 ft). The difference between these two oceans is associated with the age, acidity, and circulation of the deep water in each ocean.

Calcareous oozes are found where the production of organisms is high and dilution by other sediments is small. Microscopic, single-celled foraminifera, relatives of amoebae; small, swimming snails, the pteropods; minute, single plant-type cells called coccolithophores (fig. 4.13) all contribute their hard parts. Look at figure 4.14 to locate the deposits of calcareous oozes that are often found on the major ridge systems and in the warmer shallower areas of the South Pacific. Because there is no major mid-ocean ridge in the North Pacific, and because the calcite saturation depth is shallow, calcareous materials falling through deep water dissolve before reaching the sea floor.

(a)   (b)

FIGURE 4.13
Scanning electron micrographs of biogenous sediments: (a) Diatoms and coccolithophores. The small disks are detached coccoliths. (b) Radiolarians.

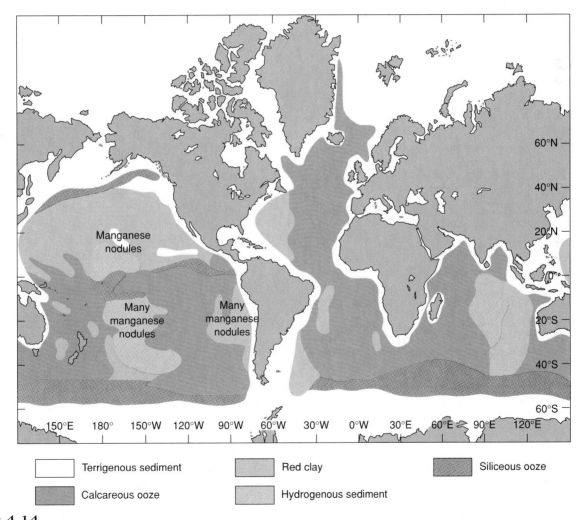

Terrigenous sediment   Red clay   Siliceous ooze

Calcareous ooze   Hydrogenous sediment

FIGURE 4.14
Distribution of the principal sediment types on the deep sea floor. Sediments are usually a mixture but are named for their major component.

Siliceous remains of organisms are found at all depths because of their slow dissolving rates at all temperatures. The principal source of siliceous oozes at colder latitudes is a single-celled plant, the diatom (fig. 4.13a). Diatoms reproduce rapidly and large populations are found in the areas that combine light, nutrients, and suitable temperatures around Antarctica

FIGURE 4.15

Classification of sediments by location of deposit. The distribution pattern is partially controlled by the sediments' proximity to source and rate of supply.

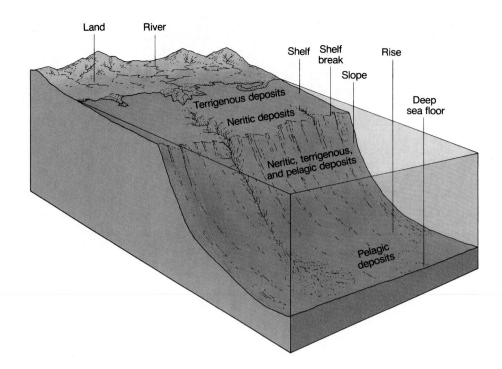

and in a band across the North Pacific; see figure 4.14. Siliceous sediments are also found in equatorial latitudes, but there they are produced by a single-celled amoeba-like animal, the radiolarian. See figure 4.13b for radiolarian ooze at high magnification and figure 4.14 for the extent of these deposits on the sea floor.

On the deep ocean floor, far away from the land, and under water that holds little marine life, **red clay,** a very fine lithogenous material, accumulates very slowly (fig. 4.12a). This fine rock powder, blown out to sea by wind and swept out of the atmosphere by rain, may remain suspended in the water for many years. The red-brown color is due to the oxidizing (or rusting) of the iron present in the sediment. Although traces of red clay are found throughout the oceans, it only becomes a substantial component of the sediment in areas where sediments from other sources do not mask its presence, such as the deep basin areas of the Pacific (fig. 4.14).

Ocean scientists classify sediments by the area in which they are deposited as well as by their source. **Terrigenous sediments** are derived from land and are found close to their land source; they contain silt, sand, gravel, and wood chips, for example. **Neritic sediments,** which come from land and sea sources, are found under the shallow waters of the continental shelf. **Pelagic sediments** are deep-sea sediments that are laid down far from the direct influence of land. This pattern of sediment deposits is shown in figure 4.15.

The patterns formed by the sediments on the sea floor reflect both the distance from their source and the processes that produce, transport, and deposit them. Coarse sediments are concentrated close to their sources; finer sediments are deposited offshore away from the currents and waves or in quiet bays and harbors. Seventy-five percent of marine sediments are terrigenous and primarily lithogenous; the majority are deposited on the continental margins, supplied by rivers and wave erosion along the coasts. River deltas and bays serve as sediment traps, preventing some of the terrigenous sediments from reaching the deep sea floor. Evaporative deposits left by ancient shallow seas are found on land and, in some cases, are found in the sea buried under more recent sediments. Marine sediments are also recycled as the earth's tectonic processes elevate old sea floor to the land, where it is eroded by the winds and the rain and returned once again to the sea.

Oceanic sediments are laid down in layers, characterized by changes in color, particle size, and kinds of particles (fig. 4.16). Layering occurs if the sediment type varies over time or if the rate at which a sediment type is supplied varies. Patterns of long and short growing seasons among marine life, for example, can be determined from the properties and thicknesses of the layers of biogenous material. Over long periods of geologic time, climatic changes such as the ice ages have altered the populations of organisms that produce sediments and so have left their record in the sediment layers.

A summary of the major sediment types is given in table 4.1.

What are the principal sources of ocean sediments?

What can you learn about the formation of manganese nodules from figure 4.12?

Relate the distribution of calcareous oozes to seafloor features.

Explain the factors responsible for the seafloor location of siliceous oozes and red clays.

Give examples of sediments classified by their place of deposit.

Explain the layering of the core shown in figure 4.16.

Figure 4.16
A deep-sea sediment core obtained by the drilling ship *Glomar Challenger*. Note the layering of the sediments.

4.9 Sediment Sizes, Sinking Rates, and Accumulation

Sediments may also be classified by particle size; see table 4.2. Familiar terms such as gravel, sand, mud, and pebble are defined by the specific size ranges of their particles. Geologists use a phi or ϕ size scale, based on the negative of the power of 2, to describe these size ranges. For example, if particle diameter is 64 millimeters, or 2^6 millimeters, then its phi size is -6; a fine silt with grains $1/128$ millimeters, or 2^{-7} millimeters, has a phi size of 7. When a sediment sample is collected it is dried and shaken through a series of sieves that have woven mesh of decreasing size (fig. 4.17). Material that passes through one sieve but not the next is said to have a diameter size range between the two phi sizes of the screens.

A sample is said to be well sorted if it is nearly uniform in particle size and poorly sorted if it is made up of many different particle sizes. If a poorly sorted sediment is moved by turbulent water and currents, the smaller particles (with larger positive phi sizes) are carried away by the water and deposited elsewhere. Size analysis is a useful tool both for classifying sediments and for determining how size distribution has changed with time. Different size particles sink at different rates; the smaller the particle is the more slowly it sinks (table 4.2). Because small particles remain in the water much longer than coarse particles, they are more likely to be displaced, contributing to the pattern of coarser particles near their source and finer particles at a greater distance. Smaller soluble particles also have time to dissolve as they slowly sink in the deep ocean.

If small particles are differently charged, they will attract each other, forming larger particles that sink more rapidly. Also, when small organisms ingest tiny particles, the particles are expelled as larger fecal pellets, which sink more rapidly.

Table 4.1 Sediments Summary

Type	Source	Areas of Significant Deposit	Examples
Lithogenous	Eroded rock	Terrigenous	Boulder–sand
	Volcanoes	Neritic	Sand–silt
	Airborne dust	Pelagic	Silt–red clay
Hydrogenous	Chemical precipitation from seawater	Neritic	Phosphates
		Pelagic	Manganese nodules
Biogenous	Living organisms	Terrigenous	Plant matter
		Neritic	Coral, shell
		Pelagic	Hard bony remains
			Oozes:
			Siliceous
			Calcareous
Cosmogenous	Outer space	Pelagic	Meteorites

Table 4.2 Sediment Size Classifications

Descriptive Name		Diameter (mm)	Phi Size[1]	Sinking Rate (cm/s)
Gravel	Boulder	>256	<−8	>4.29×10^6
	Cobble	64–256	−6 to −8	2.68×10^5 to 4.29×10^6
	Pebble	4–64	−2 to −6	1.05×10^3 to 2.68×10^5
	Granule	2–4	−1 to −2	2.62×10^2 to 1.05×10^3
Sand	Very coarse	1–2	0 to −1	65.5–262
	Coarse	0.5–1	+1 to 0	16.4–65.5
	Medium	0.25–0.5	+2 to +1	4.09–16.4
	Fine	0.125–0.25	+3 to +2	1.02–4.09
	Very fine	0.0625–0.125	+4 to +3	0.256–1.02
Mud	Silt	0.0039–0.0625	+8 to +4	9.96×10^{-4} to 2.56×10^{-1}
	Clay	<0.0039	>8	<9.96×10^{-4}

[1]Phi size = the negative of the power of 2 required to equal the particle diameter in millimeters.

FIGURE 4.17
Screens used for sorting sediment samples into different size ranges. Examples of materials sorted into phi (ϕ) sizes.

FIGURE 4.18

A geological dredge with heavy-duty chain bag. The dredge picks up loose rock and breaks off fragments from rock outcroppings on the sea floor.

These processes help to speed small particles to the sea floor, minimizing their displacement by water movements.

An average accumulation value for sediments in the deep oceans is 0.5 to 1.0 centimeters (0.2 to 0.4 in)/1000 years, while at the ocean margins the rates are considerably greater and more varied. In river estuaries, sedimentation rates can reach several meters per year; the rivers of Asia, such as the Ganges, the Yangtze, the Yellow, and the Brahmaputra, contribute more than one-quarter of the world's terrigenous marine sediments each year. In quiet bays, the deposit rate may be 500 centimeters (200 in)/1000 years, while on the continental shelves and slopes, values of 10 to 40 centimeters (4 to 16 in)/1000 years are typical, with the flat continental shelves receiving the larger amounts.

Although deep-sea sedimentation rates are extremely slow, there has been plenty of time during geologic history to accumulate the average deep-sea sediment thickness of approximately 500 to 600 meters (1600 to 2000 ft). At a rate of 0.5 centimeters (0.2 in)/1000 years, it takes only 100 million years to accumulate 500 meters of sediment.

Relate sediment size and sinking rate to the accumulation patterns found on the sea floor.

FIGURE 4.19

Vanveen (left) and orange peel (right) grab samplers are shown in open positions. Grabs take surface sediment samples.

4.10 SAMPLING THE SEDIMENTS

In order to analyze sediments, the geological oceanographer must have an actual bottom sample to examine. A variety of devices have been developed to take a sample from the sea floor for laboratory analysis. **Dredges** are net or wire baskets that are dragged across the bottom to collect loose bulk material, surface rocks, and shells; see figure 4.18. **Grab samplers** are hinged devices that are spring- or weight-loaded to snap shut when the sampler strikes the bottom. Figure 4.19 shows examples of this kind of device.

A **corer** is a hollow pipe with a sharp cutting end (fig. 4.20). The free-falling pipe is forced down into the sea floor, by its weight (fig. 4.20a and c) or, for longer cores, by a piston device within the core barrel that uses water pressure to drive the corer into the sediments (fig. 4.20b). The use of a piston corer results in a cylinder of sediment, up to 20 meters (60 ft) long containing undisturbed sediment layers (like those in fig. 4.16). Box corers are used when a large and nearly undisturbed sample of surface sediment is needed. These corers drive a rectangular metal box into the sediment, with doors that close over the bottom of the box before the sample is retrieved (fig. 4.20d). Longer cores may be obtained by drilling, as discussed in Chapter 3.

Discuss the advantages and disadvantages of sediment sampling devices.

4.11 SEABED RESOURCES

Mineral Resources

People began to exploit the materials of the seabed long ago. The ancient Greeks extended their lead and zinc mines under the sea, and 16th century Scottish miners followed veins of coal under the Firth of Forth at Culross. As technology has developed, and as people have become concerned about the depletion of onshore mineral reserves, interest in seabed minerals

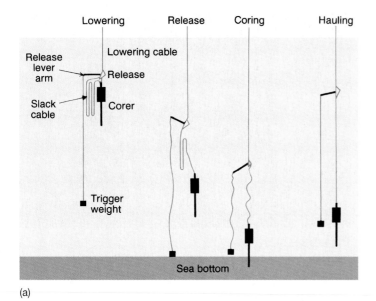

(a)

(b)

(c)

(d)

FIGURE 4.20

(a) The Phleger corer is a free-fall gravity corer. The weights help to drive the core barrel into the soft sediments. Inside the core is a plastic liner. The sediment core is removed from the corer by removing the plastic tube, which is capped to form a storage container for the core. (b) Loading a piston corer with weights to prepare it for use. (c) A gravity corer ready to be lowered. (d) A box corer is used to obtain large, undisturbed seafloor surface samples.

and mining has grown. The development of a seabed source depends largely on international markets, needs for strategic materials, and whether offshore production costs can compete with onshore costs.

The largest superficial seafloor mining operations are for sand and gravel, used in cement and concrete for buildings, for landfills, and to construct roads and artificial beaches. This is a high-bulk, low-cost material tied to the economics of transport

FIGURE 4.21
The Chevron oil-drilling platform *Hermosa* off the California coast.

and the distance to market. Annual world production is approximately 1.2 billion metric tons; the reported potential reserve is more than 800 billion tons. Sand and gravel mining is the only significant seabed mining done by the United States at this time. It is estimated that the United States has a reserve of 450 billion tons of sand off its northeast coast. Along the Gulf coast, shell deposits are mined for use in the lime and cement industries and as a gravel substitute. Iron-rich sediments are dredged in Japan with an estimated reserve of 36 million tons, and tin is recovered from sands found from northern Thailand to Indonesia.

Phosphorite, which can be mined to produce the phosphates needed for fertilizers, is found in shallow waters as phosphorite muds and sands and as nodules on the continental shelf and slope. Large deposits are known to exist off Florida, California, North Carolina, Mexico, Peru, Australia, Japan, and northwestern and southern Africa.

In the Gulf of Mexico, sulfur is present in mineable quantities. Today, the cheaper and easier recovery of sulfur waste from pollution-control equipment has replaced all but one of the area's mining operations. Millions of tons of sulfur reserves exist in the Gulf of Mexico and the Mediterranean Sea.

Coal deposits under the sea floor are mined when the coal is present in sufficient quantity and quality to make the operation worthwhile. In Japan, undersea coal deposits are reached by shafts that stretch under the sea from the land or descend from artificial islands.

Oil and gas represent 90 percent of the mineral value presently taken from the sea floor. 1988 figures show that 10.8 percent of U.S. oil and 24.6 percent of U.S. gas production came from offshore areas. Major offshore oil fields are found in the Gulf of Mexico, the Persian Gulf, the North Sea, and off the northern coast of Australia, the southern coast of California (fig. 4.21), and the coasts of the Arctic Ocean. Three new, large fields are opening in the 1990s off northeastern Newfoundland, off Louisiana, and off Norway in the North Sea. There are still many areas of the world that are relatively unexplored in the search for oil and gas, for example, the continental shelves of east and south Asia, east and northwest Africa, parts of South America, Siberia, and Antarctica. Continental margin deposits account for nearly one-third of the world's estimated gas and oil resources. Although the cost of drilling and equipping an offshore well is three to four times greater than a similar venture on land, the large size of the deposits allows offshore ventures to compete successfully. The gas and oil potential of the deeper areas of the sea floor is still relatively unknown, but the deeper the water in which the drilling must be done, the higher the cost. Though legal restraints, environmental concerns, and political uncertainties continue to slow the development of offshore deposits, petroleum exploration and development is expected to continue as the main focus of ocean mining in the near future.

Manganese nodules, the hydrogenous pelagic deposits found scattered across the world's deep ocean floors, have been the focus of intense research and development of mining and extraction techniques over the last twenty-five years. The mineral content varies from place to place, but the nodules in some areas contain 30 percent manganese, 1 percent copper, 1.25 percent nickel, and 0.25 percent cobalt, much higher concentrations than are usually found in land ores. The nodules grow at an estimated rate of 0.1 millimeters/1000 years, but they are present in huge quantities. An estimated 16 million additional tons of nodules accumulate each year.

Since the 1960s, large multinational consortia and mining corporations have spent over $600 million to locate the highest nodule concentrations and develop technologies for their collection. However, expectations of rapid development have been disappointing; some of these consortia have withdrawn completely, while others have become dormant. In the 1990s, Australia and New Zealand with twelve small South Pacific island nations have continued to support survey cruises. India has applied to the United Nations to develop manganese nodule deposits in the Indian Ocean, and Japan continues its research interests. See the following section on mining laws and treaties for a discussion of the difficulties surrounding the ownership and harvesting of the pelagic nodules.

Expeditions to the rift valleys of the East Pacific Rise near the Gulf of California, the Galápagos Ridge off Ecuador, and the Juan de Fuca and Gorda Ridges off the northwestern United States have found deposits of minerals combined with sulfur to form sulfides of zinc, iron, copper, and possibly silver, lead, chromium, gold, and platinum. Molten material from beneath the earth's crust rises along the rift valleys, fracturing and heating the rock. Seawater percolates into and through the fractured rock forming these mineral-rich deposits. Deposits may be tens of meters thick and hundreds of meters long. Too little is presently known about these deposits to know whether they might be of economic importance at some future time. There is no technology to retrieve them at this time, and like the manganese nodules these deposits are found outside national economic zones presenting ownership problems; see the next section.

What seabed resources have been identified with commercial value, and which of these are being exploited at the present time?

Mining Laws and Treaties

Because of the potential value of deep-sea minerals, specifically manganese nodules, and because the nodules are found in international waters, the developing nations of the world feel they have as much claim to this wealth as those countries that are presently technically able to retrieve the nodules. The developing nations want access to the mining technology and a share in the profits. For nearly ten years, the United Nations worked to produce the United Nations Convention on the Law of the Sea (UNCLOS). One section of this treaty regulated deep-ocean exploitation, including mining. The treaty recognized the manganese nodules as the heritage of all humankind, to be regulated by a UN seabed authority, which would license private companies to mine in tandem with a UN company. The quantities removed would be limited and the profits would be shared. The Law of the Sea Treaty was completed in April 1982. The United States and some other industrialized countries chose not to sign the treaty at that time.

During the past several years, negotiations have been underway to resolve issues related to the part of the convention that deals with seabed mining. The United States announced that it intends to sign a new agreement that changes the provisions for managing the mining of deep-sea resources, when UNCLOS came into force in 1994. The new agreement and convention was forwarded to the U.S. Senate for consideration and ratification in the fall of 1994. Much discussion will be provoked by the need to conform existing federal laws and policies with the provisions of the convention. It may take several years of deliberation before the United States ratifies the agreement and convention.

In any case, it is unlikely that any deep-sea mining will occur until well into the twenty-first century. The high costs of sea mining, low international metal market prices, and still undeveloped land sources combine to make rapid commercialization unlikely.

Explain the present situation concerning the commercial exploitation of deep-sea resources.

SUMMARY

The features of the ocean floor are as rugged as the features of the land but erode more slowly. The continental margin includes the continental shelf, shelf break, slope, and rise. Submarine canyons are major features of the continental slope and, in some cases, the continental shelf. Some canyons are associated with rivers; others are believed to have been cut by turbidity currents. Turbidity currents deposit sediments known as turbidites.

The flat abyssal plains are interrupted by scattered abyssal hills, volcanic seamounts, and flat-topped guyots. In warm shallow water, corals have grown up around the seamounts to form fringing reefs. A barrier reef is formed when a seamount subsides while the coral grows. An atoll results when the seamount's peak is fully submerged. Great, continuous volcanic mountain ranges, the mid-ocean ridges and rises, extend through all the oceans. Rift valleys run along the ridge and rise crests; fracture zones form perpendicular to the ridge system. Trenches are long, deep depressions in the ocean floor that are associated with volcanic island arcs and are found mainly in the Pacific Ocean.

Depth soundings were made first with hand lines, then with wire lines, and are now made with echo sounders.

Sediment classifications are based on their sources, areas of deposit, and particle size. Sediments formed from rock are lithogenous, those from seawater are hydrogenous, those from outside the earth's atmosphere are cosmogenous, and those from living organisms are biogenous. Sediments made up of 30 percent or more biogenous material are called oozes. Siliceous particles are less soluble than calcareous particles, which dissolve at depths below 4000 meters. Red clay is fine, land-derived material that accumulates slowly in the deep sea. Sediments arising from and deposited close to land are terrigenous; those produced by processes over the continental shelf are neritic; and pelagic sediments are found on the deep sea floor.

The sinking rate of particles is determined by their size, which is described by the phi size scale. Small particles sink more slowly than large particles. The sinking rate of particles is increased by clumping and incorporation into the fecal pellets of small marine organisms. Coarse sediments are concentrated close to shore; finer sediments are found in quiet offshore or nearshore environments. In general, sediments accumulate slowly in the deep sea and most rapidly near the continents.

Sediments are sampled with dredges, grab samplers, corers, and drilling ships.

Seabed resources include sand and gravel, used in construction and landfills. Sediments that are rich in mineral ores are mined. Phosphorite nodules have potential as fertilizer. Oil and gas are the most valuable of all seabed resources. Manganese nodules, rich in several metals, are present on the ocean floor in huge numbers. The status of manganese-nodule mining is clouded by disputes over international law with reference to mining claims and shared technology. Sulfide mineral deposits have been discovered along rift valleys; their economic importance is unknown.

KEY TERMS

sediment	abyssal hill	echo sounder/depth recorder	calcareous ooze
continental margin	fringing reef	lithogenous sediment	siliceous ooze
continental shelf	barrier reef	hydrogenous sediment	red clay
continental shelf break	atoll	carbonate	terrigenous sediment
continental slope	guyot	phosphorite	neritic sediment
submarine canyon	ridge and rise system	manganese nodule	pelagic sediment
turbidity current	rift valley	cosmogenous sediment	dredge
turbidite	island arc	biogenous sediment	grab sampler
continental rise	fathom	ooze	corer
abyssal plain			

SUGGESTED READINGS

Broadus, J. M. 1987. Seabed Materials. *Science* 235 (4791):853–60. Discussion of seabed resources.

Cruickshank, M. J. 1990. Ocean Mining Futures: Only a Matter of When. *Sea Technology* 31 (1):24–25.

Cruickshank, M. J. 1991. Ocean Mining: For the Future a Good Omen. *Sea Technology* 32 (1):38–39.

Emery, K. O. 1969. The Continental Shelves. In *Ocean Science, Readings from Scientific American* (1977), 221 (3):32–44.

Koski, R. A., W. R. Normark, J. L. Morton, and J. R. Delaney. 1982. Metal Sulfide Deposits on the Juan de Fuca Ridge. *Oceanus* 25 (3):43–48.

Menard, H. W. 1969. The Deep-Sea Floor. In *Ocean Science, Readings from Scientific American* (1977), 221 (3):55–64.

Oceanus. Fall 1983. 26 (3). Issue devoted to offshore oil and gas.

Pagano, S. 1991. Offshore Drilling, Production—New Waves of Technology. *Sea Technology* 32 (4):19–22.

Rice, A. L. 1991. Finding Bottom. *Sea Frontiers* 37 (2):28–33. History of depth measurement.

Rona, P. 1986. Mineral Deposits from Seafloor Hot Springs. *Scientific American* 254 (1):84–92.

Sea Technology. 1992. 33 (1) (Review and forecast of oil, gas, and mining activities.)

Vadus, J. R. 1989. Technology Need for Ocean Resources Utilization. *Sea Technology* 30 (8):14–25.

C H A P T E R

5

Water

Outline

Learning Objectives

After reading this chapter, you should be able to

- Describe the structure of the water molecule and how this structure affects the behavior of water.

- Describe the changes of state of water.

- Define density and explain how it is affected by changes in salinity, temperature, and pressure.

- Describe how water transmits energy in the form of heat, light, and sound.

- Discuss the practical use of sound in seawater.

- Understand the presence of salts as ions in seawater and the constant composition of seawater.

- Explain the concept of salinity.

- Explain variations in salinity along the coasts and in the surface waters of the deep sea.

- Discuss the processes that add and remove salt from seawater.

- Explain the importance of nutrients in seawater.

- Compare the distribution of dissolved oxygen and carbon dioxide in the oceans.

- Understand the importance of carbon dioxide as a buffer.

- Describe ways of extracting fresh water from seawater.

◄ Fog at Big Sur, California.

Not only is water the most common of substances on the earth's surface, but it is also uncommon in many of its properties. It makes life possible, and its properties largely determine the characteristics of the oceans, the atmosphere, and the land. Seawater is salt water, but it is much more than just salty water. Seawater contains dissolved gases, nutrient molecules, and organic substances as well as salt. To understand the oceans, it is necessary to learn something of the physical and chemical characteristics of seawater and to follow the processes that influence and regulate it.

5.1 THE WATER MOLECULE

The properties that make water such a special, useful, and essential substance result from its molecular structure. The water molecule is deceptively simple, made up of three atoms: two hydrogen atoms and one oxygen atom (H_2O). These combine to form a V-shaped molecule with the hydrogen atoms on one side and the oxygen atom at the other (fig. 5.1a). The negatively charged electrons within the molecule are distributed unequally, staying closer to the oxygen atom and giving this end a slightly negative charge. The hydrogen end of the molecule carries a slightly positive charge. Although the water molecule as a whole is neutral, its opposite sides have opposite charges. These opposing charges make water molecules attract each other, and bonds form between the positively charged side of one water molecule and the negatively charged side of another water molecule (fig. 5.1b). Any single bond is relatively weak, and as one bond is broken, another is formed. As a result, water is not a typical liquid. Compared with other liquids, it has extraordinary properties that are the consequence of attractions among water molecules.

Describe the relationship between the hydrogen and oxygen atoms in a water molecule.

How does this arrangement affect the behavior of water molecules?

5.2 CHANGES OF STATE

Water exists on the earth in three physical states: as a solid, a liquid, and a gas. When it is a solid, we refer to it as ice, and when it is a gas, we call it water vapor. Because of the bonding between the water molecules, it takes a great deal of energy to separate water molecules from the water's surface to form vapor and to separate water molecules from ice, melting it to water.

Heat is a form of energy that can be measured in **calories.** One calorie is the amount of heat needed to raise the temperature of 1 gram of water 1°C. If you are unfamiliar with the Celsius temperature scale, see Appendix C. Do not confuse heat energy with temperature. If 2 cups of water and 2 quarts of water at the same temperature are placed on the same size stove burners at the same time, it will take longer to boil the larger amount because more energy is needed to bring the larger quantity to a boil. Both amounts of water boil at the same temperature, but the amount of heat required to achieve the boiling is different in each case.

Water changes its state, between liquid and gas or liquid and solid, by the addition or loss of heat and the breaking and forming of bonds between molecules. When enough heat is added to ice at 0°C, these bonds break and the ice melts. When heat is added to liquid water, the temperature of the water increases and bonds break; the water evaporates to form vapor. When heat is removed from water vapor, the water vapor condenses and bonds form between molecules, producing a liquid, and when liquid water at 0°C loses sufficient heat, more bonds form and ice is the result. The **heat capacity** of water is the amount of heat required to raise the temperature of 1 gram of liquid water 1°C. The high heat capacity of water, compared to most substances, allows it to gain or lose large quantities of heat with little change in temperature.

FIGURE 5.1

(a) Two hydrogen atoms bond to an oxygen atom and produce a water molecule with a positive side and a negative side. The angle between the hydrogen atoms and the oxygen atoms is about 105°. (b) The positive and negative charges allow each water molecule to form bonds with other water molecules.

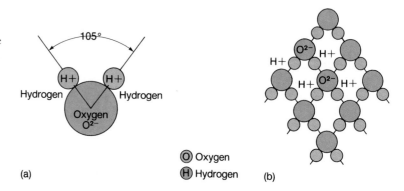

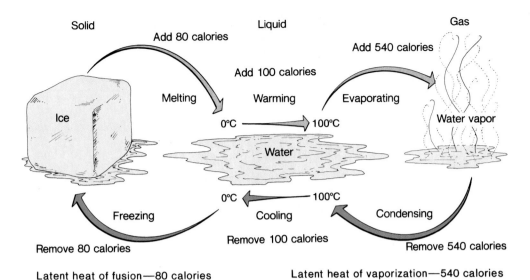

FIGURE 5.2
Heat energy must be added to convert a gram of ice to liquid water and to convert liquid water to water vapor. The same quantity of heat must be removed to reverse the process.

To change pure fresh water from a solid at 0°C to a liquid at 0°C requires the addition of 80 calories for each gram of ice (fig. 5.2). This is called latent heat of fusion; there is no change in temperature, only a change in the physical state of the water. This heat is released again in changing liquid water to ice. Ice is a very stable form of water, and it takes a lot of heat to melt it.

The change of state between liquid water and water vapor requires 540 calories of heat to convert 1 gram of water at 100°C to water vapor at 100°C. This is known as the latent heat of vaporization. When 1 gram of water vapor condenses and returns to the liquid state, 540 calories of heat are liberated. There is no change in temperature, only a change in the water's physical state. This process may also be seen in figure 5.2. Water does convert from the liquid to the vapor state at temperatures other than 100°C; for example, rain puddles evaporate and clothes dry on the clothesline. This type of change requires slightly more heat to convert the liquid water to a gas at these lower temperatures.

Adding salt to water changes its boiling and freezing points: The boiling temperature is raised and the freezing temperature is lowered. The amount of change is controlled by the amount of salt added. The rise in the boiling temperature is of little consequence since seawater does not normally boil in nature, but the lowering of the freezing point is important in the formation of sea ice. Typical seawater freezes at about –2°C.

What is heat and how is it measured?

How does heat differ from temperature?

Refer to figure 5.2 and compare the change of state between ice and water to that between water and water vapor. Consider heat requirements and bond formation.

5.3 DENSITY

Density is defined as mass per unit volume of a substance and is usually measured in grams per cubic centimeter, or g/cm^3. Less dense substances will float in more dense liquids (for example, oil on water or dry pine wood on water). Ocean water is more dense than fresh water; therefore fresh water floats on salt water.

When water is heated, energy is added and water molecules move apart; therefore the mass per cubic centimeter becomes less because there are fewer water molecules per cubic centimeter. For this reason, the density of warm water is less than that of cold water. When water is cooled, it loses heat energy, and the water molecules slow down and come closer together; there are then more water molecules, or a greater mass, per cubic centimeter. Because cold water is more dense than warm water, it sinks below the warm water.

In pure fresh water, molecules move closer and closer together as water is cooled to 4°C. At this temperature, fresh water reaches its greatest density, $1 \ g/cm^3$. As the temperature falls below 4°C, some water molecules start forming a crystal lattice (see fig. 5.1b), the water expands slightly, and the density decreases. Fresh water begins to freeze at 0°C when the water molecules no longer move enough to break the bonds with neighboring molecules. As the ice crystals form, the water molecules become widely spaced, resulting in fewer water molecules per cubic centimeter. Therefore, water at temperatures less than 4°C and ice are both less dense than water at 4°C.

When salts are dissolved in water, the density of the water increases because the salts have a greater density than water. The typical density of average ocean water at 4°C is $1.0278 \ g/cm^3$, in comparison to $1 \ g/cm^3$ for fresh water. Densities of water with and without dissolved salt are shown in table 5.1. Notice in this table that (1) if the salt content remains constant, density increases as temperature decreases and (2) if

Table 5.1 Density of Water With and Without Dissolved Salt, g/cm³

°C	No Salt	Salt 20 g/kg	25 g/kg	30 g/kg	35 g/kg
–1	ice 0.917	1.01606	1.02010	1.02413	1.02817
0	0.99984	1.01607	1.02008	1.02410	1.02813
1	0.99990	1.01605	1.02005	1.02406	1.02807
2	0.99994	1.01603	1.02001	1.02400	1.02799
3	0.99996	1.01598	1.01995	1.02393	1.02791
4	0.99997	1.01593	1.01988	1.02384	1.02781
5	0.99996	1.01586	1.01980	1.02374	1.02770
10	0.99970	1.01532	1.01920	1.02308	1.02697
15	0.99910	1.01450	1.01832	1.02215	1.02599
20	0.99820	1.01342	1.01720	1.02098	1.02478
25	0.99704	1.01210	1.01585	1.01960	1.02336
30	0.99565	1.01057	1.01428	1.01801	1.02175

temperature remains constant, density increases with increasing salt content. High densities are associated with cold salty water. Fresh water is densest at 4°C, and freshwater ice is less dense than either fresh or salty water. Chapter 7 presents a discussion of changes in seawater density with dissolved salts and temperature; see figure 7.1.

Density is also affected by pressure. Increasing the pressure increases the density of water by crowding the molecules together and reducing the volume occupied by a fixed mass of water. Pressure in the oceans increases with depth. For every 10 meters of descent, the pressure increases by about 1 atmosphere, or 14.7 lbs/in². (See Appendix C for other units of pressure.) Pressure in the deepest ocean trench, 11,000 meters (36,000 ft), is about 1100 atm. However, the compressibility of seawater is so small, decreasing the total world's ocean volume by reducing sea level about 37 meters (121 ft), that the pressure effect can be ignored except when a very accurate determination of density is required.

Define density.

Why are ice and water vapor less dense than water?

How are density, temperature, and the salt content of seawater related?

How does pressure cause changes in density?

5.4 Transmission of Heat

There are three ways in which heat energy moves through water: by **conduction,** by **convection,** and by **radiation.** Conduction is a molecular process. When heat is applied at one location, the molecules move faster due to the addition of energy; gradually, this molecular motion passes on to adjacent molecules, and the heat spreads by conduction. For example, if the bowl of a metal spoon is placed in a hot liquid, the handle soon becomes hot. Metals are excellent conductors; water is a poor conductor and transmits heat slowly in this way.

Convection is a density-driven process in which a heated fluid moves and carries its heat to a new location. Water rises when it is heated from below and its density is decreased; it sinks when it is cooled at its surface and its density is increased. Review the convection cells in Chapter 3.

Radiation is the direct transmission of heat from its energy source. The sun provides the earth's surface with radiant energy, which warms the surface waters. Some of this solar radiation is reflected from the surface and adds no heat to the water, and some is absorbed and raises the temperature of the surface water. Since warming the surface water makes it less dense, this warm water remains floating at the surface and convection does not occur. Some heat gained at the surface is conducted down, aided by the natural stirring action of the wind and currents. The oceans' surface waters lose heat to the atmosphere, where the heat is efficiently transferred upward by convection of the air.

Distinguish among the three methods of heat transmission in water.

5.5 Transmission of Light

Seawater transmits only the visible wavelengths of sunlight (fig. 5.3). About 60 percent of the entering light energy is absorbed in the first meter, and about 80 percent is gone after 10 meters (33 ft). Only 1 percent of the total light available at the surface is left in the clearest water below 150 meters (500 ft), and no sunlight penetrates below 1000 meters (3300 ft). Not all wavelengths of visible light are transmitted equally (fig. 5.4). The long wavelengths at the red end of the spectrum are absorbed rapidly, while the short wavelengths of blue-green light are transmitted to the greater depths. The light undergoes **absorption** and **scattering** by suspended particles, including silt, single-celled organisms, plants, and the water and salt molecules. This decrease in the intensity of light over distance is known as **attenuation.** The clearer the water, the greater the light penetration and the smaller the attenuation.

Color perception is due to the reflection back to our eyes of wavelengths of a particular color. The oceans usually appear blue-green because wavelengths of this color, being absorbed the least, are most available to be reflected and scattered. Coastal waters appear green or even brown or red, because this water usually contains silt from rivers as well as large numbers of microscopic organisms that add their color. Open-ocean water is often clear and blue, which indicates that it is nearly empty of both living and nonliving suspended materials. The clearer the water, the deeper the light penetration, and the less suspended matter present.

When light passes from the air into the water, it is bent by **refraction** because its speed is faster in air than in water. Because the light rays bend as they enter the water from the air, objects seen through the water's surface are not where they appear to be (fig. 5.5). Refraction is affected slightly by changes in salinity, temperature, and pressure. Devices that measure refraction of seawater (refractometers) can be used for convenient but low-precision determination of salinity.

Oceanographers use a light meter to measure the intensity of light in water, but the simplest way of measuring light attenuation in surface water is to use a **Secchi disk** (fig. 5.6). This is a white disk about 30 centimeters (12 in) in diameter, which is lowered to the depth at which it just disappears from view. The measure of this depth can be used to determine the average attenuation of light. In waters rich with living organisms or suspended silt, the Secchi disk may disappear from view at depths of 1 to 3 meters (3 to 10 ft) while in the open ocean it may be visible down to 20 to 30 meters (60 to 100 ft).

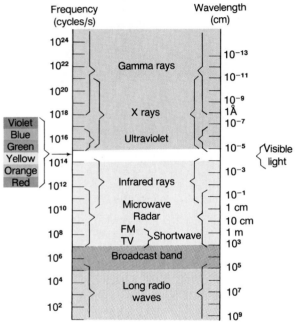

FIGURE 5.3

The electromagnetic spectrum. Visible light occupies only a small portion of the spectrum. The visible light is bounded on one side by longer infrared rays and on the other side by shorter ultraviolet rays. Water is relatively opaque to all wavelengths of this spectrum except visible light.

What portion of the electromagnetic spectrum is transmitted in seawater?

What processes affect the attentuation of light in seawater?

Why does the sea generally appear blue?

Why are light waves refracted in water and what is the result?

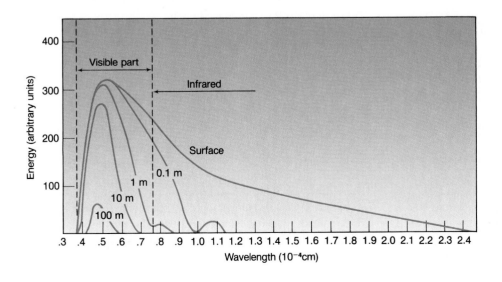

FIGURE 5.4

Total available solar energy in the sea (shown by the areas under the curves) decreases as depth increases. The longer, red wavelengths are absorbed first. The color peak shifts to the shorter, blue wavelengths as the depth increases.

Water

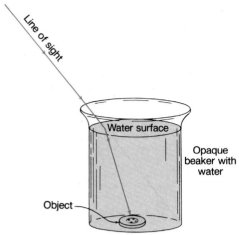

FIGURE 5.5
Objects that are not directly in one's line of sight can be seen in water due to the refraction of the light rays. The refraction is caused by the decreased speed of light in water.

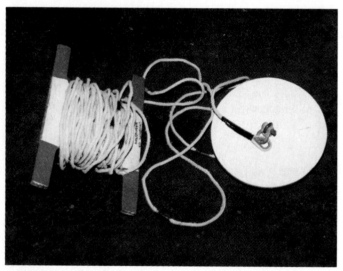

FIGURE 5.6
A Secchi disk. The line is marked with tape at meter intervals.

5.6 TRANSMISSION OF SOUND

Sound travels farther and faster in seawater than in air (average velocity in seawater is 1500 meters [5000 ft] per second as compared with 334 meters [1100 ft] per second in dry air at 20°C). The speed of sound in seawater increases with increasing temperature, pressure, and salt content, and decreases with decreasing temperature, pressure, and salt content.

Because sound is reflected back after striking an object, sound can be used to find objects, sense their shape, and determine their distance from the sound's source. If a sound signal is sent into the water and the time required for the return of the reflected sound, or echo, is measured accurately, the distance to an object may be determined. For example, if 6 seconds elapse between the outgoing sound pulse and its return after reflection, the sound has taken 3 seconds to travel to the object and 3 sec-

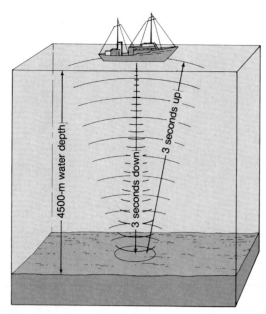

FIGURE 5.7
Traveling at an average speed of 1500 meters per second, a sound pulse leaves the ship, travels downward, strikes the bottom, and returns. In 4500 meters of water, the sound requires 3 seconds to reach the bottom and 3 seconds to return.

onds to return. Since sound travels at an average speed of 1500 meters per second in water, the object is 4500 meters away, as shown in figure 5.7.

Oceanographic vessels use echo sounders or depth recorders to direct a narrow sound beam vertically to the sea bottom. The depth recorder chart produces a continuous reading of depth as the vessel moves along its course. If high-intensity sound pulses are transmitted, sound energy can penetrate the sea floor and reflect back from layers in the sediments, providing a display of the sea floor's structure on the depth recorder chart (fig. 5.8). Section 6 in Chapter 4 discusses this topic as well.

Sonar (sound navigation and ranging) is a location system that directs a beam of sound pulses through the water and detects their echoes. Trained technicians are able to distinguish between the echoes produced by a whale, a school of fish, or a submarine. It is also possible to distinguish animals from vessels by listening to the sounds they generate, even to determine the type and class of a vessel from the sounds produced by its propellers and engines.

However, sound may be bent or refracted as it passes at an angle through water layers of differing density. When refraction occurs, targets appear displaced from their actual positions. Figure 5.9 illustrates the curvature of sound beams and how **sound shadow zones,** into which the sound does not penetrate, can form. Sound beams bend toward regions in which sound travels more slowly and away from regions in which the sound waves travel more rapidly. Understanding the behavior of sound in the oceans is crucial to naval vessels, both above and below the surface, and is the subject of much ongoing research.

At about 1000 meters (3300 ft), the combination of salt content, temperature, and pressure creates a depth zone of

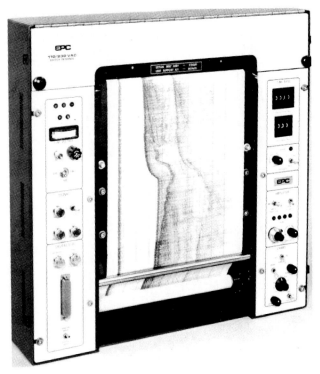

minimum velocity for sound, the **sofar** (sound fixing and ranging) **channel** (fig. 5.10). Sound waves produced in the sofar channel do not escape from it unless they are directed outward at a sharp angle. Instead, the majority of the sound energy bounces back and forth along the channel for great distances.

Based on the principle that sound speed is determined primarily by the temperature of the water, a January 1991 experiment generated a loud sound near Heard Island in the southern Indian Ocean. This experiment used the sofar channel to transmit sound to special listening stations around the world. Oceanographers intend to repeat the experiment periodically over the next ten years, and by comparing each experiment's sound travel time, learn if there are increases in sound speed that can be used to demonstrate an ocean response to global warming. In 1994 a larger, long-term Pacific Ocean experiment was planned. However, this test was delayed due to the concern that the sound signals might interfere with marine mammals. New tests to evaluate the impact of sound on marine mammals are being planned, and if the results are favorable to the researchers, the Pacific Ocean tests will be resumed.

FIGURE 5.8

A precision depth recorder displays bottom and subbottom profiles. A second image of the seafloor surface appears as the deepest layer at right. This image comes from the first seafloor echo bouncing off the sea surface to make a second round trip.

How can sound be used to measure depth?

How does sound refraction affect sonar?

Explain the effect of the sofar channel on sound.

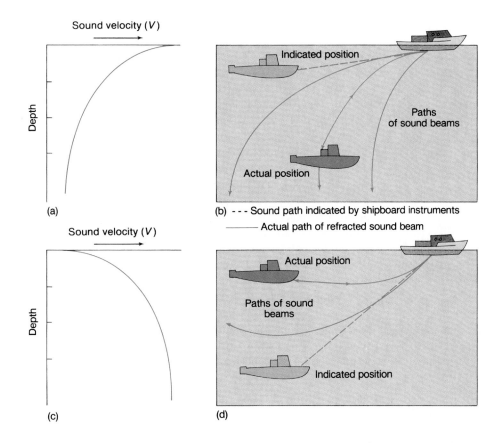

FIGURE 5.9

Sound waves change velocity and refract as they travel at an angle through water layers of different densities. The angle at which the sound beam leaves the ship indicates a target in the indicated, or ghost, position. To determine the actual position of the target the degree of refraction and change in velocity with depth must be known.

Source: Adapted from G. Neumann and W. J. Pierson, *Principles of Physical Oceanography,* page 51, Prentice-Hall, 1996.

Water

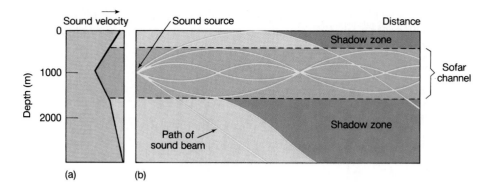

FIGURE 5.10

(a) The temperature, salinity, and pressure variation with depth combine to produce a minimum sound velocity at about 1000 meters. (b) Sound generated at this depth is trapped in a layer known as the sofar channel.

From Lawrence E. Kinsler and Austin R. Frey, *Fundamentals of Acoustics*, 2d ed. Copyright © 1962 John Wiley & Sons, New York. Reprinted by permission of John Wiley & Sons, Inc.

5.7 DISSOLVING ABILITY OF WATER

Water is an extremely good solvent. More substances—solids, liquids, and gases—dissolve in water than in any other common liquid. The dissolving ability of water is related to its molecular chemistry. For example, common table salt dissolves in water. Common salt is chemically sodium chloride (NaCl); each atom of sodium has exactly one matching atom of chlorine. When the salt is in its natural crystal form, the positively charged sodium and the negatively charged chlorine are bonded in an alternating pattern because of the strong attractive force between positive and negative charges. When sodium chloride is placed in water, the bonds between the atoms break and the resulting charged atoms are called **ions.** This reaction can be written as:

Sodium chloride → Sodium ion$^+$ + Chloride ion$^-$

or it can be written with chemical abbreviations as:

$$NaCl \rightarrow Na^+ + Cl^-$$

The positive sodium ions are attracted to the negatively charged oxygen side of the water molecules, while the negative chloride ions are attracted to the positively charged hydrogen side. The ions thus are surrounded and separated by water molecules (fig. 5.11). The ability of water molecules to separate compounds into their ions makes water an excellent solvent.

This property of water, as well as those discussed earlier, appears in table 5.2, a summary of water's properties.

What happens to soluble salts in seawater?

5.8 SALTS IN SEAWATER

Oceanographers measure the salt content of ocean water, or **salinity,** in grams of salt per kilogram of seawater (g/kg), or parts per thousand (‰). The dissolved salts are present as positively or negatively charged ions or groups of ions.

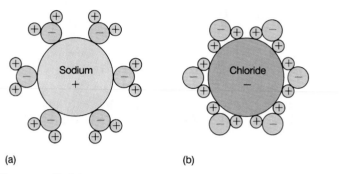

FIGURE 5.11

Salts dissolve in water because the polarity of the water molecule keeps positive salt ions separated from negative ions. (a) Sodium ions are surrounded by water molecules with their negatively charged portion attracted to the positive ion, cation. (b) Chloride ions are surrounded by water molecules with their positively charged portion attracted to the negative ion, anion.

Six ions make up more than 99 percent of the salts dissolved in seawater: sodium (Na^+), magnesium (Mg^{2+}), calcium (Ca^{2+}), potassium (K^+), chloride (Cl^-), and sulfate (SO_4^{2-}). An ion with a positive charge is a **cation:** an ion with a negative charge is an **anion.** The salts in seawater are mostly present in their ionic forms as anions and cations. Table 5.3 lists these six ions and five more known as the **major constituents** of seawater. Note that sodium and chloride ions together account for 86 percent of the salt ions present in seawater.

All the other elements dissolved in seawater are present in concentrations of less than 1 part per million and are called **trace elements.** Some of these elements are important to organisms that are able to concentrate the ions. For example, long before the presence of iodine could be determined chemically as a trace element of seawater, it was known that shellfish and seaweeds were rich sources for this element, and seaweed has been harvested commercially for iodine extraction.

Seawater is a well-mixed solution; currents at the surface and at depth, vertical mixing processes, and wave and tidal action have all helped to stir the ocean over geologic time. Because of this thorough mixing, the ionic composition of the major ions of open-ocean seawater (table 5.3, excluding bicarbonate and fluoride) is the same from place to place and from

Table 5.2 Properties of Water

Definition	Comparisons	Effects
Physical States Gas, Liquid, Solid Addition or loss of heat breaks or forms bonds between molecules to change from one state to another.	The only substance that occurs naturally in three states on the earth's surface.	Important for the hydrologic cycle and the transfer of heat between the oceans and atmosphere.
Heat Capacity Amount of heat required to raise the temperature of 1 gram of water by 1°C.	Highest of all common solids and liquids.	Prevents large variations of surface temperature in the oceans and atmosphere.
Surface Tension Elastic property of water surface.	Highest of all common liquids.	Important in cell physiology, water-surface processes, and drop formation.
Latent Heat of Fusion Heat required to change a unit mass from a solid to a liquid without changing temperature.	Highest of all common liquids and most solids.	Results in the release of heat during freezing and the absorption of heat during melting. Moderates temperature of polar seas.
Latent Heat of Vaporization Heat required to change a unit mass from a liquid to a gas without changing temperature.	Highest of all common substances.	Results in the release of heat during condensation and the absorption of heat during vaporization. Important in controlling sea surface temperature and the transfer of heat to air.
Compressibility Average pressure on total ocean volume 200 atmospheres; ocean depth decreased by 37 meters (121 ft).	Seawater is only slightly compressible, about 4×10^{-5} cm^3/g for an increase of 1 atmosphere of pressure.	Density changes only slightly with pressure. Sinking water can warm slightly due to its compressibility.
Density Mass per unit volume: grams per cubic centimeter, g/cm^3.	Density of seawater is controlled by temperature, salinity, and pressure.	Controls the ocean's vertical circulation and layering. Affects ocean temperature distribution.
Viscosity Liquid property that resists flow. Internal friction of a fluid.	Decreases with increasing temperature. Salt and pressure have little effect. Water has a low viscosity.	Some motions of water are considered friction free. Low friction dampens motion; retards sinking rate of single-celled organisms.
Dissolving Ability Dissolves solids, gases, and liquids.	Dissolves more substances than any other solvent.	Determines the physical and chemical properties of seawater and the biological processes of life-forms.
Heat Transmission Heat energy transmitted by conduction, convection, and radiation.	Molecular conduction slow; convection effective. Transparency to light allows radiant energy to penetrate seawater.	Affects density; related to vertical circulation and layering.
Light Transparency Transmits light energy.	Relatively transparent for visible wavelength light.	Allows plant life to grow in the upper layer of the sea.
Sound Transmission Transmits sound waves.	Transmits sound very well compared to other fluids and gases.	Used to determine water depth and to locate targets.
Refraction The bending of light and sound waves by density changes that affect the speed of light and sound.	Refraction increases with increasing salt content and decreases with increasing temperature.	Makes objects appear displaced when viewed by light or sound.

Table 5.3 Major Constituents of Seawater

Constituent	Symbol	g/kg in Seawater[1]	Percentage by Weight
Chloride	Cl^-	19.35	55.07
Sodium	Na^+	10.76	30.62
Sulfate	SO_4^{2-}	2.71	7.72
Magnesium	Mg^{2+}	1.29	3.68
Calcium	Ca^{2+}	0.41	1.17
Potassium	K^+	0.39	1.10
Bicarbonate	HCO_3^-	0.14	0.40
Bromide	Br^-	0.067	0.19
Strontium	Sr^{2+}	0.008	0.02
Boron	B^{3+}	0.004	0.01
Fluoride	F^-	0.001	0.01
Total		~ 35.00	99.99

(The percentages for Chloride through Potassium are bracketed together totaling 99.36)

[1]Salinity = 35‰.

From Riley and Skirrow, *Chemical Oceanography,* Vol. 1, 2d ed., p 366. Copyright © 1975 Academic Press Ltd., London. Used by permission of Academic Press.

depth to depth. That is to say, the ratio of one major ion or seawater constituent to another remains the same. The total salinity may change as fresh water is removed or gained, but the major ions exist in the same proportions. This principle may not apply along the shores, where rivers may bring in large quantities of dissolved substances or reduce the salinity to very low values. The major ions are also called conservative ions, because they do not change their ratios to each other with changes in salinity, and because they are not generally removed or added by living organisms. Certain ions present in much smaller quantities, some dissolved gases, and assorted organic molecules do change their concentrations with biological and chemical processes; these are called nonconservative ions.

What is meant by salinity?

Distinguish between the major constituents and the trace elements found in seawater.

How does the ratio of major salt ions change when open-ocean salinity changes?

5.9 DETERMINING SALINITY

The great advantage of most of the major ions being in constant proportion is that it allows one to determine salinity by measuring the concentration of one type of ion. Historically, the quantity of chloride ions has been measured to establish a sample's salinity. To do so, silver nitrate is added to the sample so that the silver combines with the chloride ion, and when the amount of silver required to react with all of the chloride ion is known, the amount of chloride is known. However, the silver also combines with bromine, iodine, and some other trace elements; the chloride concentration measured in this way has the special name chlorinity. Chlorinity (Cl ‰) and salinity (S ‰) in parts per thousand are related by the following equation:

$$\text{Salinity ‰} = 1.80655 \times \text{Chlorinity (‰)}$$

or

$$\text{S ‰} = 1.80655 \times \text{Cl ‰.}$$

Because of the ions it contains, seawater conducts electricity; the more ions there are, the greater the conductivity. This relationship makes it possible to determine salinity by using an instrument called a **salinometer.** The great advantage of the salinometer is that the measurements are made quickly and directly on the water sample with an electrical probe. It is therefore not necessary to analyze a water sample chemically except to test and calibrate the conductivity instruments. Because the conductivity of seawater is affected by both salinity and temperature, a conductivity instrument must correct for the temperature of the sample if it is calibrated to read directly in salinity units.

Why can salinity be calculated from the measurement of chloride ions?

What other method can be used to measure salinity?

5.10 OCEAN SALINITIES

In the major ocean basins, a typical kilogram of seawater is made up of 965 grams of water and 35 grams of salt. The average ocean salinity is approximately 35‰. The salinity of ocean surface water is associated with latitude. The relationship between evaporation, precipitation, and surface salinity with latitude is shown in figure 5.12. Notice the low surface salinities in the cool and rainy belts at 40–50°N and S latitude, high evaporation rates, and high surface salinities in the desert belts of the world centered on 25°N and S, and low surface salinities again in the warm, rainy tropics centered at 5°N. Sea surface salinities during the Northern Hemisphere summer are shown in figure 5.13. Refer to section 5.3 and table 5.1 for a discussion of the effect of salts on seawater density, and section 7.2 for information on the effects of temperature and salinity on ocean density.

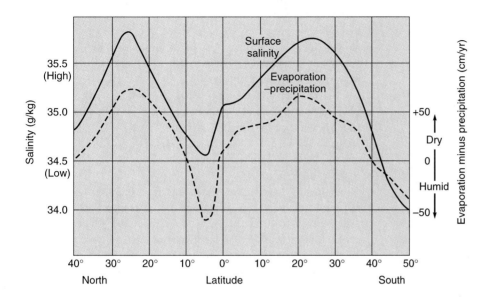

FIGURE 5.12
Mid-ocean average surface salinity values match the average changes in evaporation minus precipitation values that occur with latitude.

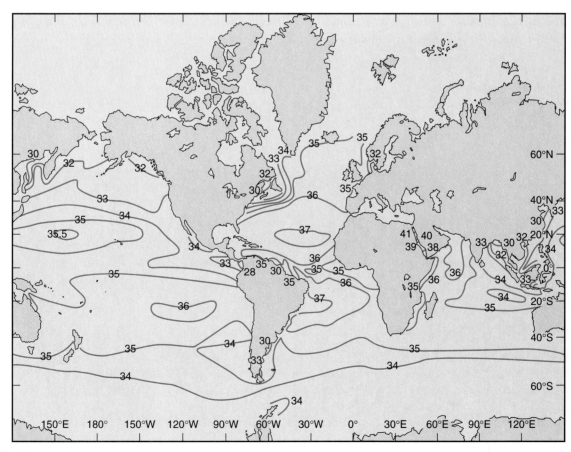

FIGURE 5.13
Average sea surface salinities in the Northern Hemisphere summer, given in parts per thousand (‰).

Water

In coastal areas of high precipitation and river inflow, surface salinities fall below the average. For example, during periods of high flow, the water of the Columbia River lowers the Pacific Ocean's surface salinity to less than 24‰ as far as 30 kilometers (20 mi) out to sea. In subtropic regions of high evaporation and low freshwater input, the surface salinities of nearly landlocked seas are well above the average: 40 to 42‰ in the Red Sea and the Persian Gulf, and 38 to 39‰ in the Mediterranean Sea. In the open ocean at these same latitudes, the surface salinity is closer to 36.5‰. Surface salinities change seasonally in polar areas, where the surface water forms sea ice in winter, leaving behind the salt and raising the salinity of the water under the ice. In summer, the water returns as a freshwater layer when the sea ice melts.

> Relate changes in open-ocean salinity to latitude and explain these changes.
>
> Why do coastal surface salinities differ from open-ocean salinities?
>
> What happens to surface salinities in the polar seas during the year?

5.11 THE SALT BALANCE

The salts of the oceans come from the land through rivers, from the chemical reactions of seawater with sediments, from the gases produced by volcanoes, and from the spreading centers of the mid-ocean ridge and rise systems. During volcanic eruptions sulfide and chloride gases are released, dissolved in rainwater, and carried to the oceans as Cl^- (chloride) and SO_4^{2-} (sulfate). River water carrying chloride and sulfate ions is acidic and erodes and dissolves the rock over which it flows, helping to liberate the positive ions, such as Na^+ (sodium), Ca^{2+} (calcium), and Mg^{2+} (magnesium), from the earth's crust. At the mid-ocean ridge systems where molten rock rises from the mantle into the crust, magma chambers are formed, and seawater becomes heated as it flows through the fractured crust. The heated water reacts chemically with the rocks of the crust, and copper, iron, manganese, zinc, potassium, and calcium are dissolved into the water. Follow the input of ions into the oceans from these sources in figure 5.14.

Although salt ions continually flow into the oceans, chemical and geologic evidence leads researchers to believe that the salt composition of the oceans has not changed for about the last 1.5 billion years. Therefore, the addition of salts must be balanced by the removal of salts if ocean salinity is to remain the same.

Salts are removed from seawater in a number of ways; some are depicted in figure 5.14. Sea spray from waves is blown ashore, depositing a film of salt on the land. Over geologic time, shallow arms of the sea may become isolated, the water evaporates, and the salts are left behind to become land deposits. Salt ions can also react with each other to form insoluble products that settle to the ocean floor. At the mid-ocean ridges, magnesium and sulfate ions are transferred from the water to form mineral deposits. Biological processes concentrate salts, which are removed if the organisms are harvested.

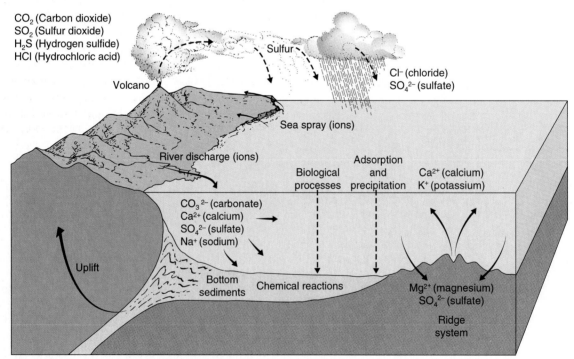

FIGURE 5.14
Processes that distribute and regulate the major constituents in seawater.

Organisms' excretion products trap ions, which are transferred to the sediments or returned to the seawater. Other biological processes remove calcium by incorporating it into shells, and silica is used to form the hard parts of certain plants and animals. These hard parts accumulate to form sediments when the organisms die. Chapter 4 discussed the sediments produced by biological processes.

Tiny clay mineral particles, weathered from rock and brought to the oceans by the rivers or the winds, bind ions and molecules to their surfaces by a process known as **adsorption.** Ions and trace metals sink with the clay particles and are eventually incorporated into sediments. The fecal pellets and skeletal remains of small organisms also act as adsorption surfaces. Although many processes remove ions from seawater and transfer them to the sediments, adsorption is the most important single process of ion removal.

Ions deposited in the sediments are trapped there, but geologic processes elevate some sediments from the sea floor to positions above sea level. Erosion then works to dissolve and wash these deposits back to the sea.

The relative abundance of the salts in the sea is due in part to the ease with which they are introduced from the earth's crust and in part to the rate at which they are removed from the seawater. Sodium is moderately abundant in fresh water, but because it reacts slowly with other substances in seawater, it remains dissolved in the ocean. Calcium is removed rapidly from seawater, forming limestone and the shells of marine organisms.

The average time that a substance remains in solution in the oceans is called its **residence time** (table 5.4). Aluminum, iron, and chromium ions react with other substances quickly and form insoluble mineral solids in the sediments. They have short residence times, in the hundreds of years. Sodium, potassium, and magnesium are very soluble, combine with other substances slowly, and so have residence times in the millions of years.

Table 5.4	Approximate Residence Time of Ions in the Oceans
Ion	**Time in Years**
Chloride	80 million
Sodium	60 million
Magnesium	10 million
Sulfate	9 million
Potassium	6 million
Calcium	1 million
Manganese	7 thousand
Aluminum	1 hundred
Iron	1 hundred

Compare processes that add and remove salts from the oceans.

What is the role of adsorption in controlling ocean salinity?

How are the abundance, solubility, and residence time of ions related?

5.12 NUTRIENTS AND ORGANICS

Ions required for plant growth are known as **nutrients;** these are the fertilizers of the oceans. As on land, plants require nitrogen and phosphorus in the forms of nitrate (NO_3^-) and phosphate (PO_4^{3-}) ions. A third nutrient required in the oceans is the silicate ion (SiO_4^-), which is needed to form the hard outer wall of the single-celled plants called diatoms and the skeletal parts of some protozoans. These three nutrients are among the dissolved substances brought to the sea by the rivers and land runoff. Nitrates and phosphates are continually recycled from decomposing organic matter. Despite their importance, nitrates, phosphates, and silicates are present in very low concentrations (table 5.5).

The concentration ratios of some nutrient ions vary, because some of these ions are closely related to the life cycles of living organisms. Populations of organisms remove nutrients during periods of growth and reproduction, temporarily reducing the amounts in solution. Later, when the population declines, decay processes and animal wastes return the ions to the seawater. Therefore nutrients are nonconservative for they do not maintain constant ratios in the way that most major salt ions do. The biological cycling of nutrients is discussed in Chapter 10, Productivity of the Oceans.

A wide variety of organic substances is present in seawater. Proteins, carbohydrates, lipids (or fats), vitamins, hormones, and their breakdown products are all present. Some are eventually reduced to their inorganic components; others are used directly by organisms and are incorporated into their systems. Another portion of the organic matter accumulates in the sediments, where over geologic time it slowly forms deposits of oil and gas.

What are nutrients and why are they important?

Table 5.5	Nutrients in Seawater	
Element	**Concentration ($\mu g/kg^1$)**	
Nitrogen (N)	500	
Phosphorus (P)	70	
Silicon (Si)	3000	

[1]Parts per billion.

Table 5.6 Abundance of Gases in Air and Seawater

Gas	Symbol	Percentage by Volume in Atmosphere	Percentage by Volume in Surface Seawater[1]	Percentage by Volume in Total Oceans
Nitrogen	N_2	78.08	48	11
Oxygen	O_2	20.99	36	6
Carbon dioxide	CO_2	0.03	15	83
Argon, helium, neon, etc.	Ar, He, Ne	0.90	1	
Totals		100.00	100	100

[1]Salinity = 36‰, temperature = 20°C.

5.13 THE GASES IN SEAWATER

The most abundant gases in the atmosphere and in the oceans are nitrogen (N_2), oxygen (O_2), and carbon dioxide (CO_2). The percentages of each of these gases in the atmosphere and in seawater are given in table 5.6. Oxygen and carbon dioxide play important roles in the ocean because they are necessary to life, and biological activities modify their concentrations at various depths. Nitrogen is not used directly by living organisms except for certain bacteria. Gases such as argon, helium, and neon are present in small amounts, but they neither interact with the ocean water nor are used by its inhabitants.

The amount of any gas that can be held in solution without causing the solution either to gain or to lose gas is the **saturation value.** The saturation value changes because it depends on the temperature, salinity, and pressure of the water. Colder water holds more dissolved gas than warmer water; less salty water holds more gas than more salty water, and water under more pressure holds more gas than water under less pressure.

In the process known as **photosynthesis,** plants use carbon dioxide to form organic molecules and produce oxygen as a by-product. Since plants require sunlight for photosynthesis, plants are confined, on the average, to the top 100 meters (330 ft) of the sea, where sufficient light is available. Therefore oxygen is produced in surface water, and carbon dioxide is consumed there. By contrast, **respiration,** which breaks down organic substances to provide energy, requires oxygen and produces carbon dioxide and energy. All living organisms respire in order to produce energy in living cells, and respiration occurs at all depths in the oceans. Decomposition, the bacterial breakdown of nonliving organic material, also requires oxygen and releases carbon dioxide. The processes of photosynthesis and respiration and their equations are discussed in more detail in Chapter 10.

Oxygen can be added to the oceans only at the surface, from exchange with the atmosphere or as a waste product of photosynthesis. Carbon dioxide also enters from the atmosphere at the sea surface, but it is available at all depths from respiration and decomposition. Figure 5.15 shows typical oxygen and carbon dioxide concentrations with depth.

If the water is quiet, the nutrients and sunlight abundant, and a large population of plants is present, oxygen values at the surface can rise above the saturation value to 150 percent or

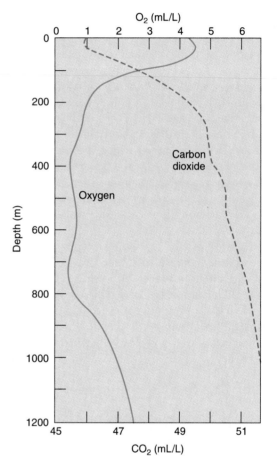

FIGURE 5.15
The distribution of O_2 and CO_2 with depth.

more. This water is **supersaturated.** Wave action tends to liberate oxygen to the atmosphere and return the condition to the saturated state.

Below the surface, the **oxygen minimum** occurs at about 800 meters (2600 ft) depth. At depths greater than this, the rates of removal of oxygen and production of carbon dioxide both fall because the population density of animals and the abundance of decaying organic matter have decreased. The slow supply of oxygen to depth by water sinking from the surface gradually increases the oxygen concentration above that found at the oxygen minimum depth.

Very low or zero concentrations of oxygen occur in the bottom waters of isolated deep basins, which have little or no exchange or replacement of water. These include the bottoms of trenches, deep basins behind a shallow entrance sill (the Black Sea, for example), and the bottoms of deep fjords. The water is trapped and becomes stagnant; respiration and decomposition use up the oxygen faster than the slow circulation of water to this depth can replace it. The bottom water becomes **anoxic,** or stripped of dissolved oxygen.

The average overall concentration of carbon dioxide throughout the oceans is not substantially affected by biological processes, but tends to remain almost constant, controlled by temperature, salinity, and pressure.

The carbon dioxide cycle and its impact on the earth's atmosphere is discussed in section 6 of Chapter 6.

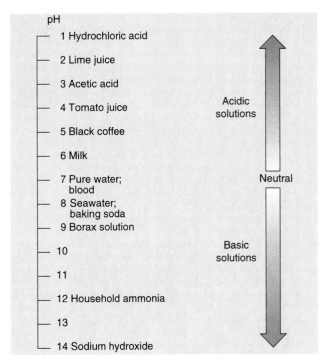

FIGURE 5.16
The pH scale.

Which two gases are biologically important in the oceans?

Explain the distribution of these gases with depth.

How is it possible for seawater to become supersaturated with oxygen?

What produces the oxygen minimum layer?

Where might you find anoxic conditions in the oceans and why?

5.14 CARBON DIOXIDE AS A BUFFER

Carbon dioxide plays an additional and extremely important role in the ocean: in water it acts as a **buffer.** A buffer prevents sudden changes in the acidity or alkalinity, or **pH,** of a solution. The buffering capacity of seawater is important to organisms requiring a steady pH for their life processes and to the chemistry of seawater, which is controlled, in part, by pH. The pH scale indicates acidity and alkalinity by measuring the concentration of hydrogen ions in a solution (fig. 5.16). A pH of 7 indicates neutrality, neither acid nor alkaline. Values between 1 and 6 are acidic while the pH values between 8 and 14 are alkaline, or basic. The pH range of seawater is between 7.5 and 8.5; an average pH value for the world's oceans is approximately 7.8.

Why is the capacity of CO_2 to buffer seawater significant?

5.15 SALT AND WATER

About 30 percent of the world's salt is extracted from seawater. In warm, dry climates, seawater is directed into shallow ponds and evaporated down to a concentrated brine solution. The process is repeated several times, until a dense brine is produced. Then the evaporation is allowed to continue until a thick, white salt deposit is left on the bottom of the pond. The salt is collected and refined to produce table salt. This technique has been recently used in southern France, Puerto Rico, and California (fig. 5.17).

In cold climates, salt has been recovered by freezing seawater in shallow ponds. The ice that forms is nearly fresh; the salts are concentrated in the brine beneath the ice. The brine is heated to remove the last of the water.

Sixty percent of the world's magnesium comes from the sea, and so does 70 percent of the bromine. There are vast amounts of dissolved minerals in the world's seawater, including 10 million tons of gold and 4 billion tons of uranium, but the concentration is very low (one part per billion or less), and the cost of extraction is too high for economic production at the present time. Dense, hot salt brines at the bottom of the Red Sea are estimated to contain minerals with values in the billions of dollars.

Desalination is the process of obtaining fresh water from salt water. The greatest drawback to desalination is the cost, which is linked to the energy required. There are several possible desalination methods; two processes use a change of state of the water (liquid to solid or liquid to vapor), and another process uses a membrane permeable to water molecules but not to salt ions.

The simplest and least energy-expensive process involving a change of state is the solar still. In this process a pond of seawater is capped by a low plastic dome. Solar radiation penetrates the dome, causing evaporation of the seawater. The evaporated water condenses on the undersurface of the dome and trickles down the curved surface to be caught in a trough,

Water

FIGURE 5.17

The southern end of San Francisco Bay has been diked into shallow ponds, where seawater is evaporated to obtain salt.

where it accumulates and flows to a freshwater reservoir. The rate of production is slow, and a very large system is needed to supply the water requirements of even a small community. Yet the costs for this type of system are low and are principally confined to construction of the facility, since the solar energy is free. This system is illustrated in figure 5.18.

When salt water is distilled by heating it to boiling, evaporation proceeds at a rapid rate and large quantities of fresh water are produced, but considerable energy is required. If water is introduced into a chamber with a reduced air pressure, the boiling occurs at a much lower temperature, with less expenditure of energy.

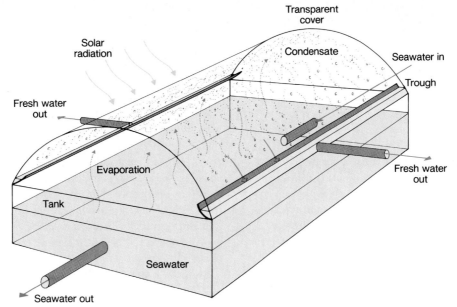

FIGURE 5.18

Solar energy is used to evaporate fresh water from seawater. Solar radiation penetrates the transparent cover of the still and causes evaporation of the seawater contained in the tank.

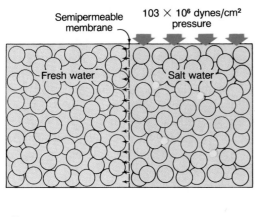

$\bigcirc$ = Water molecule

· = Salt molecule

FIGURE 5.19

Reverse osmosis. Water moves from the saltwater side to the freshwater side when pressure is applied.

In areas like Kuwait, Saudi Arabia, Morocco, Malta, Israel, the West Indies, California, and the Florida Keys, water is a limiting factor for population and industrial growth. Thousands of cubic meters of fresh water are produced daily in these areas by desalination. Around the world, nearly 4000 desalination plants produce a total of about 13 billion liters (3.4 billion gal) of fresh water from seawater each day. About 60 percent of the world's desalination capacity is located on the Arabian Peninsula. Water costs there are low, because fuel costs are low. Desalination plants in California and Texas produce fresh water at more than twice the cost of water from other sources; it is cheap enough to use for drinking water at approximately $4.00 per 1000 gallons and for certain indus-

trial applications, but far too expensive for agriculture. Small evaporation-type desalination plants are used on military and commercial vessels.

It is also possible to produce fresh water from seawater by applying pressure to seawater. If the pressure applied to the seawater exceeds 24.5 atmospheres, water molecules will move through a membrane that is permeable only to water in a process called **reverse osmosis** (fig. 5.19). Fresh water is squeezed out of the salt water. The theoretical energy requirement is about one-half that needed for the evaporative process.

Reverse osmosis is the most rapidly growing form of desalination technology. Older evaporative desalination plants are being replaced with reverse osmosis plants. Ninety-nine percent of Florida's 110 desalination plants use reverse osmosis. Santa Barbara, California has built the nations's largest reverse osmosis plant, producing 26 million liters (7 million gal) per day. Twenty-six miles off the coast from Los Angeles, Santa Catalina Island is producing 5×10^5 liters (132,000 gal) per day, about one-third of its potable water needs. Small commercial units using this process produce 400 liters (105 gal) per day; these are available for ship and summer home use. Units powered by hand pumps are available for emergency use in lifeboats and life rafts. The military uses portable reverse osmosis plants under emergency conditions.

What products are presently extracted from seawater?

Explain the operation of a solar still.

Where are most of the world's desalination plants located and why?

What is the process of reverse osmosis?

SUMMARY

The water molecule has a specific shape with oppositely charged sides. Because of the distribution of the charges, water molecules interact with each other, forming bonds between molecules. The molecule's structure is responsible for the properties of the water.

Water exists as a solid, a liquid, and a gas. Changes from one state to another require the addition or extraction of heat energy. Water has a high heat capacity; it is able to take in or give up large quantities of heat with a small change in temperature.

The density of water increases with a decrease in temperature and an increase in salt content. The effect of pressure on density is small. Less dense water floats on more dense water. In the oceans, pressure increases 1 atm for every 10 meters of depth, but the water is nearly incompressible.

Seawater transmits energy as light, heat, and sound. Heat is transmitted from the sea surface slowly downward by conduction. Convection transfers heat effectively from the sea surface to the atmosphere. Radiant energy penetrates seawater because seawater transmits visible light. The long red wavelengths of light are lost in the first 10 meters (33 ft); only the shorter wavelengths of blue-green light penetrate to depths of 150 meters (500 ft) or more. The intensity of light is decreased over distance by scattering and absorption, leading to attenuation. Light attenuation is measured either by a light meter or with a Secchi disk. Light is also refracted by water.

Sound transmission is affected by temperature, pressure, and salt content of the seawater. Echo sounders are used to measure the depth of water, and sonar is used to locate objects. Sound is refracted as it passes at an angle through the density layers of the ocean, and sound shadows are formed. The sofar channel, in which sound travels for long distances, is being used in experiments to monitor ocean temperatures and global warming.

The ability of water to dissolve substances is exceptionally good. Soluble salts are present as ions in seawater. Six major constituent ions make up 99 percent of the salt in seawater; elements present in small quantities are known as trace elements. The proportion of most major ions to each other remains the same for all waters of the open oceans. The average salinity of ocean water is 35‰. The salinity of the surface water changes with latitude and is affected by evaporation, precipitation, and the freezing and thawing of sea ice.

Ions from weathering and erosion of the earth's crust are added to the sea by rivers; other ions from the gases of volcanic eruptions are dissolved in river water and rain, and others come from seawater circulating through magma chambers at the spreading centers. Since the average salinity of the oceans remains constant, there must be an equal loss of salts. Salts are removed from seawater by sea spray, evaporite formation, and adsorption. Other mechanisms that remove salts include production of hydrogenous and biogenous sediments as well as geological uplift and activity at spreading centers. The residence time for salts in solution depends on their reactivity.

Nutrients include the nitrates, phosphates, and silicates required for plant growth. A wide variety of organic products is also present.

The saturation value of gases dissolved in seawater varies with salinity, temperature, and pressure. Carbon dioxide is added to seawater from the atmosphere at the sea surface and by respiration and decay processes at all depths; it is removed at the surface by photosynthesis. Oxygen is added only at the surface from the atmosphere and the photosynthetic process; it is depleted at all depths by respiration and decay. Seawater may be supersaturated with oxygen or it may become anoxic. Carbon dioxide levels tend to remain constant over depth. Carbon dioxide is the seawater's buffer, keeping the pH range of ocean water between 7.5 and 8.5.

Salt, magnesium, and bromine are extracted commercially from seawater. Desalination methods include change-of-state processes and reverse osmosis. The practicality of desalination is determined by cost and need.

KEY TERMS

calories	refraction	major constituent	supersaturation
heat capacity	Secchi disk	trace element	oxygen minimum
density	sonar	salinometer	anoxic
conduction	sound shadow zone	adsorption	buffer
convection	sofar channel	residence time	pH
radiation	ion	nutrients	desalination
absorption	salinity	saturation value	reverse osmosis
scattering	cation	photosynthesis	
attenuation	anion	respiration	

SUGGESTED READINGS

Alper, J. 1991. Munk's Hypothesis. *Sea Frontiers* 37 (3):38–41. The Heard Island acoustic experiment.

Baker, D. J. 1991. Toward a Global Ocean Observing System. *Oceanus* 34 (1):76–83. Methods for long-term data collection.

Bowditch, N. 1984. *American Practical Navigator*, vol. 1. U.S. Defense Mapping Agency Hydrographic Center, Washington D.C., 1414 pp. Soundings are covered in Chapter 28 and sound in Chapter 35.

Brown, N. 1991. The History of Salinometers and CTD Sensor Systems. *Oceanus* 34 (1):61–66.

Friedman, R. 1990. Salt-Free Water from the Sea. *Sea Frontiers* 36 (3):48–54. Desalination methods.

MacIntyre, F. 1970. Why the Sea Is Salt. In *Ocean Science, Readings from Scientific American* (1977) 223 (5):104–15.

Oceanus. Spring 1977. 20 (2). Issue includes articles on uses of sound in navigation, warfare, animal behavior, and seismics.

Richardson, P. L. 1991. SOFAR Floats Give a New View of Ocean Eddies. *Oceanus* 34 (1):23–31.

Stewart, W. K. 1991. High Resolution Optical and Acoustic Remote Sensing for Underwater Exploration. *Oceanus* 34 (1):10–22. Concerns sonar, optical imaging, cameras, and video.

ITEM OF INTEREST

Clouds

Clouds in their different shapes, ever changing and always moving have been watched and studied by those at sea and those on land from the beginnings of human history to the present. We see clouds huge and fluffy building up into great towers, or high and wispy spreading across the sky. We see them white in the sunshine, rosy and red at dawn and sunset, black and threatening as a storm approaches. Different kinds of clouds are placed in categories based on the work of Luke Howard, an English pharmacist. In 1803 Howard proposed the first useful way to classify clouds: stratus or layered, cumulus or puffy, cirrus or wispy, and nimbus or clouds releasing snow or rain that travels all the way to the ground. This system is the basis for the cloud types recognized by today's World Meteorological Organization (WMO); see figure 1.

Cloud Formation

Clouds form as moist air rises and cools. When the rising air is cooled below the dew point, the temperature at which water vapor begins to condense, the cloud begins to form. Water vapor condenses around condensation nuclei, forming droplets. Near the earth, in the lower troposphere, condensation nuclei may be small particles of dust, salt, or other matter; these particles are generally abundant, and condensation takes place readily. In the higher troposphere the nuclei are composed of crystallized water vapor or ice crystals and snow. If nuclei are sparse, the droplets are fewer and larger, increasing the chance of precipitation and reducing further the amount of nuclei. Many small nuclei produce abundant small droplets, increasing cloud reflectivity and absorption and decreasing incoming solar energy. In oceanic areas where precipitation is very frequent, particles are removed by the rain, and a low supply of condensation nuclei may hinder cloud formation. In these areas satellite photos show passing ships leaving cloud trails as condensation occurs on particles from the ships' exhausts.

Air temperature in the troposphere decreases with height, therefore when air with a high water content and a high dew point temperature rises and cools, the water vapor condenses at a low elevation above the earth's surface. When condensation occurs, the latent heat of vaporization (refer to table 5.2) is released to the air, warming it, and increasing its ability to rise and therefore cool further. If the air carries little water vapor with a low dew point temperature, it must rise higher to cool to the dew point. High altitude cirrus clouds are wispy because the air at high altitudes contains very little water vapor for forming ice crystals.

When temperatures are above freezing, clouds form from liquid water droplets; super-cooled water droplets form clouds when temperatures are below freezing. Clouds below 6 kilometers (19,000 ft) elevation are usually formed of water droplets, while those above 8 kilometers (26,000 ft) elevation are formed from ice crystals.

continued . . .

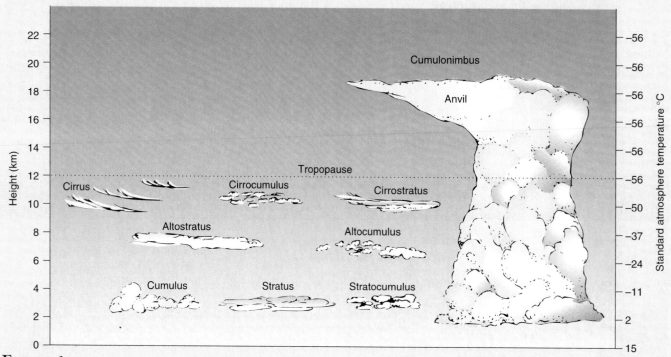

FIGURE 1

Cloud types used by the World Meteorological Organization.

Clouds are part of the earth's hydrologic cycle; refer to figure 2.13. In this figure the clouds are the visual manifestations of water droplets and ice crystals carried in the air, moving from a source of evaporation and rising air to a region of precipitation where condensed droplets become large enough to fall to the earth's surface.

Climate Effects

Clouds simultaneously heat and cool the earth. Clouds reflect the sun's energy, and they also absorb some of its incoming energy. Both processes reduce incoming energy to the earth's surface and so help to cool the earth. Clouds also intercept the sun's energy reflected and reradiated from the earth's surface, and because they are generally cooler than the earth, the clouds reradiate only part of this absorbed energy to space. This tends to trap heat and warm the earth and its atmosphere. The net effect is that clouds play a significant part in the earth's heat budget (see fig. 6.2), helping to maintain the earth's average temperature.

Recent research indicates that, on a global scale, the earth's present cloud cover provides more shielding from incoming energy than trapping of reflected and reradiated energy. At present, cloud shielding results in a total net reduction in solar energy to the earth of about 14 to 21 watts per square meter per month. It is estimated that, if there were no clouds, average earth surface temperature would increase by about 10°C.

Clouds not only affect the earth's climate, they are affected by it. If the earth's climate changed, the feedback mechanism described previously could impose a new cloud-controlled climate and heat balance on the earth. If the present reduction in solar energy due to clouds became less negative, the earth's surface would warm. If cloud feedback produced a greater reduction in available solar energy, the earth's surface would become cooler than it is at present. If the earth warmed due to the greenhouse effect (refer to Chapter 6), an increase in evaporation would increase opaque, low-altitude clouds over the ocean; these clouds would then absorb and reflect incoming solar energy back

6

The Air and the Oceans

Outline

Learning Objectives

After reading this chapter, you should be able to

■ Explain the variation of solar radiation with latitude.

■ Define the earth's heat budget.

■ Define heat capacity and compare the heat capacity of the land and oceans.

■ Understand sea surface temperature changes over the year.

■ Describe the formation of sea ice and icebergs.

■ Explain how the atmosphere moves with changes in density.

■ Explain the greenhouse effect.

■ Understand the formation of winds and the role of the Coriolis effect.

■ Show the latitude and direction of the major wind bands and name them.

■ Explain the jet stream and its relationship to surface winds.

■ Describe how the wind systems change over the year.

■ Explain wind changes along coasts and those caused by monsoon effects.

■ Distinguish the kinds of sea fog.

■ Understand how El Niño is related to changes in oceanic and atmospheric conditions.

■ Explain what causes a storm surge.

◀ Cumulus clouds over the Caribbean.

The sun's energy reaches the earth through a thin envelope of air—the atmosphere. The processes that control the atmosphere are closely related to ocean processes, and together they form much of what people call weather and climate. Clouds, winds, storms, rain, and fog are all the result of the interaction of the sun's energy, the atmosphere, and the earth's water covering. Some of these processes are more predictable than others; some are better understood than others. The complex of interactions between sun, atmosphere, and water provides the earth's daily weather, sometimes pleasant and stable, at other times severe and turbulent. This chapter presents an overview of these self-adjusting relationships as well as specific examples of their complex interactions.

6.1 DISTRIBUTION OF SOLAR RADIATION

Instantaneous solar radiation per unit surface area of the earth has its greatest intensity at the equator, moderate intensity in the middle latitudes, and least intensity at the poles. Radiation intensity is greatest between 23½°N (Tropic of Cancer) and 23½°S (Tropic of Capricorn), because only between these latitudes does sunlight strike the earth at a right angle. All other latitudes receive less instantaneous solar energy because of the decreasing angle at which the sun's rays strike the earth (fig. 6.1). The intensity of solar radiation is also reduced because radiation is absorbed by the atmosphere on its way to the earth's surface. The greater the latitude, the longer the distance through the atmosphere the sun's rays must travel (fig. 6.1). As the earth turns on its axis, the intensity of solar radiation along each latitude line changes with time, from a maximum at high noon to a minimum of zero between sunset and sunrise. In addition, a semiannual variation in the average intensity of solar radiation appears at tropical latitudes, because the sun crosses the equator twice during the year, as it moves between the Tropics of Cancer and Capricorn.

Although the intensity of instantaneous solar radiation changes with daily and annual cycles, the hours of daylight and darkness are also important in determining the amount of radiation received over a 24-hour period. At the equator both the intensity of radiation and the length of daylight change little over the year. At the poles, the high daily levels of average radiant energy reflect the long periods of daylight in summer rather than high levels of instantaneous solar radiation. The maximum daily radiation levels at the South Pole are slightly larger than those at the North Pole, because the earth is closer to the sun during the Southern Hemisphere summer. The earth's surface responds to this annual pattern of incoming solar radiation by warming the tropical areas and keeping the polar areas cold.

Why does the intensity of solar radiation vary with latitude?

Why are daily solar radiation values at the poles high in summer?

Why do Southern Hemisphere summer radiation values exceed the summer values in the Northern Hemisphere?

6.2 THE HEAT BUDGET

In order to maintain its long-term mean temperature of 16°C, the earth must reradiate as much heat back to space as it receives from the sun. The gains and losses in heat are represented in a **heat budget.** Just as incoming funds must equal outgoing funds in a balanced monetary budget, in the heat budget incoming radiation must equal outgoing radiation. If incoming funds (or deposits) are insufficient, or if outgoing funds (or withdrawals) become too great, serious financial problems arise. In the same way, if less heat were returned to space than is gained, the earth would become hotter, and if more heat were lost to space than is gained, the earth would become colder.

To follow the deposits and withdrawals in the heat budget, assume that the solar energy available to the earth and its atmosphere is 100 units. Incoming solar radiation from space contains short to long wavelengths of radiation, ultraviolet, visible, and infrared. The energy reradiated to space from the earth's surface and the heat reradiated to space from the atmosphere are assumed to be infrared radiation. Refer to figure 6.2 as you

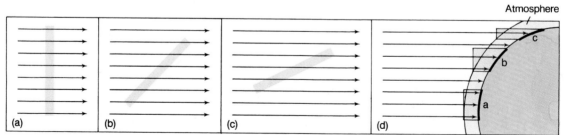

FIGURE 6.1
Areas of the earth's surface that are equal in size receive different levels of solar radiation as they become more oblique to the sun's rays (see a, b, and c). As latitude increases, the angle between the sun's rays and the earth decreases and the solar radiation received on the surface decreases (see d). Note also that the sun's rays must travel an increasing distance through the atmosphere as latitude increases.

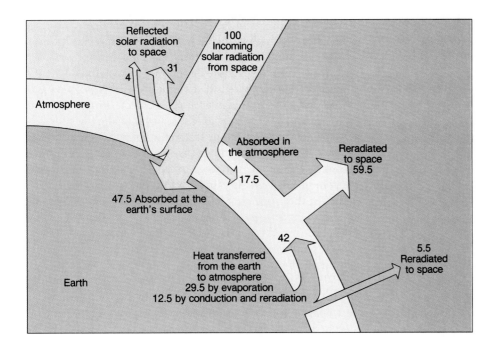

FIGURE 6.2
The earth's heat budget. Incoming solar energy
is balanced by reflected and reradiated energy.
The atmosphere's loss of heat is balanced by
heat transferred from the earth to the
atmosphere by evaporation, conduction, and
reradiation.

read this discussion. Thirty-one of these units are reflected directly back to space from the atmosphere; no heating of the earth or atmosphere occurs. Four more units are reflected back by the earth's surface through the atmosphere to outer space and also take no part in heating the earth or atmosphere. Of the remaining 65 units, 47.5 units are absorbed by the earth's surface, and 17.5 units are absorbed by the atmosphere. To balance the budget, 65 units must be returned to space, 5.5 units from the surface and 59.5 units from the atmosphere. The budget is balanced; incoming radiation balances outgoing radiation.

Closer inspection shows that although the earth's surface absorbs 47.5 units, it loses only 5.5 units to space, for a gain of 42 heat units, while the atmosphere absorbs 17.5 units and loses 59.5 units to space, for a loss of 42 units. The 42 units gained by the earth must be transferred to the atmosphere. First, 29.5 of the units are transferred by evaporation processes, which cool the surface and liberate heat when condensation of water vapor occurs in the atmosphere. Second, the remaining 12.5 units are transferred to the atmosphere by conduction or reradiation and are absorbed as heat, raising the temperature of the air. On the world average, the atmosphere is primarily heated from below by heat given off from the earth. These radiation averages do not take into account variation of heat exchange with latitude, time, or properties of the earth's surface. Changes in the heat budget related to changes in the atmosphere are discussed in section 6.5.

On the average, at latitudes less than 45°N or S each unit of surface area absorbs more radiant energy annually than it loses to space, and at latitudes greater than 45° each unit area loses more radiant energy than it absorbs. See figure 6.3. However, the surplus of energy at low latitudes sets the atmosphere and oceans in motion, and heat is transferred by winds and currents from lower to higher latitudes, maintaining a more even temperature distribution than might be expected.

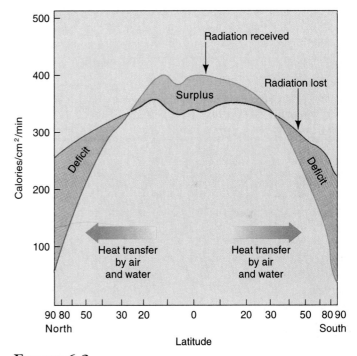

FIGURE 6.3
Comparison of incoming and outgoing radiation with latitude. A transfer of energy is required to maintain a balance.

Source: NOAA Meteorological Satellite Laboratory.

Why must incoming radiation be balanced by outgoing radiation?

Compare the effect of reflected radiation to the effect of absorbed radiation.

How is the earth's atmosphere heated?

The Air and the Oceans

6.3 EARTH SURFACE TEMPERATURES AND HEAT TRANSFER

Land and sea respond differently to solar radiation. The low heat capacity of the land results in large temperature changes as heat is gained or lost between day and night or summer and winter. The high heat capacity of the oceans allows them to absorb and release large amounts of heat with little change in temperature. Review the concept of heat capacity and refer to table 5.2. The average temperature of the sea surface is shown for the Northern Hemisphere summer in figure 6.4. At each latitude, the annual temperature changes are controlled by changes in available solar radiation and heat loss combined with the heat capacity of the surface material. The average annual temperature change for both land and mid-ocean surface water (fig. 6.5) shows how heat capacity affects sea and land differently. The maximum seasonal change in sea surface temperature, about 8 to 9°C, occurs at temperate latitudes in the Northern Hemisphere. At higher and lower latitudes, sea surface temperatures change less over the annual cycle. At polar latitudes, seasonal energy gains and losses are associated with latent heat of fusion as sea ice melts and freezes, rather than changes in the temperature of the water. At low latitudes, changes in sea surface temperatures are small because radiant energy is nearly constant over the annual cycle.

The continents undergo large annual changes in surface temperatures at high latitudes. Because of their low heat capacity the continents respond to the large annual changes of solar energy, rising in temperature during the summer and cooling in winter. In addition there is no freezing and melting of sea ice to absorb and release energy without a change in temperature. The differences between land and ocean annual surface temperature changes can be seen by looking at figure 6.6. Compare the summer (fig. 6.6a) and winter (fig. 6.6b) temperatures of land-masses and water areas at about 60°N. The contrast in seasonal temperature patterns is caused by the presence of large land-masses in the Northern Hemisphere. At 60°S, figures 6.6a and 6.6b show little seasonal difference due to the lack of land at this latitude in the Southern Hemisphere. These very different annual temperature cycles cause distinctly different weather patterns in the two hemispheres.

Where evaporation at the sea surface is high, at subtropic latitudes (20–30°N and S), there is a high rate of removal of heat energy, and large quantities of water vapor enter the atmosphere. Each gram of water evaporated and condensed in the air transfers the latent heat of vaporization—540 calories—from the earth surface to the atmosphere. The moisture-laden atmosphere moves to higher latitudes along the earth's surface as it rises and cools. The clouds that result from condensing water vapor produce high-precipitation zones at about 50 to 60°N and S and at 5°N. Therefore there is a large net transfer

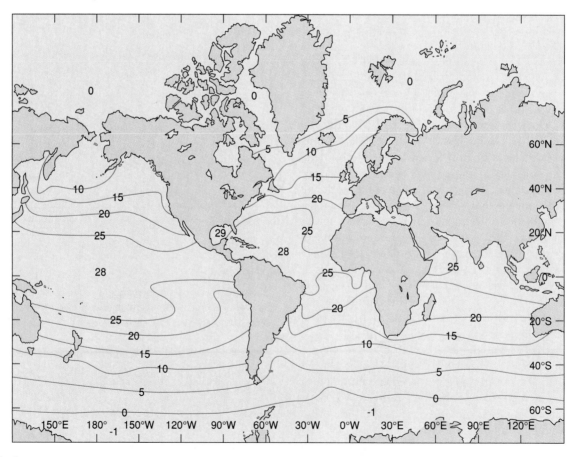

FIGURE 6.4
Sea surface temperatures during summer in the Northern Hemisphere, given in degrees Celsius.

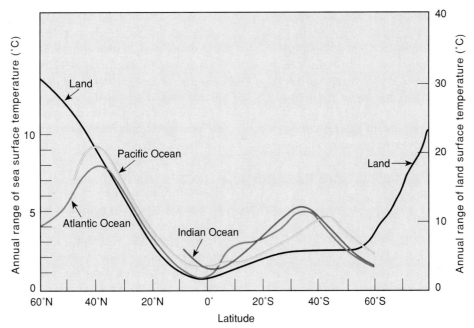

FIGURE 6.5

The annual range of mid-ocean sea surface temperatures is considerably less than the annual range of land surface temperatures at the highest latitudes. The maximum annual range of sea surface temperatures occurs at the middle latitudes.

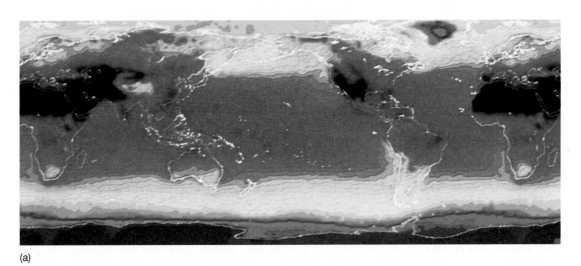

(a)

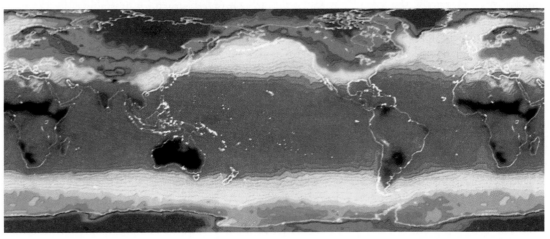

(b)

FIGURE 6.6

Meteorological satellites, such as NOAA's *TIROS*, carry high-resolution infrared sensors that measure long-wave radiation emitted from the earth's surface and atmosphere. This radiation is related to the earth's surface temperatures. Green and blue indicate temperatures below 0°C; warmer temperatures are shown in red and brown. (a) In July, the Northern Hemisphere landmasses have considerably warmer temperatures, but the ocean waters do not change dramatically from winter to summer. The oceans' warm surface water moves north and south with the change in the seasons. North-south currents along the coasts of continents are also visible. (b) In January, Siberia and Canada show surface temperatures near −30°C; at the same time, latitudes between 30° and 50° south show warm summer temperatures.

The Air and the Oceans

of heat to the atmosphere at low latitudes; the heat is carried to higher latitudes by the moving air (return to figure 6.3).

The oceans play a significant role in stabilizing the surface temperature of the earth. Their ability to store and release large quantities of heat without large changes in temperature moderates surface temperatures both seasonally and over the day and night cycles. Evaporation, conduction, and reradiation from the earth's surface all add heat to the atmosphere at the lower latitudes, which is transferred by winds and currents to the higher cooler latitudes.

Compare the heat capacity of land and water.

Explain the significance of heat capacity to the earth's surface temperatures.

Why do sea surface temperatures undergo only small annual changes at both polar and equatorial latitudes?

What is the result of evaporation at subtropic latitudes?

6.4 SEA ICE AND ICEBERGS

As seawater begins to freeze in the polar winters, a layer of slush forms, covering the ocean with a thin sheet of ice. Sheets of new **sea ice** are broken into "pancakes" by waves and wind; then as the freezing continues the pancakes move about, unite, and form floes. Ice floes move with the currents and the wind, collide with each other, and form ridges and hummocks. Some floes shift constantly, breaking apart and freezing together; others remain anchored to a landmass. As the ice forms, the heat of fusion is transferred to the cold atmosphere, and the sea water temperature remains at the freezing point.

Ice about 2 meters (6 ft) thick may be formed in one season. The thickness of the ice is limited, because the latent heat of fusion from the underlying water must be extracted through the ice by conduction, a slow process even at polar temperatures. Snow that falls on the ice surface acts as an insulator, further retarding extraction of heat from the water below.

As the ice is formed, some seawater is trapped in the voids between the ice crystals. If the ice forms slowly, most of the trapped seawater drains out and escapes; if the ice forms quickly, more salt water is trapped. As time passes, the salt water slowly escapes through the ice, and eventually the ice becomes fresh enough to drink when melted. The formation of the fresh sea ice concentrates salt in the underlying seawater, increasing its salinity and density and causing it to sink.

Sea ice is seasonal around the bays and shores of parts of the northeast United States, Canada, Russia, Scandinavia, and Alaska. The ice exists year-round in the central Arctic and around Antarctica. Sea ice covers the entire Arctic Ocean in the northern winter and pushes far out to sea around Antarctica during the Southern Hemisphere's winter, and melts back along the edges in both oceans during summer (fig. 6.7). In areas where the ice does not melt entirely during the year, new ice is added each winter, and a thickness of 3.5 to 5 meters (10 to 15 ft) can accumulate.

Castle **icebergs** are massive, irregular in shape, and float with about 12 percent of their mass above the sea surface (fig. 6.8). They are formed by glaciers, large masses of fresh-water ice, that begin inland in the snows of central Greenland, Antarctica, and Alaska, and inch their way toward the sea. The forward movement, the melting at the base of the glacier where it meets the ocean, and wave and wind action cause blocks of ice to break off and float out to sea.

The icebergs produced in the Arctic drift south with the currents as far as New England and the busy shipping lanes of the North Atlantic. It was one of these icebergs, probably from a glacier in Greenland, that sank the *Titanic* with a loss of 1517 lives on its maiden voyage in 1912. Alaskan icebergs are usually released in narrow channels and bays; they do not readily escape into the open ocean. Flat, tabular icebergs produced from the broad continental ice sheets of Antarctica tend to stay close to the polar continent, caught by the circling currents, although they have been known to reach latitudes of 40°S. In late 1987, a large tabular berg, 155 kilometers (96 mi) long and 230 meters (755 ft) thick, with a surface area about the size of Delaware, broke from Antarctica and drifted 2000 kilometers (1250 mi) along the Antarctic coast. Two years later, it grounded and broke into three pieces. The volume of ice in this berg was estimated to have been enough to provide everyone on earth with two glasses of water daily for about 2000 years.

How is sea ice formed?

What limits the thickness of seasonal sea ice?

How are icebergs formed, and where are they most commonly found?

6.5 THE ATMOSPHERE

The atmosphere extends 90 kilometers (54 mi) above the earth and is made up of a series of layers; see figure 6.9. In the first layer, or **troposphere,** where the air is warmed by heat reradiated and conducted from the earth's surface and by evaporation of water vapor and its condensation, the temperature decreases with altitude. Precipitation, evaporation, wind systems, and clouds are all found in the troposphere. In the next layer, or **stratosphere, ozone,** an unstable form of oxygen, absorbs ultraviolet radiation and the temperature increases with increasing elevation. Above the stratosphere are the mesosphere and the thermosphere.

Air is a mixture of gases, as listed in table 6.1. This mixture called air becomes less dense when it is warmed, when its water vapor content is increased, and when the atmospheric pressure decreases. Air becomes denser when it is cooled, when its water vapor content is decreased, and when the atmospheric pressure increases. Because water vapor is a low-density gas, its presence decreases the density of air.

Atmospheric pressure is the force with which a column of overlying air presses on an area of the earth's surface. The average atmospheric pressure at sea level is 1013.25 millibars (1 bar = 1×10^6 dynes/cm^2), or 14.7 lbs/in^2. Regions of air with a density greater than average are known as **high-pressure zones,** and regions with a density less than average are **low-pressure zones.**

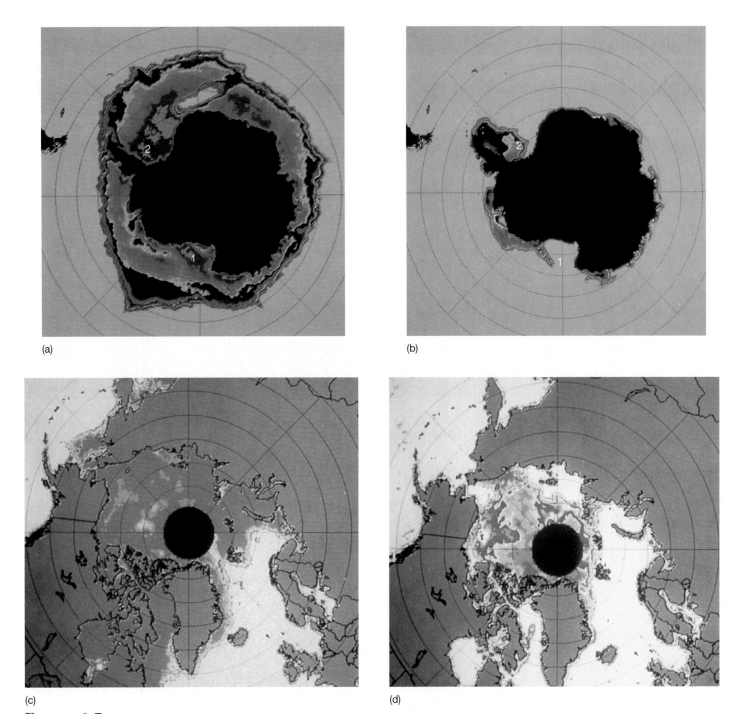

(a)

(b)

(c)

(d)

Figure 6.7

Sea ice around the Arctic and Antarctica. Data from NASA's *NIMBUS-5* satellite was used to construct seasonal maps of sea ice coverage at high latitudes. The Antarctic ice extends more than 100 kilometers from the land into the Ross (1) and Weddel (2) seas during the winter (a) and is much reduced during the summer (b). The Arctic Ocean is filled with ice in the winter (c) but by the next fall the ice cover is much less (d). Short-term motions of the ice boundaries can be calculated from repeated satellite images. These show that the ice floes move up to 50 kilometers (30 mi) per day. Purple indicates high ice concentration, and blue indicates open water.

The Air and the Oceans

FIGURE 6.8

A castle berg drifting in the
Atlantic Ocean just north
of the Grand Banks.

FIGURE 6.9

The structure of the atmosphere.

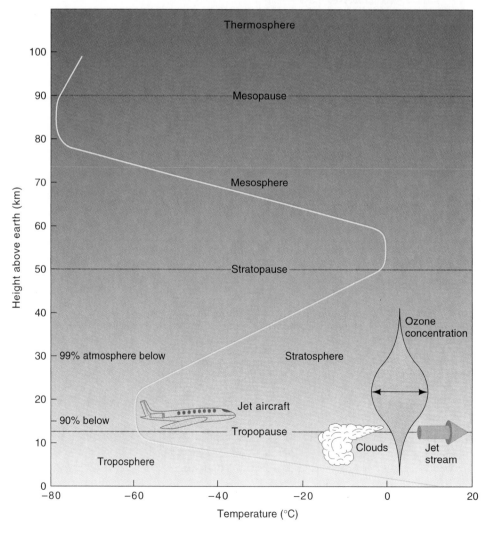

Table 6.1 The Composition of Dry Air

Gas	Percentage of Mixture by Volume
Nitrogen	78.08
Oxygen	20.95
Argon	0.93
Carbon dioxide	0.03
Neon	1.8×10^{-3}
Helium	5×10^{-4}
Krypton	1×10^{-4}
Xenon	1×10^{-5}
Hydrogen	$< 1 \times 10^{-5}$
	~100%

What is air?

What processes regulate the density of air?

How are density and pressure zones related?

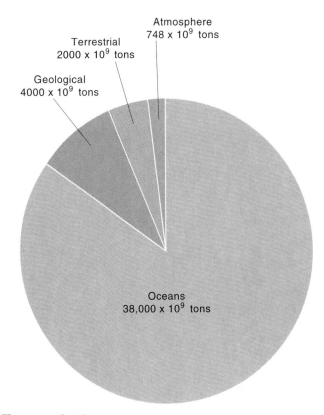

FIGURE 6.10

World carbon dioxide distribution. Values are in billions of tons.

6.6 CHANGES IN ATMOSPHERIC GASES

The atmosphere has changed continuously over the last 4.6 billion years, but today it is changing measurably and rapidly as a result of human activities. The rising level of carbon dioxide (CO_2) is of increasing concern. There are three active reservoirs for carbon dioxide (CO_2): the atmosphere, the oceans, and the terrestrial system; in addition there is the inactive reservoir of the earth's crust. The oceans store the largest amount of CO_2, and the atmosphere has the smallest amount; see figure 6.10. The atmosphere is the link between the other reservoirs, and the oceans play a major part in determining the atmosphere's CO_2 concentration by physical (mixing and circulation), chemical, and biological means.

Carbon dioxide is one of several gases that are transparent to incoming short-wave radiation but absorb outgoing, long-wave radiation from the earth's surface. An increase in the atmosphere's carbon dioxide level also increases the absorption of incoming infrared radiation by the atmosphere and decreases the infrared reaching the earth's surface. This increase in the atmosphere's CO_2 content allows less transfer of long-wave energy from the earth's surface to space and increases the transfer of this energy from the earth's surface to the atmosphere. The net result of increased absorption of infrared in the atmosphere due to increasing CO_2 is the warming of the atmosphere. The retention of the outgoing radiation keeps the earth warm by what is commonly known as the **greenhouse effect.** Human activities have recently added an excess of CO_2 from fossil fuels to the atmosphere, an excess currently estimated at 6 billion tons per year.

As a part of the effort to compare current CO_2 levels with those of the recent past, scientists have sampled the gases trapped in the polar ice and glaciers during the past 160,000 years. Bubbles in the ice contain samples of the air present at the time the ice was formed and allow comparisons of its composition before and after the Industrial Revolution. Since 1850, the concentration of carbon dioxide in the atmosphere has increased from 280 parts per million to 350 parts per million. Recently, the average rate of increase has been 1.3 parts per million per year.

In addition, for more than thirty-five years, scientists have been recording CO_2's annual cycle and the steady increase in its concentration (fig. 6.11). In the Northern Hemisphere, CO_2 decreases in the spring and summer when plants increase photosynthesis and remove CO_2 faster than it is contributed by respiration and decay. In the autumn and winter, plants lose their leaves, photosynthetic activity is reduced, decay processes release CO_2, and atmospheric CO_2 increases. The effects of deforestation, conversion of forest land to agriculture (often accompanied by burning), the burning of fossil fuels in homes and industries, and the growth of human populations are responsible for the increasing values that have been superimposed on the natural cycle. If this trend were to continue unstopped, the concentration of carbon dioxide would double over historic levels some time before the middle of the next century.

Global warming of 2 to 4°C has been predicted by some in response to such an increase in CO_2 levels. A change in temperature of this magnitude in a few decades is comparable to that which has occurred since the last ice age, over 10,000 years ago. It is difficult to predict the extent of the effects brought

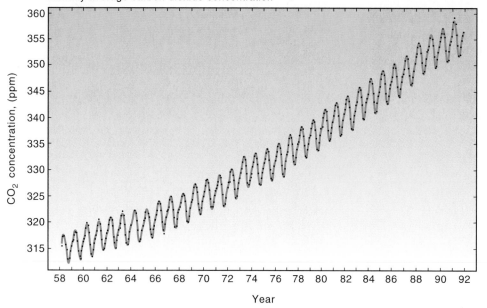

Mauna Loa Observatory, Hawaii
Monthly average carbon dioxide concentration

FIGURE 6.11

Concentration of atmospheric carbon dioxide in parts per million of dry air versus time in years as observed at the Mauna Loa Observatory, Hawaii. The dots indicate monthly average concentrations.

about by such a change, because many other uncertain factors are involved. However, if warming does occur, it is expected to affect the high latitudes more than the low latitudes, causing melting of polar land ice. That, along with increasing seawater temperature and volume, might lead to a sea-level rise of about 1 meter. Another possibility is that the polar ice caps may expand rather than diminish because of increased evaporation caused by the warming, which causes increased precipitation at high latitudes.

In the stratosphere, ozone absorbs much of the ultraviolet radiation from the sun, protecting life-forms on land and at the sea surface. A decrease in the atmosphere's ozone allows more ultraviolet radiation to pass through the atmosphere and increases this radiation reaching the earth's surface. Significant reduction of ozone in the stratosphere has been occurring over Antarctica since the late 1970s. The Arctic winters are not as cold as those of the Antarctic, and ozone loss over the Arctic is less. It is estimated that the average global loss of ozone since 1978 is about 3 percent; at polar latitudes the loss is estimated at about 8 percent per decade. Studies done in 1990 in the Southern Ocean around Antarctica indicated that springtime increases in ultraviolet radiation are reducing the annual production of single-celled plants at the sea surface. The average reduction is estimated to be 2 to 4 percent of normal. These single-celled plants are at the base of the food chains that support the Antarctic's marine life.

The most widely accepted theory of ozone destruction is related to the release of chlorine into the atmosphere. Chlorine is a product of vulcanism, but it is commonly released in increased quantities as a component of chlorofluorocarbons (CFCs) used for refrigeration, air-conditioning, in solvents, and for production of insulating foams.

The greenhouse effect, the ozone loss, and global warming are complex problems. For example, the microscopic plants at the sea surface produce dimethyl sulfide (DMS) gas. Current es-

timates are that the oceans emit about 60 million tons of DMS, or 30 million tons of sulfur, to the atmosphere each year. During their time in the atmosphere, sulfur compounds influence the amount of incoming radiation by reflecting radiation back into space before it can reach the earth's surface. They are also thought to control the density of marine clouds and increase the clouds' reflective properties. Denser, more reflective clouds reduce incoming radiation and decrease the heating of the earth's surface, counteracting the effect of the increasing CO_2 levels. In addition, the 1991 volcanic eruptions of Mount Pinatubo in the Philippines and Mount Unzen in Japan released large quantities of sulfur dioxide to the atmosphere; the resulting sulfur compounds also act to decrease incoming radiation and promote cooling.

The complexity of these interconnected systems makes it very difficult to accurately predict the rate and extent of anticipated changes. However, despite the complexity, there is a consensus that the increasing CO_2 content of the atmosphere is producing a warming trend.

What is the greenhouse effect?

How is carbon dioxide related to the greenhouse effect?

Explain the Northern Hemisphere's annual carbon dioxide cycle.

How have carbon dioxide levels in the atmosphere changed since the Industrial Revolution, and how do we know?

What is the significance of the loss of ozone from the stratosphere?

Why is it difficult to predict the rate of global warming and its effects? Give examples.

6.7 THE ATMOSPHERE IN MOTION

The air moves because at one place less-dense air rises away from the earth while in another place more-dense air sinks toward the earth. Between these areas, the air that flows horizontally along the earth's surface is the wind. This process is shown in figure 6.12. Note that there are really two horizontal air flows, or wind levels, moving in opposite directions: one at the earth's surface and one aloft. This pattern of circulating air is a convection cell that is based on vertical air movements driven by changes in the air's density.

Imagine an earth heated and cooled as described in sections 6.2 and 6.3, but with no continents and no rotation. Due to the unequal distribution of the sun's heat over the earth's surface, large amounts of heat and water vapor are transferred to the atmosphere around the equator. The warm air rises and cools, so that the water vapor condenses, producing precipitation. Equatorial regions are known for their warm wet climate. The dry air aloft flows toward the poles where its density causes it to sink, producing a cold, dry climate. The air then moves over the water back toward the equator, increasing its water vapor content and gaining heat. Such air movement is shown in figure 6.13.

In this system, the winds at the earth's surface in the Northern Hemisphere blow from north to south, and the upper winds blow from south to north; in the Southern Hemisphere the reverse is true. It is important to remember that winds are named for the direction *from which they blow*. A north wind blows from north to south, while a south wind blows from south to north.

How are winds formed?

Explain how the air would circulate in the Northern and Southern Hemispheres if there was no land and the earth did not rotate.

How are the winds named?

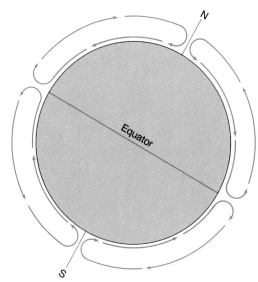

FIGURE 6.13

Heating at the equator and cooling at the poles produces a single large convection cell in each hemisphere on a nonrotating water-covered earth model.

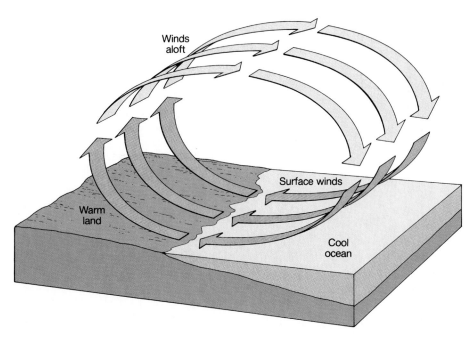

FIGURE 6.12

A convection cell is formed in the atmosphere when air is warmed at one location and cooled at another.

FIGURE 6.14

(a) Air moving toward the equator passes from a latitude of lower eastward speed to a latitude of higher eastward speed. The result is deflection to the right of the initial wind direction in the Northern Hemisphere and deflection to the left in the Southern Hemisphere. There is no deflection at the equator.
(b) Because air moving northward from the equator to point A carries with it its initial eastward velocity, the air is deflected to the right of the initial wind direction in the Northern Hemisphere. In the Southern Hemisphere, air moving southward to point B is deflected to the left. (c) The deflection of eastward-moving and westward-moving air. There is no deflection at the equator.

a = Initial north wind velocity

b = Initial eastward velocity minus the eastward velocity of the earth at a lower latitude

c = Resultant velocity of the wind—Northern Hemisphere

a = Initial south wind velocity

b = Initial eastward velocity minus the eastward velocity of the earth at a higher latitude

c = Resultant velocity of the wind—Northern Hemisphere

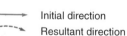

Initial direction

Resultant direction

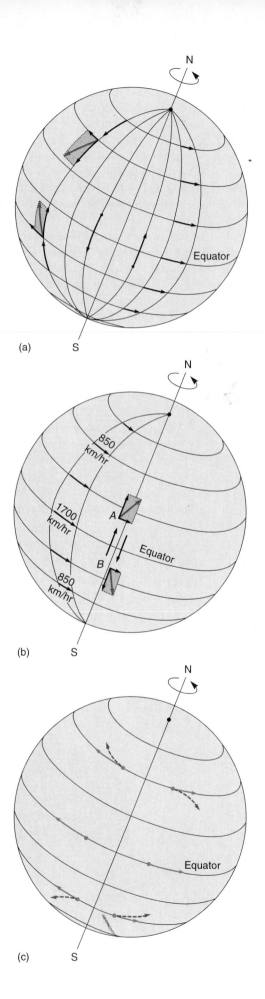

(a)

(b)

(c)

6.8 THE EFFECT OF ROTATION

Adding rotation to the system just described changes the atmospheric circulation. Although each of us is moving with the earth in its daily rotation, we are unaware of the motion. However, if we could step out onto a platform fixed in space, we could trace a parcel of surface air moving straight south from the North Pole, while the earth turned to the east under it. If we were to trace this same parcel of moving air relative to the rotating earth, however, the air would appear to be deflected from its straight-line movement to the south, moving instead to the right or the west. See figure 6.14. This deflection is caused by an apparent force, not a true force since no true force is exerted on the wind, and the deflection is caused by measuring the path of the moving air relative to the rotating earth. The deflection of the moving air relative to the earth's surface is called the **Coriolis effect,** after Gaspard Gustave de Coriolis (1792–1843), who mathematically solved the problem of deflection in frictionless motion when the motion is referred to a rotating body.

If the same experiment were performed with a parcel of air flowing north from the South Pole, the wind would be deflected to the left or west of the direction of motion (fig. 6.14). At the equator no deflection occurs; the Coriolis effect is zero.

The Coriolis effect is equally important in determining the relative motion of ocean currents. *Both moving air and moving water are deflected relative to their direction of motion, to the right in the Northern Hemisphere and to the left in the Southern Hemisphere.*

Why is the Coriolis effect not a true force?

How does the earth's rotation affect wind direction?

6.9 THE WIND BANDS

When the water-covered earth rotates, air continues to rise at the equator and flows toward the north and south poles. As it moves, it is deflected to the right in the Northern Hemisphere and to the left in the Southern Hemisphere. This deflection short-circuits the single hemispheric convection cells of the earlier stationary model. The deflected air sinks at 30°N and 30°S; it then moves over the water, either back toward the equator or toward 60°N and 60°S. The air that cools and sinks over the poles then moves to lower latitudes, warming and picking up water vapor; at 60°N and 60°S it rises again. The result is three convection cells (fig. 6.15) in each hemisphere for the rotating earth.

FIGURE 6.15

The circulation of the earth's atmosphere results in a six-band surface wind system on a rotating water-covered earth.

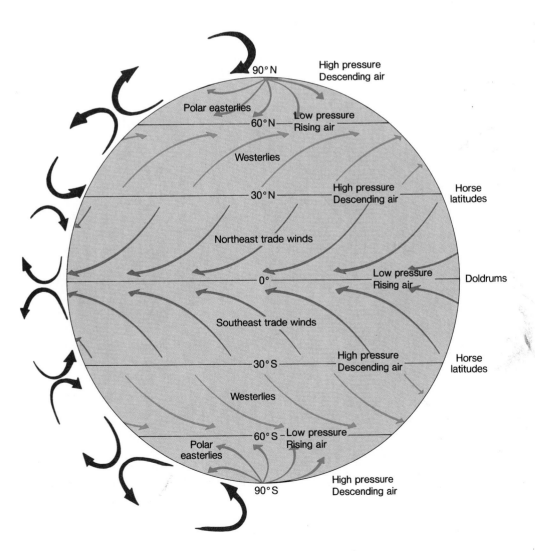

123

The surface winds between 30°N and 30°S are deflected relative to the earth, blowing from the north and east in the Northern Hemisphere and from the south and east in the Southern Hemisphere. These are the **trade winds:** the northeast trade winds north of the equator, and the southeast trade winds south of the equator. Between 30°N and 60°N, the deflected surface flow produces winds that blow from the south and west, while between 30°S and 60°S, they blow from the north and west. In both hemispheres these winds are called the **westerlies.** Between 60°N and the North Pole, the winds blow from the north and east, while between 60°S and the South Pole, they blow from the south and east. In both cases they are called the **polar easterlies.** The six surface wind bands are shown in figure 6.15.

At 0° and at 60°N and S, moist low-density air rises; these are areas of low atmospheric pressure, zones of clouds and rain. Zones of high-density descending air at 30° and 90°N and S are areas of high atmospheric pressure, zones of low precipitation and clear skies. Air flows over the earth's surface from atmospheric high pressure to atmospheric low pressure. In the zones of vertical motion, between the wind belts, the surface winds are unsteady. Such areas were troublesome to the early sailors, who depended on steady winds for propulsion. The area of rising air at the equator is known as the **doldrums,** and the high-pressure areas at 30°N and S are known as the **horse latitudes.** In all these areas, sailing ships could find themselves becalmed for days. The origin of the word "doldrums" is obscure, but the horse latitudes are said to have gotten their name from the stories of ships carrying horses that were thrown overboard when the ships were becalmed and the freshwater supply became insufficient for both the sailors and the animals.

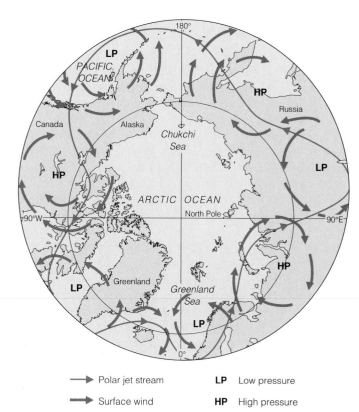

FIGURE 6.16
The polar jet stream circles the earth in the Northern Hemisphere above the boundary between the polar easterlies and the westerly winds. It is deflected north and south by the alternating air-pressure cells of the northern temperate zone.

Locate the trade winds, the westerlies, and the polar easterlies with latitude.

Describe the properties of the zones between the wind bands.

What and where are the doldrums and the horse latitudes?

6.10   THE JET STREAMS

The weather is generated by changes in the atmosphere, but annual heating cycles, wind patterns, and seasonal changes in the earth's surface temperature do not paint a complete picture of day-to-day weather. The winds of the upper troposphere have considerable effect on short-term weather patterns. The polar **jet stream** is the best known of the strong winds that lie above the boundary between the polar easterlies and the westerlies. This wind flows rapidly (in excess of 300 kilometers per hour) eastward and can oscillate more than 2000 kilometers (1250 mi) north to south during the winter as pressure systems change (fig. 6.16). The oscillations

in the jet stream flow are caused by strong high-pressure systems to the south and intense low-pressure systems to the north. The westerlies move the wave form of the stream and its associated high- and low-pressure systems eastward around the earth. In winter, jet stream velocities are higher and displacements are greater. The slow eastward migration of this system constantly alters the weather patterns as they move over the earth's surface.

Above the zone between the trades and westerlies, there are also upper-atmosphere winds, the subtropical jets. In the equatorial regions, the zone where the northeast and southeast trade winds come together is known as the intertropical convergence zone (or doldrums). This zone oscillates north and south with the seasonal cycle of solar heating. Thunderstorms and great cumulus cloud systems formed in the rising air of the intertropical convergence generally migrate westward.

Why is the flight time of high-altitude jet aircraft usually shorter when flying eastward rather than westward across the United States?

How is the jet stream related to the earth's surface winds?

6.11 SEASONAL WIND CHANGES

Unlike our model earth that was one worldwide ocean, on the real earth large blocks of continents interrupt the oceans, and the land surfaces change temperature differently than the oceans. Throughout the year ocean and land surfaces remain warm at the equatorial latitudes and the oceans remain cold at polar latitudes, but the land goes through large seasonal temperature changes at higher latitudes. In the middle latitudes, the land is warmer than the ocean during the warm summer months (refer to fig. 6.6). The density of air over the warm land is reduced, creating a low-pressure area, while the

air density over the cold sea increases producing a high-pressure area (fig. 6.17). Low-pressure zones at 60°N and 0° tend to combine over the land, breaking the high-pressure belt at 30°N into several discrete high-pressure cells located over the oceans during the Northern Hemisphere's summer. In the winter the land becomes much colder than the water, and the air pressure is low over the water and high over the land, creating a low-pressure zone over the ocean. Over the land, the polar high-pressure zone spreads toward the high-pressure zone at 30°N, breaking the low-pressure belt centered on 60°N into separate low-pressure cells over the warmer ocean water.

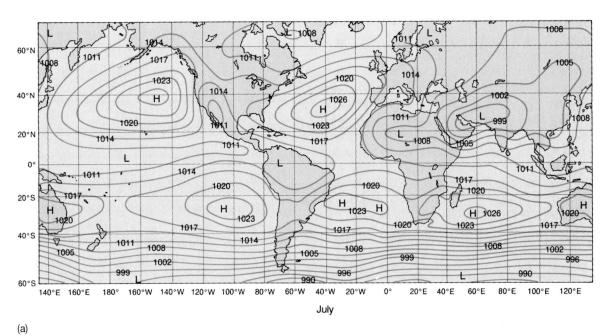

(a)

July

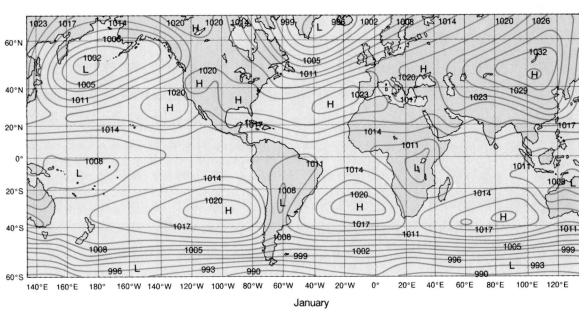

(b)

FIGURE 6.17

Mean sea-level pressures expressed in millibars for (a) July and (b) January.

The Air and the Oceans

During the Northern Hemisphere's summer, the cool air in the high-pressure cells over the central part of the North Atlantic and North Pacific Oceans descends and flows outward toward the low-pressure areas. As the descending air moves outward from the center of a high-pressure system, it is deflected to the right, producing winds that spiral in a clockwise direction around the high (fig. 6.18). On the northern side of these high-pressure cells are the westerlies, and on the southern side are the northeasterlies; the eastern side of the cell has northerly winds, and the western side has southerly ones. In the winter, the air circulates counterclockwise about the low-pressure cell over the northern oceans, and the prevailing wind directions are those of the polar easterlies and the westerlies. Figure 6.19 shows the first global measurements of ocean wind data obtained by satellite.

The strength of the wind is governed by the change in pressure with distance, or the pressure gradient. Along the Pacific coast of the United States, the northerly winds cool the coastal areas in the summer, and the southerly winds warm them in the winter. The New England coast receives warm, moist air from the low latitudes in the summer and cold air from the high latitudes in the winter. Although these seasonal changes modify the wind and pressure belts of the water-covered model of the earth, the wind and pressure belts of the model are still identifiable in the northern latitudes when the atmospheric pressures are averaged over long time periods. Rotation of the airflow about high- and low-pressure air cells is reversed in the Southern Hemisphere due to the Coriolis effect.

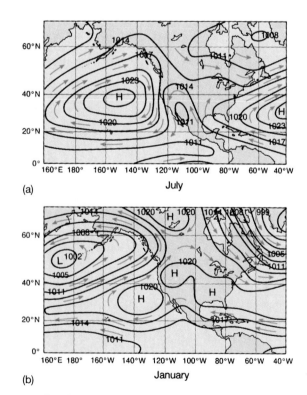

FIGURE 6.18

(a) Winds flow clockwise about a Northern Hemisphere high-pressure cell in the summer and (b) counterclockwise about a low-pressure cell in the winter.

At the middle latitudes in the Southern Hemisphere there is little land to separate the water; therefore, the water temperature predominates and there is little seasonal effect (refer to figures 6.5 and 6.6). The atmospheric pressure and wind pattern created by surface temperatures in the Southern Hemisphere changes little over the annual cycle and is very similar to that developed for the water-covered model (see fig. 6.15).

Why do high- and low-pressure systems alternate through the temperate latitudes of the Northern Hemisphere?

Why do the winds circulate clockwise around a high-pressure zone in the Northern Hemisphere?

What happens to winds around a low-pressure zone in the same hemisphere?

How does the lack of land in the Southern Hemisphere affect the seasonal wind pattern?

6.12 LAND EFFECTS ON WINDS

The difference in temperature between land and water produces large-scale and small-scale effects in coastal areas. In India and Asia during the summer, the air rises over the hot land, creating a low-pressure system. The rising air is replaced by warm, moist air carried on the southwest winds from the Indian Ocean. As this onshore airflow rises over the elevated land, the moisture condenses, producing a steady, heavy rainfall. This is the wet or summer **monsoon** (fig. 6.20a). In the winter, a high-pressure cell forms over the land, and northeast winds carry the dry, cool air southward from the Asian mainland and out over the Indian Ocean. This movement produces cool, dry weather over India known as the dry or winter monsoon (fig. 6.20b).

The same effect is seen on a smaller scale along a coastal area or along the shore of a large lake. During the day, the land is warmed faster than the water and the air rises over the land. The air over the water moves in to replace it, creating an **onshore** breeze. At night, the land cools rapidly, and the water becomes warmer than the land. Air rises over the water, and the air from the land replaces it, this time creating an **offshore** breeze (fig. 6.21). The onshore breeze reaches its peak in the afternoon, when the temperature difference between land and water is at its maximum. The offshore breeze is strongest in the late night and early morning hours. Fishing boats relying only on their sails used this daily wind cycle, leaving harbor early in the morning and returning in the late afternoon or early evening.

As the winds sweep across the oceans, they are forced to rise to continue moving across the land, as shown in figure 6.22. The upward deflection cools the air, causing rain on the windward side of islands and mountains; on the leeward (or sheltered) side, there is a low-precipitation zone, sometimes called a **rain shadow.** For example, the windward sides of the Hawaiian mountains have high precipitation and lush vegetation, while the leeward sides are much drier. On the west coast

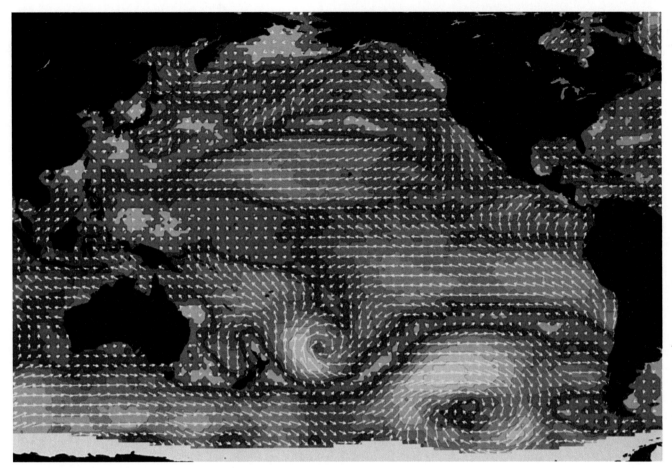

FIGURE 6.19

Pacific Ocean wind distributions. The *SEASAT* satellite obtained the first global measurements of ocean wind data during three days in 1978. Arrows point in the direction the winds blow; longer arrows indicate greater wind speed. Light winds (less than 4 m/s or 9 mph) are colored blue, and strong winds (greater than 14 m/s or 31 mph) are colored yellow. Wind speeds and directions were calculated from radar reflections from the small wind-driven waves that roughened the sea surface. The approximate accuracy is ± 2m/s and ± 20°. Winds blowing from the west in the Southern Hemisphere blow almost continuously around the entire Southern Ocean except near the tip of South America. Intense storms are shown in the South Pacific. The strong and constant southeast trade winds are shown to their north. Summer in the Northern Hemisphere is characterized by a large high-pressure cell in the North Pacific. Circulation about this cell produces the winds along the west coast of the U.S. that induce upwelling. Winds and squalls are formed at the boundary between the northeast and southeast trade winds, the doldrums.

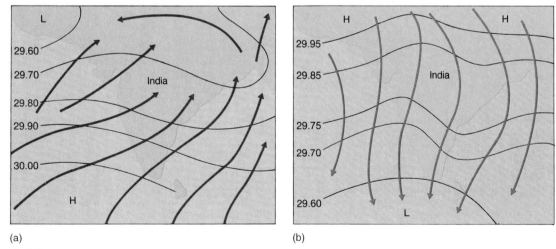

(a)   (b)

FIGURE 6.20

The seasonal reversal in wind patterns associated with (a) the summer (wet) monsoon and (b) the winter (dry) monsoon. The isolines of pressure are given in inches of mercury.

FIGURE 6.21

Differences between day and night land-sea temperatures produce an onshore breeze during the day and an offshore breeze at night.

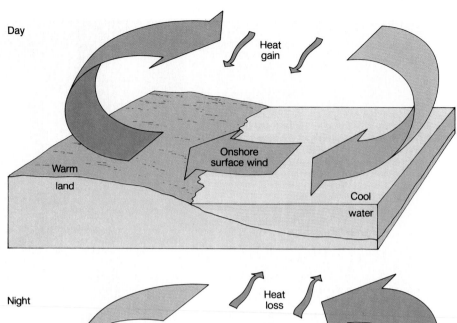

Day

Heat gain

Warm land

Onshore surface wind

Cool water

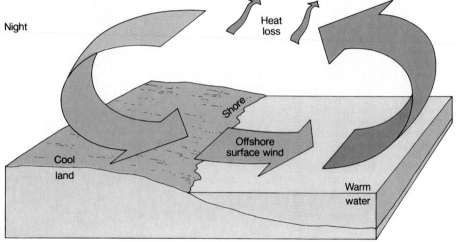

Night

Heat loss

Shore

Cool land

Offshore surface wind

Warm water

FIGURE 6.22

Moist air rising over the land loses its moisture on the mountains' windward side. The descending air on the leeward side is dry.

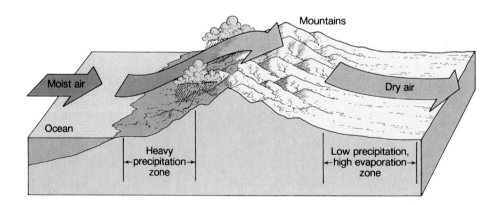

Mountains

Moist air

Dry air

Ocean

Heavy precipitation zone

Low precipitation, high evaporation zone

of Washington state, the westerlies rise to clear the Olympic Mountains and produce the Olympic rain forest, with a rainfall of as much as 5 meters (200 in) per year.

What produces the monsoon?

Why do local coastal winds frequently change direction from day to night?

Explain the effect of land elevation on precipitation.

6.13 FOG

When clouds form close to the ground, they are called **fog.** There are three basic types of fog, each formed differently. (1) The most common fog at sea is **advective fog,** formed when warm air saturated with water vapor moves over colder water. In northern California, Oregon, and Washington, when off-shore warm, moist air moves over the cold coastal water, a coastal fog results, cooling the city of San Francisco (fig. 6.23) and providing moisture to the redwood forests and grassy coastal cliffs. (2) Streamers of **sea smoke** rise from the sea sur-

FIGURE 6.23
Advective fog obscures the Golden Gate Bridge at the entrance to San Francisco Bay.

face on a cold winter day. Dry, cold air from the land or from the polar ice pack moves out over the warmer water, which warms the cold air from below. This warmed air picks up water vapor from the sea surface, rises rapidly, and forms ribbons of fog. (3) **Radiative fog** is the result of warm days and cold nights. If the air holds enough moisture during the day, it condenses as the earth cools at night, forming the low-lying thick, white fogs that are often found in river valleys and occasionally in bays and inlets along the coast. Such fogs usually disappear during the morning as the sun gradually warms the air, changing the water droplets back to water vapor.

Compare the direction of heat transfer between advective fog and sea smoke.

6.14 EL NIÑO

El Niño, part of the El Niño-Southern Oscillation or ENSO, is a weather aberration that has importance to oceanography. It begins with the breaking down of the normal systems that sup-

port the trade winds, the high-pressure system over the area of Easter Island and the low-pressure area over Indonesia. The trade winds first strengthen, causing an abnormal accumulation of warm surface water near Indonesia; as the trade winds weaken, the warm surface water surges eastward across the Pacific carrying its overlying low-pressure storm systems with it. The result is warm surface water and associated wet weather off the west coast of South America in a region that normally has cold surface water and a dry climate (fig. 6.24). Meanwhile, the Indonesian area loses its normal high precipitation. This event appears to occur on a 4- to 7-year cycle.

El Niño is named for the Christ Child due to its frequent coincidence with the Christmas season in the waters off Peru and Ecuador. The intrusion of warm surface water along this coast causes the death of the usual populations of cold-water organisms and upsets the food chains on which fish, marine mammals, and seabirds depend. The heavy rains cause flooding and landslides along the normally dry coast lands. In other parts of the world, the 1982–83 El Niño brought severe drought in Australia and the Sahel of Africa, and frequent typhoons and heavy rains throughout Polynesia.

El Niño has a counterpart called **La Niña** (the girl) that occurs when stronger than normal trade winds lower the

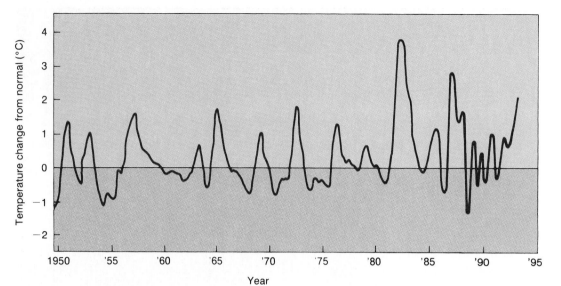

FIGURE 6.24

Sea surface temperature changes at the Peru coast, approximately 8°S. El Niño events (above 0 on the chart) and La Niña events (below 0) form an alternating pattern. Normal sea surface temperature is marked by 0. Note the lack of La Niña events between 1975 and 1987.

normal surface temperatures in the eastern tropical Pacific, while it is warmer than normal to the west. This helps establish dry conditions over the west coast areas of South America, while rainfall and flooding increase in India, Burma, and Thailand.

Other weather changes brought by these events include the failure of the wet monsoon in Asia, displacement of major storms from their usual tracks, abnormal waves forming on the jet stream, and drought conditions in the central United States and Europe. During early 1995 a persistent southerly displacement of the jet stream, which may be related to El Niño, brought unusually heavy rainstorms into California causing flooding and much property loss. The ability to forecast these events and so regulate fisheries, predict agricultural droughts, and prepare for severe weather conditions has great economic and commercial importance around the world. To make these long-range predictions, researchers watch for atmospheric pressure shifts and monitor the intensity of tropical cyclones associated with accumulating warm surface water in the western Pacific. They use satellite observations and computer modeling to recognize patterns and help them with their forecasts.

An experimental computer model correctly predicted the 1986–87 El Niño a year before its onset, and forecast the 1991–92 El Niño two years in advance. The 1991–92 episode brought a mild winter to the northern United States while inundating the southeast states with heavy rains. This statistical model did not predict the significant (1–2°C) tropical Pacific surface water warming of 1993, but other researchers using a different model system were able to do so.

How does El Niño begin?

What is the result of El Niño along the west coast of South America?

How does La Niña differ from El Niño?

Why is the prediction of El Niño important?

6.15   INTENSE OCEAN STORMS

Intense storms known as **hurricanes** are born over tropical oceans when surface water temperature exceeds 27°C. On either side of the equator very strong low-pressure cells pump large amounts of energy from the sea surface into the atmosphere as warm moist marine air rises rapidly and condenses to form clouds and precipitation. The large amounts of heat energy liberated from the condensing water vapor fuel the storm winds, raising them to destructive levels, up to 300 kilometers per hour (160 knots). A major hurricane contains energy exceeding that of a large nuclear explosion; fortunately, the energy is released much more slowly. The energy generated by a hurricane is about 3×10^{12} watt-hours per day, the equivalent of one million tons of TNT.

Hurricanes can form on either side of the equator but not at the equator, because the Coriolis effect is necessary to create their spiraling winds. When a storm of this type is formed in the western Pacific Ocean it is called a **typhoon** instead of a hurricane; both hurricanes and typhoons may be called cyclones. Areas that give rise to hurricanes and typhoons and their storm tracks are shown in figure 6.25.

There are often periods of excessive high water along a coast associated with these intense ocean storms and the changes in atmospheric pressure and wind that accompany them. These episodes are known as **storm surges** or **storm tides.** The low atmospheric pressure in the hurricane's or typhoon's center allows the sea surface to rise up into a dome or hill, while the sea surface is depressed farther away from the center where the atmospheric pressure is greater.

The strong winds circulating about the center of the storm plus the speed of the storm over the ocean create a wind-driven surface current. When the storm comes ashore, the wind pushes the water toward the coast; the elevated sea level, the large waves, and the continued strong winds retard the runoff of water from the land and increase the severity of the flooding. The effect of the high water is intensified when the coastal area is shallow and the adjacent land is low. If the storm tide arrives at the same time as the high tide, the disaster is magnified.

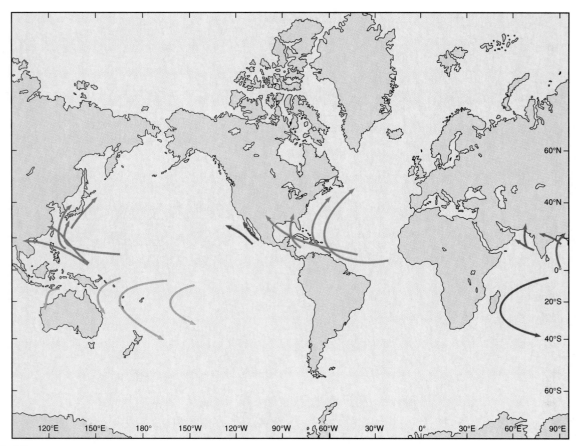

FIGURE 6.25

Tracks of hurricanes and typhoons. Colored arrows indicate areas in which these storms occur.

FIGURE 6.26

Storm tide damage caused by Hurricane Hugo, Pawley's Island, South Carolina, September, 1989.

Along shallow areas of the East and Gulf coasts of the United States, storm surges have caused great damage. Hurricane Hugo came ashore with a storm tide of 5 meters (16.5 ft) in South Carolina in 1989, destroying much property along the coast (fig. 6.26) and in the city of Charleston. In August 1992 Hurricane Andrew (fig. 6.27) struck Florida, but the coastal high-water damage was much less than that along the Carolina coast. Hurricane Andrew, traveling a short distance between the Bahamas and Florida, did not have enough time to build up energy to produce a large storm tide and high waves or drive water onshore; in addition Florida's extremely flat coast reduced the formation of high waves, and some of Andrew's wave energy was absorbed by the offshore reefs and coastal mangrove swamps. However, the damage caused by its onshore winds made Andrew the most destructive hurricane on record in the United States. An estimated 300,000 people lost their lives when storm surges struck the coast of Bangladesh in 1970; another 10,000 were killed in 1985, and 139,000 in 1991. The expanding populations of this area live and farm in the low-lying delta of the Ganges River, and there is no way to evacuate thousands of people when the storm tides cover hundreds of square miles. These storm tides or storm surges occur not only in the Indian Ocean and the Gulf of Mexico, but also in the North Sea and along any shallow coast under intense storm conditions.

What is the source of energy for intense tropical storms?

Where do hurricanes begin, and in what direction are they most likely to travel?

How is sea level affected by these storms?

Why are storm tides (storm surges) so destructive?

The Air and the Oceans

FIGURE 6.27

A satellite image of Hurricane Andrew approaching land south of Miami, August 24, 1992. Maximum sustained winds were 220 kilometers per hour (138 mi/hr).

SUMMARY

The intensity of solar radiation over the earth's surface varies with the latitude and the season. Incoming radiation and outgoing radiation are equal when averaged over the whole earth. Reflection, reradiation, evaporation, conduction, and absorption keep the heat budget in balance.

Earth surface temperatures change with latitude and seasonal changes in solar radiation as well as with the heat capacity of the earth's surface materials, sea ice formation, and evaporation. The oceans gain and lose large quantities of heat, but their temperature changes very little because of the oceans' high heat capacity. Heat that is mixed downward reduces the temperature change at the surface.

Sea ice is formed in the extreme cold of polar latitudes. The process is self-insulating, so that large thicknesses of ice do not form each winter. Icebergs are formed from glaciers that break off into the sea.

The atmosphere is made up of layers. Weather occurs in the layer closest to the earth's surface, the troposphere. Air is a mixture of gases. The density of air is controlled by the air's temperature, pressure, and water vapor content. High-pressure and low-pressure zones are related to air density.

Concern that the earth's climate may be changing is related to changes in the earth's atmosphere. Increasing carbon dioxide concentration, due to the burning of fossil fuels, is leading to predictions of global warming. The ozone layer is being depleted, increasing the ultraviolet radiation reaching the earth's surface, and reducing populations of single-celled plants in Antarctic seas.

Winds are the horizontal motion of air in the atmospheric convection cells produced by heating at the earth's surface. Winds are named for the direction from which they blow.

Because of the Coriolis effect, winds are deflected to their right in the Northern Hemisphere and to their left in the Southern Hemisphere. This action produces a three-celled wind system in each hemisphere that results in the surface wind bands of the trade winds, the westerlies, and the polar easterlies. Zones of rising air occur at 0° and 60°N and S; these are low-pressure areas of clouds and rain. Zones of descending air at 30° and 90°N and S are high-pressure areas of clear skies and low precipitation. Surface winds are unsteady and unreliable in the zones of the doldrums and the horse latitudes. The polar jet stream is a high-speed upper-atmosphere wind that marks the boundary between the westerlies and the polar easterlies in the Northern Hemisphere.

Seasonal atmospheric pressure changes modify these wind bands and cause coastal winds to change direction seasonally in the Northern Hemisphere.

Differences in temperature between land and water produce the monsoon effect. This seasonal reversal in wind pattern causes the wet and dry monsoons of the Indian Ocean; a similar daily reversal causes the onshore and offshore winds of coastal areas. Winds from the ocean rising to cross the land also produce heavy rainfall.

Fog occurs when water vapor condenses to form liquid droplets. Advective fog occurs when warm, water-saturated air passes over cold water; sea smoke occurs when dry, cold air moves over warm water; and radiative fog occurs when warm, moist air is cooled at night.

In some years, warm tropical surface water moves eastward across the Pacific and accumulates along the west coast of the Americas. This phenomenon, El Niño, is associated with changes in atmospheric pressure and wind direction in the equatorial Pacific Ocean. Colder-than-normal surface water temperatures and associated meteorological phenomena known as La Niña, appear to alternate with these El Niño events.

In the tropics, rapid heat transfer to the atmosphere from the sea surface produces intense ocean storms. A rising sea surface under a storm's low-pressure center, combined with storm winds and waves, produces severe coastal flooding known as a storm tide or storm surge.

KEY TERMS

heat budget	low-pressure zone	horse latitudes	advective fog
sea ice	greenhouse effect	jet stream	sea smoke
iceberg	Coriolis effect	monsoon	radiative fog
troposphere	trade winds	onshore	El Niño
stratosphere	westerlies	offshore	La Niña
ozone	polar easterlies	rain shadow	hurricane/typhoon
atmospheric pressure	doldrums	fog	storm tide/storm surge
high-pressure zone			

SUGGESTED READINGS

Berner, R. A., and A. C. Lasaga. 1989. Modeling the Geochemical Carbon Cycle. *Scientific American* 260 (3):74–81.

Houghton, R. A., and G. M. Woodwell. 1989. Global Climate Change. *Scientific American* 260 (4):36–44.

Jensen, J. J., and J. Hovermale. 1990/91. Numerical Air/Sea Environmental Prediction. *Oceanus* 34 (4):40–49. Role of data in predicting environmental change.

Jones, P. D., and T. M. L. Wigley. 1990. Global Warming Trends. *Scientific American* 263 (2):84–91.

Kerr, R. A. 1993. El Niño Metamorphosis Throws Forecasters. *Science* 262 (5134):656–57.

Kretschmer, J. 1990. When the Winds Blow: A Mythology of Gust and Squall. *Sea Frontiers* 36 (3):40–43.

Leetma, A. 1989. The Interplay of El Niño and La Niña. *Oceanus* 32 (2):30–34.

Olson, D. B. 1990. Monsoons and the Arabian Sea. *Sea Frontiers* 36 (1):34–41.

Post, W. M., R. Peng, W. R. Emanuel, A. W. King, V. H. Dale, and D. L. DeAngelis. 1990. The Global Carbon Cycle. *American Scientist* 78:310–26.

Smith, F. G. W. 1992. Hurricane Special: An Inside Look at Planet's Powerhouse. *Sea Frontiers* 38 (6):28–31.

Takahushi, T. 1989. The Carbon Dioxide Puzzle. *Oceanus* 32 (2):22–29.

Webster, P. J. 1981. Monsoons. *Scientific American* 245 (2):108–18.

Weiner, J. 1989. Glacier Bubbles Are Telling Us What Was in Ice Age Air. *Smithsonian* 20 (2):78–87.

Williams, J. M., F. Doehring, and I. W. Duedall. 1993. Heavy Weather in Florida. *Oceanus* 36 (1):19–26.

Zimmer, C. 1994. Good News and Bad News. *Discover* 15 (5):26–28. Update on greenhouse gases.

The Air and the Oceans

7

Circulation Patterns and Ocean Currents

Outline

Learning Objectives

After reading this chapter, you should be able to

- Understand the effects of changing surface temperatures and salinities.

- Know how temperature and salinity influence density with depth.

- Describe how surface water densities vary with latitude.

- Explain density-driven circulation and relate it to surface changes at different latitudes.

- Compare the processes of divergence and convergence.

- Discuss the layered structure of the Atlantic Ocean and compare it with the Pacific and Indian Oceans.

- Describe how temperature and salinity are measured at sea.

- Relate the effects of surface winds to the movement of water and the ocean's surface currents.

- Explain the processes that produce the oceanic gyres.

- Describe the major ocean surface current systems.

- Identify major areas of coastal upwelling and downwelling.

- Understand the role of eddies.

- Understand how ocean currents are measured.

- Describe how energy is produced from changes in seawater temperature and from energy flow.

◀ The sun reflecting from ocean ripples.

I f we could cut vertically through the ocean and remove a slice of ocean water in the same way we might cut a slice of cake, we would find that, like a cake, the ocean is a layered system. The layers are invisible to us, but they can be detected by measuring the water's changing salt content and temperature and by calculating the density of water from the surface to the ocean floor. This layered structure is a dynamic response to processes that occur at the surface: the gain and loss of heat, the evaporation and addition of water, the freezing and thawing of sea ice, and the movement of water in response to winds. This chapter explores the structure of the oceans below the surface and follows the currents as they flow, merge, and move away from each other. It examines both horizontal and vertical circulation and the links between them.

7.1 DENSITY-DRIVEN CIRCULATION

Variations in temperature, salinity, and pressure combine to control the density of seawater. Many combinations of different salinities and temperatures produce the same density; this is shown in the lines of constant density displayed in figure 7.1. This diagram is related to the density values in table 5.1 of Chapter 5.

In many places, the oceans have a well-mixed surface layer approximately 100 meters (300 ft) deep. In this layer, the increasing pressure with depth causes the density to increase very slightly. Below this surface layer, density changes rapidly between 100 meters and about 1000 meters (300–3000 ft) because the temperature decreases rapidly. Also in this layer (between 100 and 1000 meters), the salinity may increase, decrease, or remain unchanged; it has a smaller influence on the density over this depth. Below 1000 meters, both temperature and salinity change slowly with depth, and the density increases slightly down to the sea floor because of the increasing pressure. Figure 7.2 shows examples of these temperature and salinity changes with depth.

At the ocean surface, density is controlled by the earth's uneven heating and evaporation-precipitation patterns. Variations in surface densities control the vertical movement of the water. Use figure 7.3 with the following discussion. The warm, low-salinity surface water at the equator is less dense than the surface water at latitudes 30°N and 30°S, which is slightly cooler and has a higher salinity because of evaporation and diminished rainfall. Therefore, surface water produced at 30° latitudes is denser than that produced at the equator, and it sinks below the equatorial water. The combined salinity and temperature of surface water at 50 to 60°N or S produce water that is denser than either the equatorial or the 30° latitude surface water, and so this water sinks below the equatorial and 30°

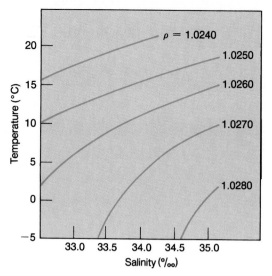

FIGURE 7.1
The density of seawater, measured in grams per cubic centimeter, is abbreviated as ρ (rho) and varies with temperature and salinity. The pressure factor is not included. Many combinations of salinity and temperature produce the same density. Low densities are at the upper left and high densities at the lower right.

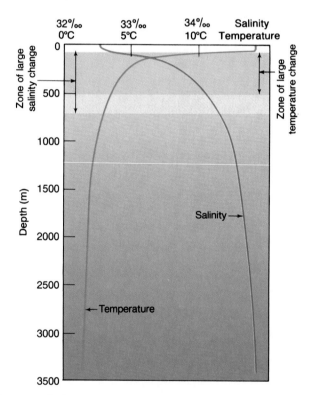

FIGURE 7.2
Temperature and salinity values change with depth in seawater. Rapid changes in temperature and salinity with depth occur in the near-surface layer (50–700 meters). (Based on data from the northeast Pacific Ocean.)

waters. Winter conditions in the polar regions lower the surface water temperature and, if sea ice forms, increase the surface salinity. This results in dense surface water that sinks toward

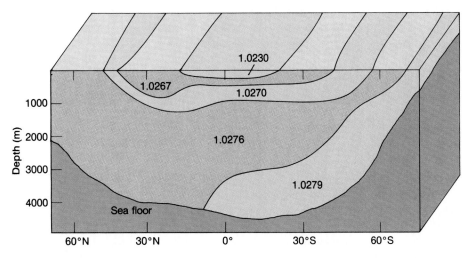

FIGURE 7.3

Water of different densities is distributed over depth. The density of each layer is determined by the salinity and temperature where each layer meets the surface.

the ocean floor at polar latitudes. The cold, dense water of the Arctic is confined by the seafloor topography to these high northern latitudes; similar water formed around Antarctica moves northward into the Southern Hemisphere oceans. The thickness and horizontal extent of each density layer is related to the rate at which the water is formed and the size of the surface region over which it is formed.

What are the effects of changing the salinity and temperature of surface water?

Explain how water samples with different salinities and temperatures have the same density.

How do temperature and salinity change with ocean depth?

Why does water density change more rapidly with depth in the upper 1000 meters than near the sea floor?

Explain why surface water densities vary with latitude.

7.2 THERMOHALINE CIRCULATION

If a surface process forms more-dense water on top of less-dense water, the water column is unstable, and as the denser water sinks, the less-dense water rises. This convective motion is called **overturn.** If the water column has the same density over depth, it has neutral stability; it does not overturn, and the water is easily mixed by wind, waves, and currents. If the water column has an increasing density with depth, it is stable, does not overturn, and is more difficult to mix vertically.

Vertical circulation driven by surface changes in temperature and salinity is known as **thermohaline circulation.** Thermohaline circulation ensures an eventual top-to-bottom exchange of water throughout the oceans. If dense surface water sinks at one location, the water it displaces at depth must rise elsewhere. An excellent example of such circulation occurs

in the Weddell Sea of Antarctica, where the winter cooling and freezing produce dense surface water that sinks to the deep sea floor. This water descending along the coast of Antarctica is the densest water found in the open oceans.

In the middle latitudes, the surface water's temperature changes with the seasons (fig. 7.4). During the summer, the surface water warms and the water column is stable, but in the fall the surface water cools, increases its density, and begins to overturn. Winter storms and winter cooling continue the mixing process. The shallow, warm, low-density surface water formed during the previous summer is lost, and the upper portion of the water column becomes neutral to greater depths. Spring brings surface warming, and the water column becomes stable and remains so through the summer.

Seasonal changes in temperature are more important than seasonal changes in salinity in altering density in the open ocean. For example, in the Atlantic Ocean the surface water at 30°N and S has a high salinity, but it is warm year-round so that there is no water sinking to great depths. On the other hand, the surface water from 50° to 60° in the North Atlantic has a lower salinity but is cold, especially during the winter. This cold water sinks and flows below the saltier but warmer surface water.

Close to shore, however, the salinity of seawater may outweigh the temperature in controlling the density. This is particularly true along coasts that receive large amounts of freshwater runoff. Here even cold (0° to 1°C), fresh river water does not sink when it meets the salt water but remains at the surface as a freshwater lid. In polar regions, when the sea ice melts, a layer of fresh water forms at the surface and slowly dilutes the surface seawater. When sea ice forms, the extraction of water to form the ice increases the salinity of the surface water, and this denser water sinks. In regions of high evaporation, the salinity of water in marginal seas rises, increasing the density of the water. Such water sinks to depth as it flows into the adjacent ocean. This occurs in the Red Sea and the Mediterranean Sea, and is discussed in Chapter 9, section 9.

Circulation Patterns and Ocean Currents

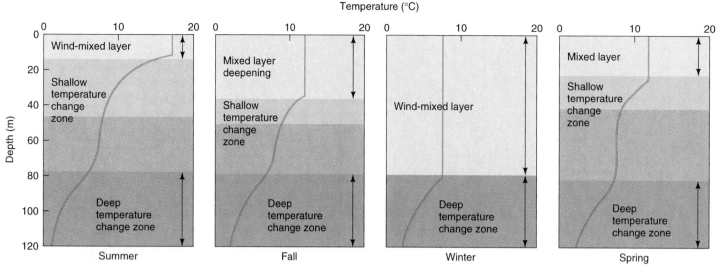

FIGURE 7.4

The surface layer temperature structure varies over the year. In the absence of strong winds and wave action in the summer, solar heating produces a shallow temperature change zone. During the fall and winter the surface cooling and storm conditions cause mixing and vertical overturn, which eliminate the shallow temperature change zone and produce a deep wind-mixed layer. In spring the shallow temperature change zone re-forms.

What are the effects of stable, unstable, and neutral water column conditions?

Why is oceanic vertical circulation called thermohaline circulation?

How do seasonal climate changes control surface overturn at middle latitudes?

Compare temperature and salinity control of surface density in the open ocean and along the coast.

7.3 Convergence and Divergence

When dense water from the surface sinks and reaches a level at which it is denser than the water above but less dense than the water below, it spreads horizontally as more water descends behind it. At the surface, water moves horizontally into the region where sinking is occurring; this is known as a surface **convergence.** The dense water that has descended displaces water upward, and surface water moves away from the area of rising water, completing the cycle and forming a surface **divergence.** Oceanographers refer to the upward vertical motion associated with a divergence zone as **upwelling;** downward vertical motion at a convergence zone is known as **downwelling.** Other important forces acting at the sea surface that cause water to converge and diverge include wind-driven water flows moving toward or away from each other or toward or away from the land. Note that these wind-driven convergences and divergences are not caused by thermohaline circulation.

A generalized north-to-south section of the middle Atlantic Ocean in figure 7.5 displays the relationships between thermohaline circulation and surface convergences and divergences. Notice that high latitude surface water is denser and sinks deeper than lower latitude surface water. These open-ocean zones of convergence and divergence are shown in figure 7.6. The **tropical convergence** is at the equator; it extends across the oceans centered under the doldrums. The two **subtropical convergences** are at approximately 30° to 40°N and S; the Arctic and Antarctic convergences occur at about 50°N and S, and the coastal Antarctic convergence is at 60°S. There are three major oceanic divergence zones: two tropical divergences and the Antarctic divergence. Downwellings or surface convergence zones transport oxygen-rich surface water to depth, and upwellings or divergences deliver water rich in nutrients to the surface, where these fertilizers promote the production of plants that supply the food chain supporting the remainder of the oceans' organisms.

This vertical motion of the water flows at rates of 0.1 to 1.5 meters (0.3 to 5 ft) per day, compared to fast-flowing ocean surface currents that reach speeds of 1.5 meters per second. Horizontal movement at depth due to thermohaline circulation is about 0.01 centimeters (0.004 in) per second. Water caught in this slow but relentlessly moving cycle may spend 1000 years at depth before it again rises to the surface.

Compare the processes of divergence and convergence.

If a convergence zone is present, why must there also be a divergence zone?

Using figures 7.5 and 7.6 follow the water sinking at the Arctic convergence and identify the area in which it resurfaces.

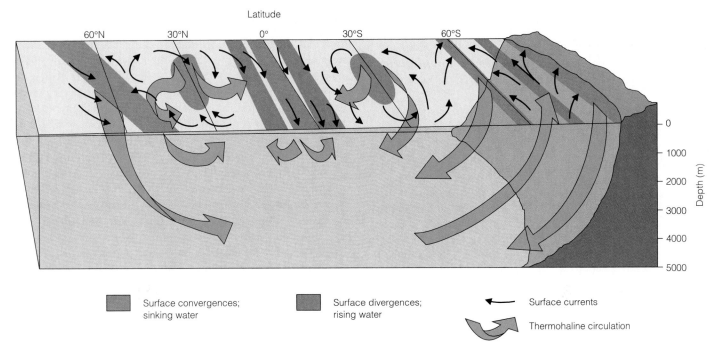

FIGURE 7.5

A north-south midsection of the Atlantic Ocean shows surface zones of convergence and divergence. Thermohaline and wind-driven circulation associated with these zones produces vertical circulation patterns.

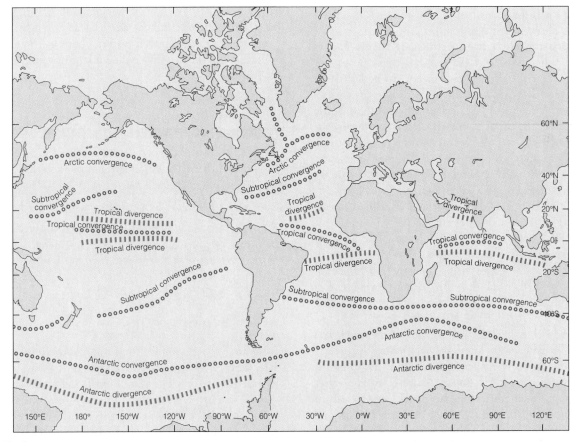

FIGURE 7.6

The principal zones of surface convergence and divergence associated with wind-driven and thermohaline circulation.

Circulation Patterns and Ocean Currents

7.4 THE LAYERED OCEANS

Thermohaline circulation and the deep-water flows associated with convergences and divergences determine the distribution of temperature, salinity, and density with depth in the world's oceans.

Figure 7.7 compares average salinity and temperature distributions for the Atlantic, Pacific, and Indian Oceans.

The Atlantic Ocean

At the surface of the North Atlantic, water from high northern latitudes moves southward, while water from low latitudes moves northward along the coast of North America and then east across the Atlantic. These waters converge in areas of cool temperatures and high precipitation at approximately 60°N. The resulting mixed water has a salinity of about 34.9‰ and a temperature of 2° to 4°C. This water, known as **North Atlantic deep water,** sinks and moves southward (see fig. 7.7a and b). Above this water, at about 30°N, a shallow lens of very salty (36.5‰), very warm (25°C) surface water remains trapped by the circular movement of the oceanic surface currents (to be discussed later in section 7). Between this surface water and the North Atlantic deep water lies water of intermediate temperature (10°–13°C) and salinity (35–37‰). Very warm, high-salinity water from the Mediterranean Sea is added to the North Atlantic through the Strait of Gibraltar. This water finds its own density at approximately 1000 meters (3000 ft) depth in the North Atlantic.

Near the equator, **Antarctic intermediate water,** warmer (5°C) and less salty (34.4‰) than the North Atlantic deep water, remains above the denser water below, while less dense but high-salinity surface waters lie above it. Along the edge of Antarctica, very cold (–0.5°C), salty (34.8‰), and dense water is produced at the surface by sea ice formation during the Southern Hemisphere's winter. It sinks to the ocean bottom, because this **Antarctic bottom water** is the densest water produced in the world's oceans. After it descends to the ocean floor, it begins to move northward. When it meets North Atlantic deep water, it creeps beneath it and continues northward along the coast of South America (fig. 7.7a and b). Antarctic bottom water forms in small quantities and does not accumulate enough height to be able to flow over the mid-ocean ridge system and into the basins along the African coast. It is therefore confined to the deep basins on the west side of the South Atlantic and has been found as far north as the equator.

The North Atlantic deep water trapped between the Antarctic bottom and intermediate waters rises to the surface in the area of the 60°S divergence. As it reaches the surface it splits; part moves northward as **South Atlantic surface water** and Antarctic intermediate water; part moves southward toward Antarctica. A mixture of North Atlantic deep water and Antarctic bottom water becomes the circumpolar water for the Southern Ocean around Antarctica.

The Pacific Ocean

In the Pacific high latitude, waters sinking from relatively small areas of surface convergence lose their identities rapidly, making the layers difficult to distinguish (fig. 7.7c and d). Antarctic bottom water forms in small amounts along the Pacific rim of Antarctica but is lost in the great volume of the Pacific Ocean. The deeper water of the South Pacific Ocean has its source in the circumpolar flow of water around Antarctica. Because the North Pacific is isolated from the Arctic Ocean, only a small amount of water comparable to North Atlantic deep water can be formed. In the extreme western North Pacific, convergence of the southward-flowing cold water from the Bering Sea and the northward-moving water from the equatorial latitudes produces a small volume of water that sinks to mid-depths (fig. 7.7c). There is no dramatic source of deep water similar to that found in the North Atlantic. Warm, salty surface water is found at subtropic latitudes in each hemisphere, and Antarctic intermediate water is produced in small quantities, but its influence is small (fig. 7.7c). Deep-water movements in the Pacific are sluggish, and conditions are very uniform below 2000 meters (6600 ft).

The Indian Ocean

Since the Indian Ocean is principally an ocean of the Southern Hemisphere, there is no counterpart of the North Atlantic deep water (fig. 7.7e and f). Small amounts of Antarctic bottom water are soon lost, and the deeper waters tend to be supplied from Antarctic circumpolar water. There is a small amount of Antarctic intermediate water and a lens of warm, salty water at the surface.

> Identify the water layers of the Atlantic Ocean. What is the origin of each? In which direction does each flow?
>
> Why is the layering of the Pacific Ocean less dramatic than the layering of the Atlantic Ocean?
>
> In what ways is the Indian Ocean similar to the South Atlantic Ocean?

7.5 MEASURING WATER PROPERTIES

Collecting water samples for chemical analysis requires specially designed bottles that are attached to a wire rope or cable, shown in figure 7.8. The water bottles are opened and fastened to the cable at predetermined intervals, and the water bottles are lowered to the desired depths. A small weight is then placed around the cable; it slides down the cable and when it strikes the uppermost water bottle it causes the water bottle to close,

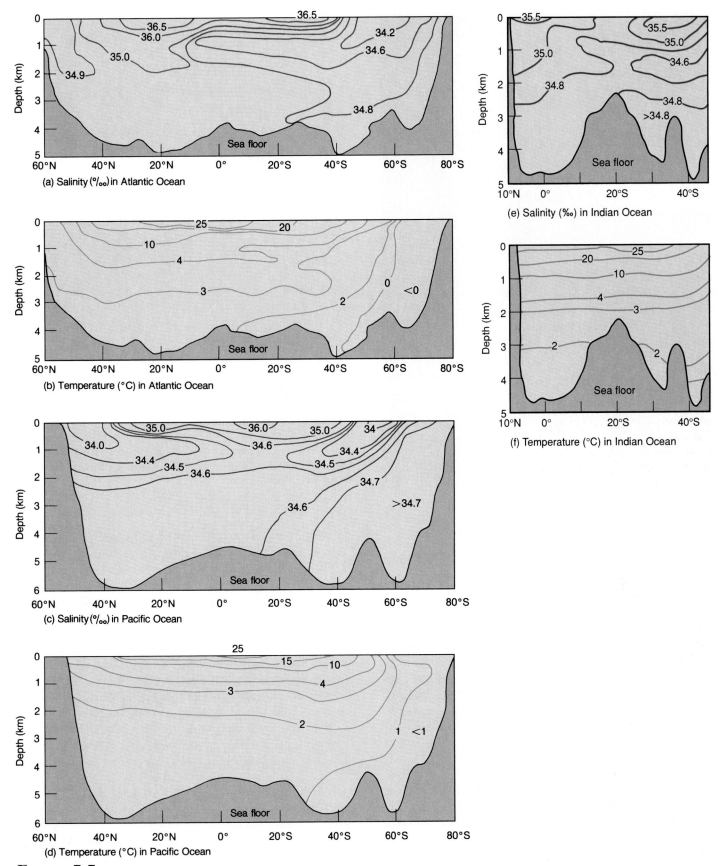

FIGURE 7.7

Mid-ocean salinity and temperature profiles of the Atlantic, Pacific, and Indian Oceans. Temperature and salinity patterns caused by thermohaline circulation are shown as functions of depth and latitude.

FIGURE 7.8

Hanging a water bottle. The cable is measured as it passes over the meter wheel.

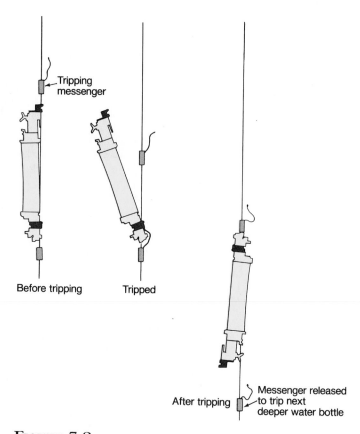

FIGURE 7.9

The reversing of a messenger-activated water bottle equipped with thermometers, and the release of a second messenger to activate the next bottle.

capturing a water sample. This action releases another weight to slide down and strike the next bottle on the cable (fig. 7.9). This sequence continues, one bottle at a time, until the last bottle has turned over.

Some water bottles close by turning over; others stay upright but have spring-loaded valves that seal the bottle. Small-volume samples of about 1 liter (0.26 gal) are used if routine chemical analyses are required. The material used to construct the water bottle is important when dealing with trace materials. Metal water bottles are a source of contamination in experiments with trace metals; plastics are a problem if hydrocarbon and organic compounds are being detected.

When all the water bottles have turned over and have taken their samples, the cable is winched up; the water bottles are removed and placed in a rack on shipboard. The water samples are carefully drained off into storage bottles and are taken for analysis to a chemical laboratory on ship or on land.

Water temperatures are obtained by using a specialized mercury thermometer, the **deep-sea reversing thermometer (DSRT),** shown in figure 7.10. These thermometers are attached in groups of two or three to water sampling bottles and are lowered with them into the sea. When the desired depth has been reached, the thermometer is allowed approximately 5 minutes to adjust to the water temperature, then the weight used to close the water bottle and turn it over also turns over or reverses the thermometers. When the thermometer is reversed, it locks in the reading of temperature at that depth by isolating the mercury that registers the temperature from the mercury reservoir (fig. 7.10b). A small auxiliary thermometer that gives the shipboard temperature at the time of reading is used for correcting the temperature reading from depth. Reversing thermometers are usually used in pairs so that their readings can be compared. Their accuracy is about ± 0.01°C.

Modern electronic instruments known as CTDs, for conductivity-temperature-depth sensors, record salinity by directly measuring the electrical conductivity and temperature of the seawater. The data are returned to the ship as an electronic

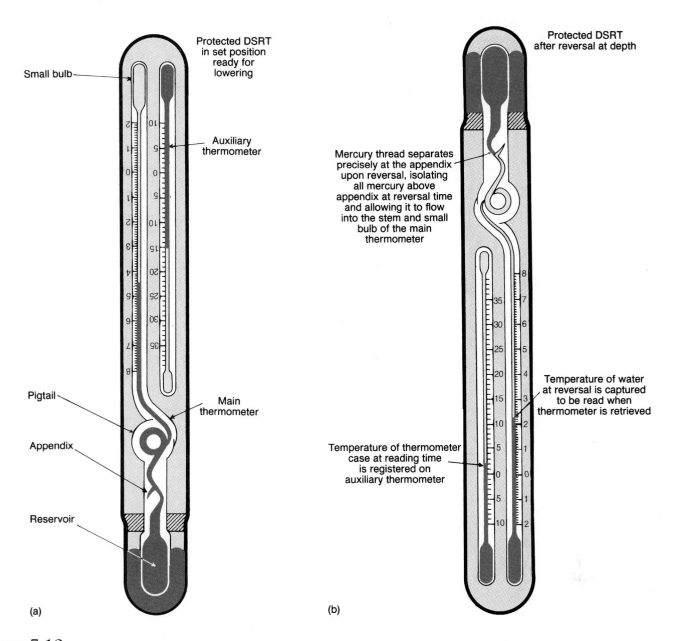

Small bulb

Protected DSRT in set position ready for lowering

Auxiliary thermometer

Pigtail

Main thermometer

Appendix

Reservoir

(a)

Protected DSRT after reversal at depth

Mercury thread separates precisely at the appendix upon reversal, isolating all mercury above appendix at reversal time and allowing it to flow into the stem and small bulb of the main thermometer

Temperature of water at reversal is captured to be read when thermometer is retrieved

Temperature of thermometer case at reading time is registered on auxiliary thermometer

(b)

FIGURE 7.10

(a) A deep-sea reversing thermometer (DSRT) in set position as it is lowered into the sea. (b) A deep-sea reversing thermometer in reversed position, with the water temperature recorded.

signal that may be fed directly into the computer, recorded on a chart, or made available as numeric data. Large amounts of data can be acquired in this way, while water bottles and DSRTs collect only one sample at each depth for later analysis. CTD sensors can be left in place, suspended from the vessel or a buoy, to monitor salinity and temperature changes with time, and the collected data can be stored in the instrument or transmitted to ship and land stations by satellite. Although large quantities of data are collected easily and quickly, these instru-

ments do not measure as many water properties as can be obtained from a water sample in the laboratory.

Electronic thermometers respond very rapidly to temperature changes with an accuracy that is comparable to that of the reversing thermometers. They can be used to record temperature fluctuations over time at a single location, or they can be lowered through the water, yielding a continuous temperature profile with depth. Readings may be recorded within the instrument or they may be fed directly to a computer to produce a graph of temperature with depth. These thermometers record five or more readings each second; reversing thermometers obtain one reading at each water bottle depth for the hour or so required to lower and retrieve them. The new electronic instruments increase the precision and speed of the measurements, but they must still be calibrated against control data collected by the old, reliable water bottles and reversing thermometers.

How is an uncontaminated water sample for laboratory analysis obtained from a specific depth?

Why is a deep-sea reversing thermometer reversed?

What are the advantages of electronic sampling devices? Are there any disadvantages?

7.6 WIND-DRIVEN WATER MOTION

When the winds blow over the oceans, they set the surface water in motion, driving the large-scale surface currents in nearly constant patterns. Refer to Chapter 6 for a discussion of the earth's surface winds. Because the friction between the water and the earth's surface is small, the moving water is deflected by the Coriolis effect, to the right of the wind direction in the

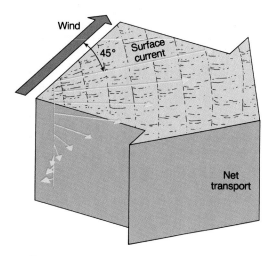

FIGURE 7.11

Water is set in motion by the wind. The direction and speed of flow change with depth to form the Ekman spiral. This change with depth is a result of the earth's rotation and the inability of water, due to low friction, to transmit a driving force downward with 100 percent efficiency. The surface current is 45° to the right of the wind, while the net transport over the wind-driven column is 90° to the right of the wind in the Northern Hemisphere.

Northern Hemisphere and to the left in the Southern Hemisphere. In the open sea, the deflection of the surface current is at a 45° angle to the wind direction, as shown in figure 7.11. The combined effect of the wind on the surface and the deflection of the water creates the large-scale pattern of the wind-driven surface currents in the open oceans. Once set in motion, these currents are further modified by the interaction that occurs among wind-driven currents, zones of converging and diverging water, and landmasses.

The water just below the surface receives less energy and moves more slowly than the water at the surface, and it is further deflected to the right (Northern Hemisphere) or left (Southern Hemisphere) of the surface water. The same is true for the water below, down to approximately 150 meters (500 ft). The overall result is a spiral in which deeper water set in motion by the wind moves more slowly and with greater deflection than the water above, and the much-reduced current at the bottom moves exactly opposite to the surface (fig. 7.11). This spiral motion is called the **Ekman spiral,** after the Swedish physicist V. Walfrid Ekman, who developed its mathematical relationship in 1902. Over the depth of the spiral, the average flow of all the water set in motion by the wind, or the net flow known as **Ekman transport,** moves 90° to the right or left of the surface wind, depending on the hemisphere in which the motion takes place. Ekman transport, therefore, differs from the surface current, which moves at an angle of 45° to the wind direction.

If a north wind blows across the sea surface, what direction does the surface current flow in the Northern Hemisphere and in the Southern Hemisphere?

What direction is the net flow of all wind-driven water in the question above?

7.7 CURRENT FLOW AND GYRES

In the open ocean between the westerlies and the trade winds (review section 6.9 and fig. 6.15), Ekman transport directs the surface water toward 30° latitudes N and S forming surface convergences that extend across the oceans in each hemisphere. As the surface water accumulates along these convergences, the sea level in each ocean is raised, and gravity causes the water to slide downhill and away from the convergence zones. As a result, the wind-driven water tends to move parallel to the convergence zones and along their elevation contours in east-west directions.

Refer to figure 7.12 as you read the following description. The water under the westerlies moves away from an ocean's western shore or boundary and toward its eastern shore or boundary; the water under the trade winds moves the water away from an ocean's eastern boundary and toward its western boundary. North of the equator, when water moving with the trade winds accumulates at the land boundary on the west side of an ocean, the water flows north and parallel to the land toward the region from which the water moves eastward in the latitudes of the westerlies. Water moving with the westerlies and accumulating at the land boundary on the east side of a Northern Hemisphere ocean flows south, parallel to the land, toward the region from which the water moves westward. This produces

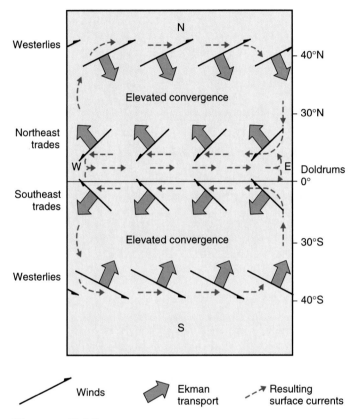

Westerlies

40°N

Elevated convergence

30°N

Northeast trades

W E Doldrums
0°

Southeast trades

Elevated convergence

30°S

Westerlies

40°S

N

S

Winds — Ekman transport — Resulting surface currents

FIGURE 7.12
Wind-driven transport and resulting surface currents in an ocean bounded by land to the east and to the west. The currents form large oceanic gyres that rotate clockwise in the Northern Hemisphere and counterclockwise in the Southern Hemisphere.

a series of interconnecting surface currents moving in a circular, clockwise path centered on 30°N latitude. In the Southern Hemisphere the water will move from east to west with the trade winds, then south to the latitude of the westerlies where it will move west to east with the westerlies, and then north back into the trade wind latitudes. South of the equator the path is circular and counterclockwise, centered on 30°S.

Such a circular flow forms a mound of converging surface water at its center. This water is driven toward the center by the Coriolis effect and can result in an increase in the water's elevation by a meter (3 ft) or more. There is a direct relationship between the speed of the flow and the force of the Coriolis deflection. When the elevation of the mound is such that the outward, gravity-driven downhill flow equals the inward Coriolis effect, a balance is achieved. When this balance is in place the water flow is no longer deflected inward, but instead the water moves around the mound parallel to the mound's contours (fig. 7.13). These large, circular, wind-driven, oceanic flows are called **gyres.**

Wind-driven open-ocean surface currents move at speeds that are about one one-hundredth of the average wind speed above the sea surface, between 0.1 and 0.5 meters per second (0.3–1.5 ft/s or 0.25–1 knot). The speed of the surface flow around a large ocean gyre is not necessarily the same on all sides. In the North Atlantic and North Pacific the currents flowing from low to high latitudes on the western sides of the gyres tend to be narrower and stronger than those on the eastern side. This is known as the **western intensification** of the currents. The increased flow on the western sides of the gyres is caused by (1) the increasing Coriolis effect with increasing latitude, (2) the change in wind strength and direction with increasing latitude, and (3) the friction between ocean currents and landmasses. The intensification of flow is most conspicuous in the Northern Hemisphere. In the Southern Hemisphere the intensification is obscured in the Atlantic by currents deflected by the coast of Brazil (see section 8), and in the Pacific a portion of the westward-driven water escapes through the islands of Indonesia. Also the southern tips of South America and Africa deflect the circumpolar current flow around Antarctica into the eastern sides of the gyres in the southern oceans (see section 8).

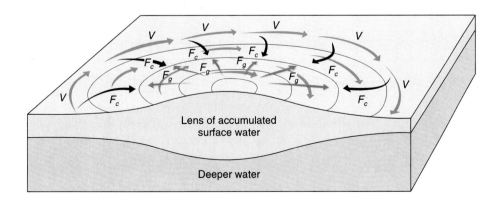

Lens of accumulated surface water

Deeper water

FIGURE 7.13
Currents flow (V) around a gyre when F_c, the inward Ekman transport due to the Coriolis effect, is balanced by F_g, the outward force due to gravity. The example is of a clockwise gyre in the Northern Hemisphere creating a convergence in the gyre's center.

Circulation Patterns and Ocean Currents

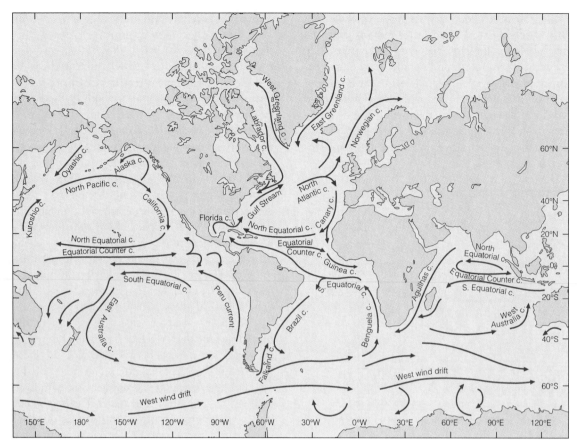

FIGURE 7.14
The major surface currents of the world's oceans.

Why is the sea surface elevated in the center of the major current gyres of the world's oceans?

Why do the current gyres rotate clockwise in the Northern Hemisphere and counterclockwise in the Southern Hemisphere?

Draw diagrams showing the direction of winds and currents in a Northern Hemisphere gyre and in a Southern Hemisphere gyre.

What factors obscure western intensification in the Southern Hemisphere?

7.8 OCEAN SURFACE CURRENTS

The major surface currents that result from the processes previously described are shown in figure 7.14. Use this figure to follow the current patterns described in this section.

The Pacific Ocean

In the North Pacific, the northeast trade winds push the water toward the west and northwest; this is the *North Equatorial Current*. The westerlies create the *North Pacific Current* or *North Pacific Drift* moving from west to east. A current's direction or set is the direction in which it flows. This is an easterly current. Note that the trade winds move the water away from Central and South America and pile it up against Asia, while the westerlies move the water away from Asia and push it against the west coast of North America. The water that accumulates in one area must flow toward areas from which the water has been removed. This movement forms two north-south currents: the *California Current*, moving south along the western coast of North America, and the *Kuroshio Current*, moving north along the east coast of Japan. The Kuroshio and California currents are part of a continuous flow or circular motion centered around 30°N latitude. This circular, clockwise flow of water is called the North Pacific gyre. Other major North Pacific currents include the *Oyashio Current*, driven by the polar easterlies, and the *Alaska Current*, fed by water from the North Pacific Current and moving in a counterclockwise gyre in the Gulf of Alaska.

In the South Pacific Ocean, the southeast trade winds move the water to the left of the wind and westward, forming the *South Equatorial Current*. At a higher latitude, the westerly winds push the water to the east, where at these southern latitudes it moves almost continuously around the earth as the *West Wind Drift*. The tips of South America and Africa deflect a portion of this flow northward on the east side of both the South Pacific and South Atlantic Oceans. As in the North Pacific, currents between the South Equatorial Current and the West Wind Drift complete the gyre. The *Peru Current* or

Humboldt Current flows north along the coast of South America, and the *East Australia Current* moves south on the west side of the ocean. These four currents form the counterclockwise South Pacific gyre.

The North Pacific and South Pacific gyres are formed not on either side of 0° but on either side of 5°N, because the doldrum belt is displaced northward due to the unequal heating between the Northern and Southern Hemispheres. Also between the North and South Equatorial currents, under the doldrums, there is a surface current moving down slope west to east, the *Equatorial Countercurrent*. This current helps to return surface water accumulated against the coast of Asia by the Equatorial Currents. See figures 7.12 and 7.14.

The Atlantic Ocean

The North Atlantic westerly winds move the water eastward as the *North Atlantic Current* or *North Atlantic Drift*. The northeast trade winds push the water to the west, forming the *North Equatorial Current*. The north-south currents are the *Gulf Stream*, moving north along the coast of North America, and the *Canary Current*, flowing south on the eastern side of the North Atlantic. The Gulf Stream is fed by the *Florida Current* and the North Equatorial Current. The North Atlantic gyre rotates clockwise. The polar easterlies provide the force for the *Labrador* and *East Greenland Currents* that balance water flowing into the Arctic Ocean from the *Norwegian Current*.

In some areas of the Gulf Stream the current speed may reach 1.5 meters per second (5 ft/s or 3 knots); its volume transport rate is 55×10^6 m³/s, more than 500 times the flow of the Amazon River. The Florida Current, exiting the Gulf of Mexico through the narrow channel between Florida and Cuba, may exceed 1.5 meters per second.

The gyre currents of the central North Atlantic isolate a lens of clear, warm surface water 1000 meters (3300 ft) deep. This is the Sargasso Sea, bounded by the Gulf Stream on the west, the North Atlantic Current on the north, the Mid-Atlantic Ridge rising upward from the sea floor on the east, and the North Equatorial Current on the south. This region is famous for its floating mats of *Sargassum*, a brown seaweed collected by the water motion and held within the gyre. Except for the floating *Sargassum*, with its rich and specialized community of small plants and animals, the area is biologically a near desert because of the downwelling and nutrient-poor surface water associated with the subtropical convergence.

In the South Atlantic, the westerlies continue the West Wind Drift. The southeast trade winds move the water to the west, but the bulge of Brazil splits the *South Equatorial Current*. Much of this flow is deflected northward over the top of South America, into the Caribbean Sea, and eventually into the Gulf of Mexico, where it exits as the Florida Current to feed the Gulf Stream. A portion of the South Equatorial Current slides south of the Brazilian bulge along the western side of the South Atlantic to form the *Brazil Current*. The *Benguela Current* moves northward up the African coast, completing the South Atlantic gyre that rotates counterclockwise.

Because much of the South Equatorial Current is deflected across the equator, the Equatorial Countercurrent appears only weakly in the eastern tropical Atlantic. The northward movement of South Atlantic water across the equator means that there is a net flow of surface water from the Southern Hemisphere to the Northern Hemisphere. This flow is balanced by a flow of water at depth from the Northern Hemisphere to the Southern Hemisphere, the North Atlantic deep water discussed in section 4. Again, the equatorial currents are displaced north of the equator, although not as markedly as in the Pacific Ocean.

The Indian Ocean

This ocean is mainly a Southern Hemisphere ocean. The southeast trade winds push the water to the west, creating the *South Equatorial Current*. The Southern Hemisphere westerlies still move the water eastward in the West Wind Drift. The gyre is completed by the *West Australia Current* moving northward and the *Agulhas Current* moving southward along the coast of Africa. In this Southern Hemisphere ocean the currents move to the left of the wind direction, and the gyre rotates counterclockwise. The northeast trade winds strengthen the *North Equatorial Current* during the dry monsoon season, and the *Equatorial Countercurrent* is reduced. The wet monsoon season winds strengthen the Equatorial Countercurrent and reduce the North Equatorial Current. The strong seasonal monsoon controls the surface flow of the Northern Hemisphere portion of the Indian Ocean. This strong seasonal shift is unlike anything found in either the Atlantic or the Pacific.

The surface currents and the deep-water flows of all the oceans have been and continue to be interconnected and dependent on each other. In high-latitude convergence zones wind-driven water is modified by surface cooling and the formation of sea ice, thus producing dense water that sinks, flows horizontally, and rises in another ocean at surface divergences. The present day flow of ocean water within and between oceans, at surface and at depth is shown in figure 7.15.

In which direction do the major ocean current gyres flow in the Northern Hemisphere and in the Southern Hemisphere?

Follow the major ocean gyres and identify the major currents in the North Atlantic, South Atlantic, North Pacific, South Pacific, and Indian oceans.

Why are the equatorial countercurrents located under the doldrums?

Which is the stronger current and why: (a) the Kuroshio or the California? (b) the Gulf Stream or the Canary?

How does deep-ocean circulation compensate for the surface flow across the equator in the Atlantic Ocean?

Explain the role of the monsoon on the currents of the northern Indian Ocean.

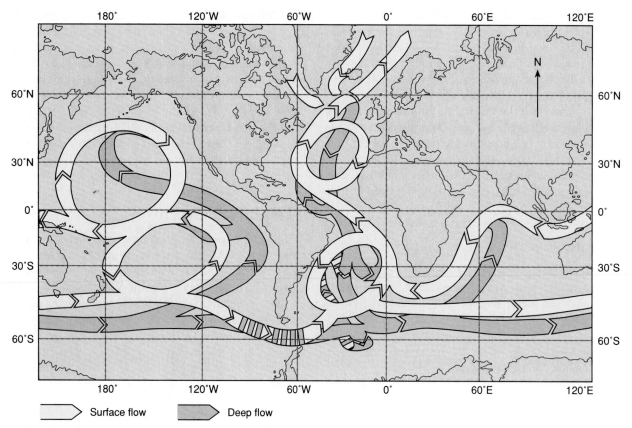

| Surface flow | | Deep flow |

FIGURE 7.15

The present flow of ocean water within and between oceans circulates at the surface and at depth. Cold surface water sinks to the sea floor in the North Atlantic Ocean, then flows south to be further cooled by the Antarctic bottom water formed in the Weddell Sea at 40°W 65°S. This deep water moves eastward around Antarctica, feeding into the surface layer of the Indian Ocean and also into the deep basins of the Pacific Ocean. A return flow of surface water from the Pacific and Indian Oceans flows north to replace the surface water in the North Atlantic Ocean.

7.9 COASTAL UPWELLING AND DOWNWELLING

Wind-driven Ekman transport can drive surface water toward or away from a coast. In coastal areas where the trade winds move the surface waters away from the western side of the continents, upwelling occurs nearly continuously throughout the year. These zones of upwelling are the product of the winds, not of thermohaline circulation. The rich fishing grounds along the west coasts of Africa and South America are supported by this coastal upwelling of nutrient-rich water that stimulates the growth of marine plants.

Seasonal downwellings and upwellings also occur along ocean coasts, for example, North America from central California to Vancouver Island. The wind pattern in this region changes from southerly in winter to northerly in summer, causing changes in the coastal Ekman transport. Remember that open-ocean Ekman transport moves at an angle of 90° to the right or left of the wind direction depending on the hemisphere.

Along this coast the wind blows from the north in the summer, and the net movement of the water is offshore, result-

ing in an upwelling along the coast (fig. 7.16a). In winter the winds blow from the south, and the net movement of the water is onshore against the coast, producing a downwelling in the same region (fig. 7.16b). The summer upwelling produces cold coastal water at San Francisco and helps cause its frequent summer fogs. Because of the lack of land to alter the wind patterns at middle latitudes in the Southern Hemisphere, this type of seasonal upwelling is less common. Refer to figure 6.17.

Convergences and divergences of surface currents produce not only downwellings and upwellings, but also mixing between waters from different geographic areas. Waters carried by the surface currents converge and form new water mixtures with intermediate values of temperature and salinity. These new water mixtures then sink to their density levels, move horizontally, blending and sharing properties with adjacent water, and eventually rise to the surface at a new location. (This is the thermohaline or density-driven circulation discussed earlier in this chapter.) Thermohaline circulation and wind-driven surface currents are closely related, so closely related that it is difficult to assign a priority of importance to one process over the other.

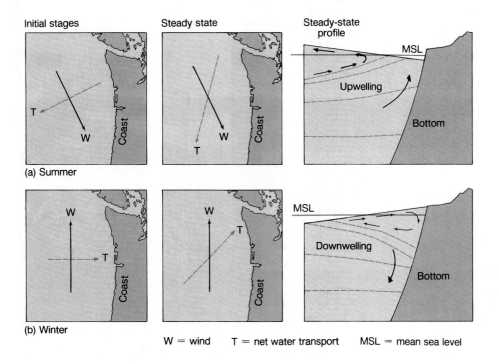

Initial stages   Steady state   Steady-state profile

MSL

Upwelling

Bottom

(a) Summer

MSL

Downwelling

Bottom

(b) Winter

W = wind   T = net water transport   MSL = mean sea level

FIGURE 7.16

Initially, the Ekman transport is 90° to the right of the wind along the northwest coast of North America. As water is transported (a) away from the shore in summer, or (b) toward the shore in winter, a sea surface slope is produced. This slope creates a gravitational force that alters the direction of the Ekman transport to produce a balance of forces and a new steady-state direction of the current.

Why is there a year-round upwelling along the west coasts of South America and Africa?

Explain the seasonal upwelling and downwelling along the west coast of the United States.

7.10 EDDIES

As the Gulf Stream moves away from the coast, it develops a meandering path. At times, the western edge of the Gulf Stream develops indentations that are filled by cold water from the Labrador Current side. These indentations pinch off and become **eddies,** or packets of water moving with a circular motion. These eddies are displaced to the east and south of the current boundary and transfer cold water into warm water. Likewise, bulges at the western edge of the Gulf Stream are filled with warm water. When these bulges are cut off, they drift to the west and north of the Gulf Stream, into cold water. This process is illustrated in figure 7.17. Figure 7.18 is a satellite view of Gulf Stream eddies as they form along the boundary of the current. Current eddies dissipate the energy of flowing water by friction, preventing the current from increasing its speed as wind energy tries to drive it faster and faster. Daily and weekly current patterns differ greatly from the long-term average current flows shown on most ocean current charts, because these charts do not include the presence of eddies, meanders, and other short-term variations.

Large and small eddies exist in all parts of the oceans and at all depths; they range from tens to several hundred kilometers in diameter. Each eddy contains water with specific chemical and physical properties and maintains its identity as it wanders through the oceans, eventually mixing with the surrounding water.

What are eddies and where are they found?

How is warm water transferred into cold water and vice versa along the edges of the Gulf Stream?

7.11 MEASURING THE CURRENTS

Direct measurements of currents are made by (1) following a parcel of the moving water or (2) measuring the speed and direction of the water as it passes a point. Moving water may be followed with buoys designed to float at predetermined depths and signal their positions to the research vessel. Surface water may be labeled with buoys (fig. 7.19) or with dye that can be photographed from the air. Buoy positions may also be tracked by satellite, and a series of pictures or position fixes may be used to calculate the speed and direction of the current.

Current meters include a rotor to measure the current's speed and a vane to measure its direction (fig. 7.20). If the

    Circulation Patterns and Ocean Currents

FIGURE 7.17

The western boundary of the Gulf Stream is defined by sharp changes in current velocity and direction. Meanders form at this boundary after the Gulf Stream leaves the U.S. coast at Cape Hatteras. The amplitude of the meanders increases as they move downstream (a and b). Eventually the current flow pinches off the meander (c). The current boundary re-forms, and isolated rotating eddies of warm water (W) wander into the cold water, while cells of cold water (C) drift into the warm water (d).

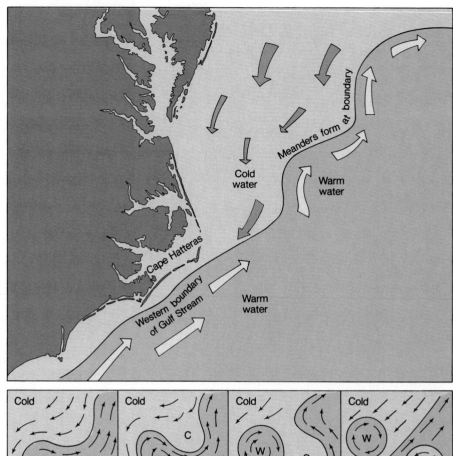

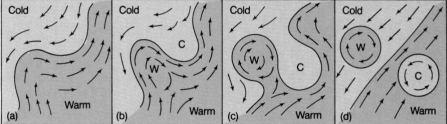

FIGURE 7.18

A satellite image of the sea surface reveals the warm (orange and yellow) and cold (green and blue) eddies that spin off the Gulf Stream. (Reddish blue areas at the top are the coldest waters.) These eddies may stir the water column right down to the ocean floor, kicking up blizzards of sediment.

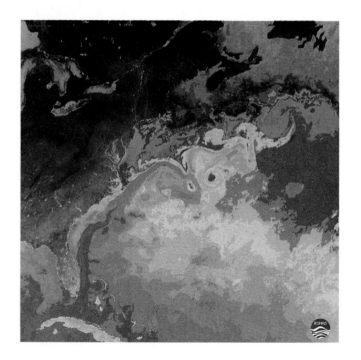

FIGURE 7.19
Surface buoys (red) carrying instruments are deployed from the research vessel *Thomas G. Thompson*.

(a)

(b)

FIGURE 7.20
(a) An internally recording Aanderaa current meter. The vane orients the meter to the current while the rotor determines current speed. (b) Inside the protective case the instrument records coded information from its external sensors. The speed and direction of the current, water temperature, pressure, and conductivity of the water are recorded on magnetic tape. A clock controls the frequency at which these data are recorded. A magnetic tape is retrieved when the current meter is picked up from its taut wire mooring.

Circulation Patterns and Ocean Currents

current meter is lowered from a stationary vessel, the measurements can be returned to the ship by a cable or can be stored by the meter for reading upon retrieval. If the current meter is attached to an independent bottom-moored buoy system, the signals can be transmitted to the ship by radio or stored in the current meter.

To measure a current passing a location, the current meter must be held fixed in space. Although a vessel can be moored and held steady in shallow water, it is usually impossible to do so in the open sea. The solution is to attach the current meter to an anchored buoy system (fig. 7.21) that is entirely submerged and so not affected by winds or waves. The floats and meters are retrieved by sending a sound signal to a special acoustical link that detaches the float and meter from the anchor. The anchor is left behind, and the buoyed equipment returns to the surface for recovery.

What are the two ways of measuring current flow?

What measurements are necessary to define a current?

7.12 ENERGY SOURCES

Because of the transparency and heat capacity of water, large amounts of solar energy can be stored in the ocean. Temperature and salinity variations with depth represent potential sources of energy and may be alternatives to the consumption of fossil fuels. Two possible techniques for extracting this energy are discussed below.

Thermal Energy

Ocean thermal energy conversion or **OTEC** depends on the difference in temperature between ocean surface water and water at 600 to 1000 meters (2000 to 3300 ft) depth. There are two different types of OTEC systems: (1) closed cycle, which uses a contained working fluid with a low boiling point, such as ammonia or Freon, and (2) open cycle, which directly converts seawater to steam.

In a closed system (fig. 7.22a) the warm surface water is passed over the evaporator chamber containing the ammonia or Freon, and this fluid is vaporized by the heat derived from the warm seawater. The vapor builds up pressure in a closed system, and this gas under pressure is used to spin a turbine, which generates power. After the pressure has been released, the working fluid is passed to a condenser, where it is cooled by cold water pumped up from depth. Cooling the fluid returns it to its liquid state, and it is pumped back to the evaporator to repeat the cycle.

In the open system (fig. 7.22b), warm seawater is converted to steam in a low-pressure vacuum chamber, and the steam is used as the working fluid. Since less than 0.5 percent of the incoming water is turned into steam, large quantities of warm water must be used. The steam passes through a turbine and is condensed by using cold water from depth to cool the condenser. The condensation product is desalinated water, another source of revenue in regions lacking both power and fresh water, for example the Caribbean and Pacific island nations.

OTEC plants using either system can be located on shore, offshore (fig. 7.22c), or on a ship that moves from place to place. However, OTEC requires at least a 20°C difference in temperature between surface and depth to generate power in excess of

---

FIGURE 7.21

Taut wire moorage. (a) Recovery is accomplished by retrieving the surface buoy and hauling in the wire. If the surface buoy is lost, it is possible to grapple for the ground wire. (b) In this system, a sound signal disconnects the anchor, and the equipment floats to the surface.

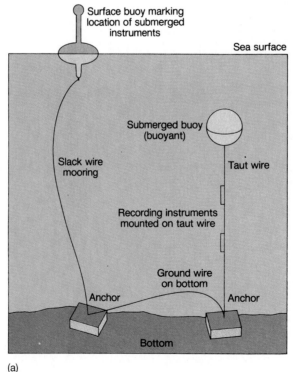
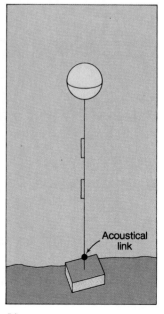

(a)

(b)

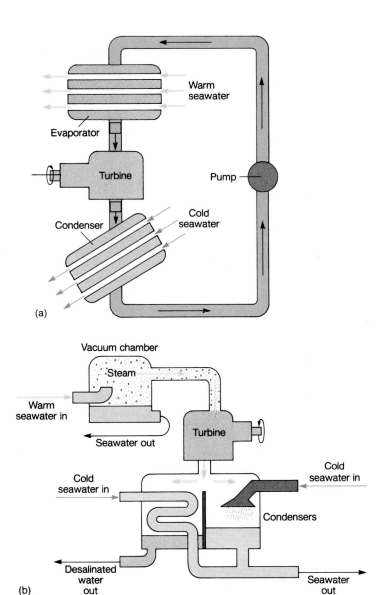

(a)

(b)

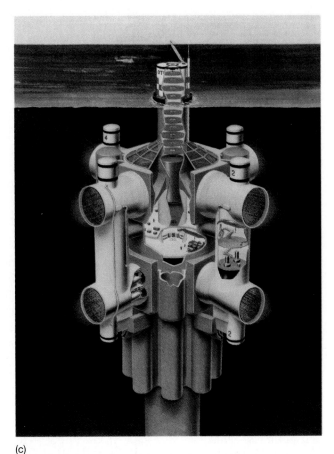

(c)

FIGURE 7.22

(a) The simplified working system of an OTEC closed-system electrical generator. (b) The simplified working system of an OTEC open-system electrical generator. (c) The proposed Lockheed OTEC Spar, designed for generating electrical energy from the ocean temperature gradient.

that needed to run its own pumps. This means that the most efficient power plants would be located at latitudes between approximately 25°N and 25°S where seawater temperature differences over depth average 22°C. Avery and Wu (1994) estimated that the area of the world's oceans meeting OTEC requirements is 60×10^6 square kilometers (23×10^6 mi²). If OTEC facilities were uniformly spaced throughout this oceanic area on a 32 kilometer (20 mi) spacing and if each power system generated 200 megawatts (MW) of useful power, the total power generated would exceed 10×10^6 MW. The total U.S. electricity generating capacity in 1987 was 165×10^3 MW.

This region of the earth does not have the highest population density or the greatest industrial base requiring power. However, the electricity generated by each open-ocean OTEC station could be used on-site to produce hydrogen, ammonia, or methanol which could be stored and transported to land bases as a fuel to power electrical generation. Islands and land areas within the tropical ocean and having access to cold water at depth could have OTEC plants that are land based and supply electrical power directly.

Engineers studying the potential and feasibility of OTEC estimate that about 0.2 MW of usable power can be extracted per km² of tropical ocean surface, which is about 0.07 percent of the average absorbed solar energy. They consider the process to be safe and environmentally benign. However, the pumping of cold water from depth to the sea surface to cool the OTEC condensers brings nutrients from depth which are liberated into the nutrient-poor tropical surface water. This artificial upwelling of cold nutrient water can change the productivity of the surface water, stimulating marine plant growth and interfering with the life process of those surface animals that are dependent on tropical temperatures.

The Natural Energy Laboratory of Hawaii at Keahole Point is building a land-based, open-system OTEC plant. This pilot project began producing power in 1992; in 1993 tests it produced net power of 50 kW from a gross production of 190 kW. It is the world's only system pumping cold water from depth at this time. The water is pumped through a polyethylene pipe, 1600 meters (1 mi) long and 1 meter (3 ft) in diameter, from 850 meters depth (2800 ft); the warm water is extracted by a shorter pipe from 25 meters (80 ft) below the surface. The cost of the plant is high, $30 million, about ten times the cost of a conventional plant burning coal or oil; however, OTEC fuel is free. Maintenance costs are estimated at about the same as or less than that for a fossil fuel plant and much less than for a nuclear plant. This land-based OTEC operation also requires that large volumes of cold, nutrient-rich water be pumped to the surface. Researchers in Hawaii believe that using this nutrient-rich water for fish and seaweed farming is one of the biggest incentives for commercial development of OTEC in their region. This water from depth may be a positive OTEC by-product for sea farmers, but it is also a hazard to tropical organisms and destructive to coral reefs and their associated species.

Taiwan, facing serious problems in supplying sufficient energy to maintain its growing economy as well as increasing public pressure against nuclear and hydroelectric projects, is exploring a land-based OTEC plant. Along Taiwan's east coast, there is nearshore deep water with a year-round temperature difference of 20 to 22°C between surface and depth. Planning is stressing the multiproduct approach, not only power production but a resource for fresh water, cooling for refrigeration and air-conditioning, and deep water rich in nutrients, which can be used as an artificial upwelling to promote fish and seaweed farming.

Harnessing the Currents

The massive oceanic surface currents of the world are untapped reservoirs of energy. Their total energy has been estimated at 2.8×10^{14} (280 trillion) watt-hours. If the total energy of a current was removed by conversion to electric power, that current would cease to exist; but only a small portion of any ocean current's energy could be harnessed, owing to the current's size and distribution. Harnessing the energy from these open-ocean currents would require the use of turbine-driven generators anchored in place in the current stream. Large turbine blades would be driven by the moving water, just as windmill blades are moved by the wind; these blades could turn the generators to harness the energy of the water flow.

The Florida Current and the Gulf Stream are swift and continuous currents moving close to shore in areas where there is a demand for power. If ocean currents are developed as energy sources, these are among the most likely to be tapped. But most of the wind-driven oceanic currents generally move too slowly or are found too far from where the power is needed. The cost of constructing, mooring, and maintaining current-driven power-generating devices in the open sea makes them noncompetitive with other sources of power at this time.

What is the difference between a closed-cycle and an open-cycle OTEC system?

In what areas of the world is OTEC a suitable energy resource?

Why is energy unlikely to be derived from open-ocean currents in the foreseeable future?

Why can energy produced by either OTEC or ocean currents be considered solar energy?

SUMMARY

The surface water changes its density with changes in salinity and temperature that are keyed to latitude. The densities at depth are more homogeneous.

If the density increases with depth, the water column is stable; unstable water columns overturn and return to a stable distribution. Neutrally stable water columns are mixed vertically by winds and waves. Vertical circulation that is driven by changes in surface density is known as thermohaline circulation. In the open ocean, temperature is more important than salinity in determining the surface density. Salinity is the more important factor in areas that are close to shore and freshwater discharge from the land.

Water sinks and downwelling occurs at the convergence of surface currents; water rises at zones of surface current divergence. There are major oceanic areas of convergence and divergence located at specific latitudes.

The oceans are layered systems. In the Atlantic Ocean, waters of differing densities are formed at the surface at different latitudes. They sink and flow north or south at differing depths. The water layers of the Pacific Ocean lose their identity in the large volume of this ocean; their movements are sluggish. The water layers of the Indian Ocean are also less distinct than those of the Atlantic, with no water that corresponds to that formed in the northern latitudes.

Water samples are taken with water bottles that turn over and close at the sampling depth. The temperature of the water at the depth of a water bottle sampling station is taken with a deep-sea reversing thermometer. Electronic instruments read salinity and temperature directly.

Winds push the surface water 45° to the right of their direction in the Northern Hemisphere and 45° to the left in the Southern Hemisphere. Water below the surface receives less energy, moves more slowly, and is deflected still further from the wind. Averaged over the depth of the motion, Ekman transport is 90° to the right (Northern Hemisphere) or left (Southern Hemisphere) of the surface wind.

Large surface current gyres occur in each ocean. Southern Hemisphere gyres rotate counterclockwise; Northern Hemisphere gyres rotate clockwise. The currents of the northern Indian Ocean change with the seasonal monsoon.

The elevated water surface in the center of a gyre is caused by the balance between a component of the force of gravity and the inward force driving Ekman transport. The surface water in the Sargasso Sea is isolated by this current flow. Ocean currents transport very large volumes of water that flow faster when the current is forced to flow through a narrow space. Currents on the western side of Northern Hemisphere oceans are stronger and narrower than currents on the eastern side. Eddies occur at all depths, wander long distances, and gradually lose their identity.

Seasonal upwellings and downwellings occur in coastal areas that have changing wind patterns and an alternating coastal flow of water onshore and offshore due to the Ekman transport. Thermohaline circulation and wind-driven surface currents are closely related and occur on local and world scales.

A variety of instruments are available to measure currents, either by following the water or by measuring the water's speed and direction as it moves past a fixed point.

Ocean thermal energy conversion (OTEC) is a method of extracting energy from the oceans; land-based plants are under construction in Hawaii and planned for Taiwan. Currents also represent a possible source of energy, although costs make them noncompetitive at present.

KEY TERMS

overturn
thermohaline circulation
convergence
divergence
upwelling
downwelling

tropical convergence
subtropical convergence
North Atlantic deep water
Antarctic intermediate water
Antarctic bottom water
South Atlantic surface water

deep-sea reversing
 thermometer (DSRT)
Ekman spiral
Ekman transport
gyre
western intensification

eddies
current meter
ocean thermal energy
 conversion (OTEC)

SUGGESTED READINGS

Avery, W. H., and C. Wu. 1994. *Renewable Energy from the Ocean—A Guide to OTEC.* Oxford University Press, New York. 474 pp.

Bowditch, N. 1984. *American Practical Navigator,* Vol. 1. U.S. Defense Mapping Agency Hydrographic Center, Washington, D.C., 1414 pp. Ocean currents are covered in Chapters 31 and 32.

Huyghe, P. 1990. The Storm Down Below. *Discover* 11 (11):71–76. Deep-ocean storms and eddies.

International OTEC/DOWA Association Conference Report. 1994. *Sea Technology* 35 (5):53–54.

Kunzig, R. 1991. Where the Water Goes. *Discover* 12 (8):26, 37. World current measurement project and equipment.

Lacombe, H. 1990. Water, Salt, Heat and Wind in the Med. *Oceanus* 33 (1):26–36.

Loupe, D. 1991. The Food Factor. *Sea Frontiers* 37 (2):22–27. Projects associated with OTEC plant.

MacLeish, W. H. 1989. The Blue God, Tracing the Mighty Gulf Stream. *Smithsonian* 19 (11):44–59.

_____. 1989. Painting a Portrait of the Stream from Miles Above and Below. *Smithsonian* 19 (2):42–55. More about the Gulf Stream.

Richardson, P. L. 1991. SOFAR Floats Give a New View of Ocean Eddies. *Oceanus* 34 (1):23–31.

_____. 1993. Tracking Ocean Eddies. *American Scientist* 81 (3):261–72.

Thomas, A., and D. L. Hillis. 1989. Hybrid Cycle Adds Potable Water to OTEC Areas in Short Supply. *Sea Technology* 30 (10):33–37.

Vadus, J. R. 1989. Technology Need for Ocean Resources Utilization. *Sea Technology* 30 (8):14–25.

Whitworth, III, T. 1988. The Antarctic Circumpolar Current. *Oceanus* 31 (2):53–58.

The Great Sneaker Spill

In May 1990 a severe storm in the North Pacific caused the Korean container ship, *Hansa Carrier*, to lose overboard twenty-one deck-cargo containers, each approximately forty feet long (fig. 1). Among the items lost were 39,466 pairs of NIKE brand athletic shoes on their way to the United States. Six months to a year later these began washing up along the beaches of Washington, Oregon, and British Columbia (fig. 2). They were wearable after washing and having the barnacles and the oil removed, but the shoes of a pair had not been tied together for shipping, and pairs did not come ashore together. As beach residents recovered the shoes (some with a retail value of $100 a pair), swap meets were held in coastal communities to match the pairs.

FIGURE 2
Beached at last.

FIGURE 1
After the storm the *Hansa Carrier* docked in Seattle.

In May of 1991 oceanographer Curtis Ebbesmeyer of Seattle, Washington read a news article on the beached NIKES. He was intrigued and realized that 78,932 shoes was a very large number of drifting objects compared to the 33,869 drift bottles used in a 1956–59 study of North Pacific currents. He contacted Steve McLeod, an Oregon artist and shoe collector who had information on locations and dates for some 1600 shoes that had been found between northern California and the Queen Charlotte Islands in British Columbia. Additional beachcombers were asked for information, and Ebbesmeyer mapped the times and locations where batches of 100 or more shoes had been found (fig. 3). Next Ebbesmeyer visited Jim Ingraham at NOAA's National Marine Fisheries Service's offices in Seattle to study his computer model of Pacific Ocean currents and wind systems north of 30°N latitude. Using the spill date (May 27, 1990), the spill location (161°W; 48°N), and the dates of the first shoe landings on Vancouver Island and Washington State beaches between Thanksgiving and Christmas 1990, Ebbesmeyer found that the shoe drift rates agreed with the computer model's predicted currents.

News of Ebbesmeyer and Ingraham's interest in the shoe spill reached an Oregon news reporter and was then circulated nationally. Readers sent letters describing their own shoe finds. Reports of single shoes were valuable,

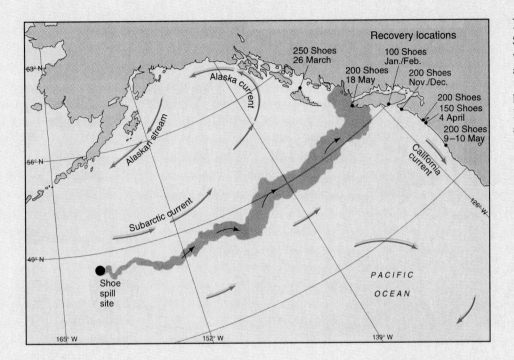

FIGURE 3
Site where 80,000 Nike shoes washed overboard on May 27, 1990, and dates and locations where 1300 shoes were discovered by beachcombers (dots at upper right). Drift of the shoes is simulated with a computer model (colored plume).

because each shoe had within it a NIKE purchase order number which could be traced to a specific cargo container. Ebbesmeyer was able to determine from these numbers that only four of the five shoe-containers broke open, so that only 61,820 shoes were set afloat.

The computer model and previous experiments with satellite-tracked drifters showed that there would have been little scattering of the shoes as the ocean currents carried them eastward and approximately 1500 miles from the spill site to shore, but the shoes were found scattered from California to northern British Columbia. The north-south scattering is related to coastal currents that flow northward in winter carrying the shoes to the Queen Charlotte Islands and southward in spring and summer bringing the shoes to Oregon and California. In the spring of 1992 three sneakers from one of the containers were found at Pololu, at the north end of the island of Hawaii, indicating that they were making their way across the Pacific.

A Japanese film crew made a video about the shoes for Japanese national television, and Ebbesmeyer asked for information concerning any shoes that might arrive in Japan. Because it takes about four and a half years for an object to drift completely around the North Pacific Current gyre, the shoes were expected to arrive on Japanese beaches during 1994–95.

Other Scenarios

Ebbesmeyer was interested in seeing where the shoes might have gone if they had been lost on the same date but under different conditions in other years. The computer allowed simulations for May 27 of each year from 1946 to 1991. Figure 4 shows the wide variation in model-predicted drift routes. If the shoes had been lost in 1951, they would have traveled in the loop of the Alaska Current. If they had been lost in 1982, they would have been carried far to the north during the very strong El Niño of 1982–83, and if lost in 1973 they would have come ashore at the Columbia River.

Tub Toys Come Ashore

A similar situation occurred in January 1992 when twelve cargo containers were lost from another vessel in the North Pacific at 180°W, 45°N. One of these containers held 29,000 small, floatable, bathtub toys. Plastic blue turtles, yellow ducks, red beavers, and green frogs began arriving on beaches near Sitka, Alaska in November of 1992. Advertisements asking for news of toy strandings were placed in local newspapers and the Canadian lighthouse keepers newsletter. Beachcombers reported a total of about 400 of these toys.

continued . . .

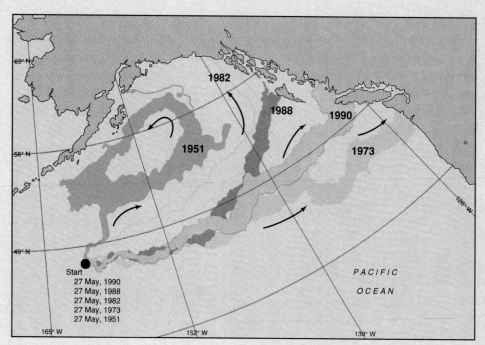

FIGURE 4
Projected drift tracks for the sneakers for the years 1951, 1973, 1982, 1988, and 1990 based on computer modeling of ocean currents and weather.

None of the toys have been found south of 60°N latitude, suggesting that once the toys reached the vicinity of Sitka, they drifted to the North. This time the computer model showed that if the toys continue to float with the Alaska Current, they will move with the Alaska stream on through the Aleutian Islands into the Bering Sea. Some may continue through the Bering Sea into the Arctic Ocean to the vicinity of Point Barrow, and from there drift north of Siberia with the Arctic pack ice. Eventually some plastic turtles, ducks, beavers, or frogs may come to rest on the coast of western Europe. If the toys turn south they may merge with the Kuroshio Current and be carried past the location where they were spilled. 🐚

C H A P T E R

8

Waves and Tides

Outline

Learning Objectives

After reading this chapter, you should be able to

- Recognize the parts and properties of waves.

- Compare deep-water and shallow-water waves.

- Follow the development of waves from creation to breaking.

- Describe interactions between waves.

- List characteristics of special wave types.

- Describe tides as observed along the coast.

- Understand the forces causing tides.

- Explain the modifications that produce observed tides.

- Give examples of energy from waves and tides.

◀ Storm waves along the coast. Big Sur, California.

We have all seen water waves. Gentle ripples move across a pond, and breaking waves provide exciting surf at the beach. Storm waves crash against our coasts and force commercial vessels to slow their speed and lengthen their sailing time. Special waves associated with earthquakes have killed thousands of coastal inhabitants and severely damaged their cities and towns. Surfers search for the perfect wave, and ancient peoples navigated by the patterns that waves form.

The tides, best known as the rise and fall of the sea around the edge of the land, are also waves, the longest of the water waves on this planet. Far out at sea, tidal changes go unnoticed; along the shores and beaches, the tides govern many of our commercial and recreational activities.

8.1   How a Wave Begins

If you stand on the beach looking out across a perfectly flat surface of water and throw a stone into the water, the stone strikes the water and displaces, or pushes aside, the water surface. As the stone sinks, the displaced water flows back into the space left behind, and as this water rushes back from all sides, the water at the center is forced upward. The elevated water falls back, causing a depression below the surface, which is refilled, starting another cycle. This process produces a series of waves that move outward and away from the point of disturbance along the water's surface.

FIGURE 8.1
Wind-generated storm waves at sea.

If the stone thrown into the water is small, the waves are small, and the force that causes the water to return toward its undisturbed surface level is the **surface tension** of the water surface (described further in table 5.2, Chapter 5). All very small water waves are affected by surface tension.

The most common force for creating water waves is the wind. As the wind blows across a smooth water surface, the friction between the air and the water stretches the surface, resulting in small wrinkles or ripples. The surface becomes rougher, and it is easier for the wind to grip the roughened water surface and add more energy, increasing the size of the waves. As the waves become larger, the force returning the water to its level state changes from surface tension to gravity (fig. 8.1).

What is required to create a wave at the sea surface?

Why are surface tension and gravity necessary to form a wave?

Why are ripples important in the creation of waves?

8.2   Anatomy of a Wave

The highest part of the wave that is elevated above the undisturbed sea surface is called the **crest;** the lowest part that is depressed below this surface is called the **trough** (fig. 8.2). The distance between two successive crests or two successive troughs is the length of the wave, or its **wavelength.** The **wave height** is the vertical distance from the top of the crest to the bottom of the trough. Sometimes the term **amplitude** is used; the amplitude is equal to one-half the wave height, or the distance from either the crest or the trough to the undisturbed water level. A wave is characterized not only by its length and height (or amplitude), but also by its **period.** The period is the time required for two successive crests or two successive troughs to pass a point in space. If you are standing on a piling and start a stopwatch as the crest of a wave passes and then stop the stopwatch as the crest of the next wave passes, you have measured the period of the wave.

Sketch a wave and label its parts.

What is the relationship between the height and amplitude of a wave?

Explain the period of a wave.

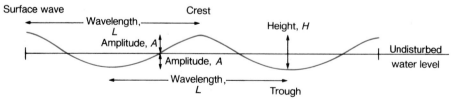

FIGURE 8.2
A profile of surface waves.

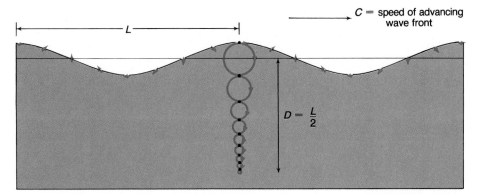

C = speed of advancing wave front

FIGURE 8.3

The moving wave form sets the water particles in motion (note the arrows in the diagram). The diameter of a water particle's orbit at the surface is determined by wave height. Below the surface, the diameter decreases and orbital motion ceases at a depth (D) equal to one-half the wavelength.

8.3 WAVE MOTION AND WAVE SPEED

As a wave form moves across the water surface, it sets particles of water in motion. Out at sea, where the waves move the water surface quietly up and down, the water is not moving toward the shore. Such an ocean wave does not represent a flow of water but instead represents a flow of motion or energy from its origin to its eventual dissipation, which may occur at sea or against the land.

As the wave crest approaches, the surface water particles rise and move forward. Immediately under the crest the particles have stopped rising and are moving forward at the speed of the crest. When the crest passes, the particles begin to fall and to slow in their forward motion. As the trough advances, the particles slow their falling rate and start to move backward, until at the bottom of the trough they move only backward. As the remainder of the trough passes, the water particles begin to slow their backward speed and start to rise again. This motion (rising, moving forward, falling, reversing direction, and rising again) creates a circular path, or **orbit,** for the water particles. Follow this motion in figure 8.3.

It is this orbital motion of the water particles that causes a floating object to bob, or move up and down, forward and backward, as the waves pass under it. This motion affects a fishing boat, swimmer, sea gull, or any other floating object on the seaward side of the surf zone. The surface water particles trace an orbit with a diameter equal to the height of the wave. This same type of motion is present in the water particles below the surface, but as less energy of motion is present, the orbits become smaller and smaller with depth. At a depth equal to one-half the wavelength, the orbital motion has decreased to almost zero (fig. 8.3). Submarines dive during rough weather for a quiet ride, since the wave motion does not extend far below the surface.

A wave's speed across the sea surface may be computed using the wave's length and period. The speed of a surface-water wave (C) is equal to the wavelength (L) divided by its period (T):

Speed = Wavelength/Wave period or $C = L/T$

Once a wave has been created, the speed at which the wave moves may change, but *its period remains the same* (period is determined by the generating force). The oceanographer at sea determines the wave period (T) by direct measurement and calculates the wave speed (C) and the wavelength (L) using equations from simple wave theory; see the next section.

Explain the relationship between flow of water and wave motion in waves seaward of the surf.

What is the maximum diameter of a wave's water particle orbit?

How does the orbit of a wave change with depth?

How are wavelength and wave period related to the speed of a wave?

If the wave period does not change, what happens to the speed of a wave if the wavelength increases or decreases?

8.4 DEEP-WATER WAVES

"Deep-water" has a precise meaning for the oceanographer studying waves. To be a **deep-water wave,** the water must be deeper than one-half the wave's length. Under this condition, the orbits of the waves do not reach the sea floor.

The wavelength in meters (L) of a large deep-water wave is related to gravity (g, 9.81 m/s^2) and the square of the wave period in seconds (T). Using simple wave theory it can be shown that

$$L = (g/2\pi) \, T^2 = 1.56 \, T^2.$$

This equation demonstrates that long-period waves also have long wavelengths and when used with the equation, $C = L/T$, it shows the relationship between the wave speed in meters per second (C) and the period in seconds (T):

$$C = 1.56 \, T.$$

Most waves observed at sea are caused by the wind, and they travel in a direction determined by the wind. These waves are formed in local storm centers or by the steady winds of the trade and westerly wind belts. In an active storm area covering thousands of square kilometers, the winds are not steady but turbulent, varying in strength and direction. Storm-area winds flow in a circular pattern about the low-pressure **storm center,** creating waves that move outward and away from the storm in all directions. The storm center may also move across the sea surface, following the waves and increasing their heights, supplying energy for a longer time and over a longer distance.

In the storm area, the sea surface appears as a jumble and confusion of waves of all heights, lengths, and periods. Small

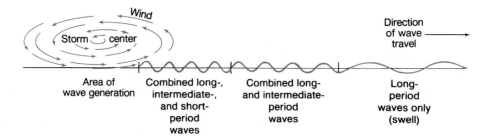

FIGURE 8.4
Dispersion. The longer waves travel faster than the shorter waves. Waves are shown moving in only one direction in the diagram.

waves ride the backs of larger waves. This turmoil of mixed waves is called a "sea" by sailors. The turbulent winds of the storm area transfer energy at different intensities and rates to the sea surface, resulting in waves with a variety of periods and heights.

Among the waves escaping from a storm center are some with long periods and long wavelengths. These waves have a greater speed than waves with short periods and short wavelengths. The faster, longer waves gradually move through and ahead of the shorter, slower waves; waves with different periods travel at different speeds. This process is called **sorting** or **dispersion.** Groups of these faster waves move as **wave trains,** or packets of similar waves with approximately the same period and speed. Near the storm center the waves are not yet sorted, but farther away from the storm the faster, long-period wave trains are ahead of the slower, short-period waves. This process is shown in figure 8.4.

When the longer waves are sorted, they appear as a regular pattern of crests and troughs moving across the sea surface. These long, uniform waves, called **swell,** lose their energy very slowly. Groups of large, long-period waves have been traced across the entire length of the Pacific Ocean from Antarctica to Alaska.

Careful observation of a group of waves or a wave train from a stone thrown into the water shows that waves constantly form on the inside of the train as it moves across the water. As each wave joins the inside of the ring, a wave is lost from the outside. The outside wave's energy is lost in advancing the wave form into undisturbed water. Therefore, each individual wave in the group moves faster than the leading edge of the wave train. The wave train moves at a **group speed** that is one-half the speed of the individual wave.

Group speed = 1/2 Wave speed = Speed of energy transport

or

$$V = C/2.$$

Define a deep-water wave.

Why do storm centers produce waves of varying sizes?

Why do waves appear more regular when they are far away from a storm center?

Compare the motion of a single wave with that of its group as it escapes a storm center.

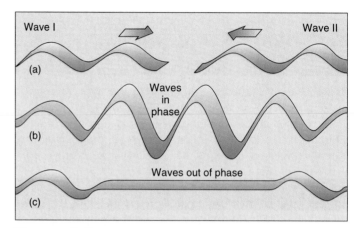

FIGURE 8.5
Constructive and destructive interference between similar wave series. (a) Waves approach each other. (b) If the crests of the approaching waves coincide, the height of the combined waves increases, and the waves are in phase. (c) If the crests of one wave and the troughs of the other wave coincide, the waves cancel, and the waves are out of phase.

8.5   WAVE INTERACTION

When wave trains from one storm meet other similar trains of swell moving away from other storm centers, they pass through each other and continue on. If the crests or the troughs of the two different wave trains are in phase or coincide, the profiles are additive and the amplitude of the crests and troughs increases. This is known as **constructive interference.** When the crests of one wave train coincide with the troughs of another, the waves are out of phase, and the wave trains cancel each other due to **destructive interference.** Figure 8.5 shows constructive and destructive interference between waves traveling in either the same or opposite direction. However, wave trains may intersect at any angle, and many possible interference patterns may result. If the two wave trains intersect each other sharply as at a right angle, then a checkerboard pattern can be formed (fig. 8.6). Two or more intersecting wave trains can phase together and suddenly develop large waves unrelated to local storms. If these waves become too high, they may break, lose some of their energy, and create new, smaller waves.

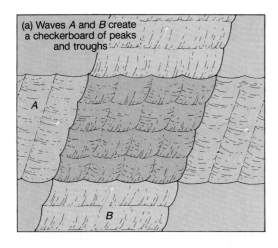

(a) Waves *A* and *B* create a checkerboard of peaks and troughs

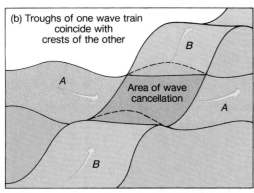

(b) Troughs of one wave train coincide with crests of the other

A

Area of wave cancellation

B

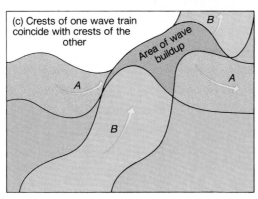

(c) Crests of one wave train coincide with crests of the other

Area of wave buildup

A

B

FIGURE 8.6

Wave trains form interference patterns where they meet. (a) Waves meeting at right angles. (b) Destructive interference, wave crests and wave troughs meet and cancel each other. (c) Constructive interference, wave crests and wave troughs coincide, building higher waves.

If two identical waves are moving in the same direction, how does the water surface appear if the crests and troughs are (a) in phase, or (b) out of phase?

Which of the situations in the above question is an example of (a) constructive interference, (b) destructive interference?

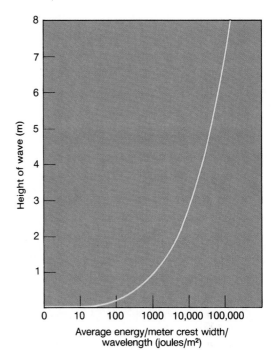

FIGURE 8.7

Wave energy increases rapidly with increases in wave height. Average wave energy is calculated per unit width of crest and averaged over the wavelength (L) and the depth ($\frac{L}{2}$).

8.6 WAVE HEIGHT

Three important factors control wind-wave height: (1) wind speed (how fast the wind is blowing), (2) wind duration (how long the wind blows), and (3) **fetch** (the distance over water that the wind blows in a single direction). If the wind speed is very small, large waves are not produced, no matter how long the wind blows over an unlimited fetch. If the wind speed is great, but it blows for only a few minutes, no high waves are produced despite unlimited wind strength and fetch. Also, if very strong winds blow for a long period over a very short fetch, no high waves form. Only when all three of these factors combine do spectacular wind waves occur at sea.

The average energy of a wave is directly related to the square of the wave height (fig. 8.7). Oceanographers can relate the height and period of storm-formed wind waves to wind speed, duration of wind blow, and fetch. In the open sea, where fetch and duration of blow are not limiting, wind speed alone determines the significant wave height of wind waves (table 8.1). The significant wave height is defined as the wave height at the beginning of the one-third highest waves in a series of observations. For example, if a height record of 120 successive wave heights is made and the wave heights are arranged in order, from the lowest to the highest, the fortieth wave from the highest wave is the one with the significant wave height.

The best-documented giant wave at sea occurred in 1933 when the USS *Ramapo*, a Navy tanker, encountered a severe storm in the Pacific Ocean. The wave, measured against the ship's superstructure by the officer on watch, was 34.2 meters

Table 8.1 The Relationship between Wind Speed and Wave Height

Average Wind Speed knots	m/s	Significant Wave Height meters	Significant Wave Period seconds	Significant Wave Speed m/s	Maximum Wave Height meters	Minimum Fetch km	Minimum Wind Duration hours
10	5.1	1.22	5.5	8.58	2.19	129	11
20	10.2	2.44	7.3	11.39	4.39	240	17
30	15.3	5.79	12.5	19.50	10.43	1017	37
40	20.4	14.33	18.0	28.00	25.79	2590	65

Table 8.2 Universal Sea State Code

Sea State Code	Description	Average Wave Heights
SS0	Sea like a mirror; wind less than one knot.	0
SS1	A smooth sea; ripples, no foam; very light winds, 1–3 knots, not felt on face.	0–0.3 m 0–1 ft
SS2	A slight sea; small wavelets; winds light to gentle, 4–6 knots, felt on face; light flags wave.	0.3–0.6 m 1–2 ft
SS3	A moderate sea; large wavelets, crests begin to break; winds gentle to moderate, 7–10 knots; light flags fully extend.	0.6–1.2 m 2–4 ft
SS4	A rough sea; moderate waves, many crests break, whitecaps, some wind-blown spray; winds moderate to strong breeze, 11–27 knots; wind whistles in the rigging.	1.2–2.4 m 4–8 ft
SS5	A very rough sea; waves heap up, forming foam streaks and spindrift; winds moderate to fresh gale, 28–40 knots; wind affects walking.	2.4–4.0 m 8–13 ft
SS6	A high sea; sea begins to roll, forming very definite foam streaks and considerable spray; winds a strong gale, 41–47 knots; loose gear and light canvas may be blown about or ripped.	4.0–6.1 m 13–20 ft
SS7	A very high sea; very high, steep waves with wind-driven overhanging crests; sea surface whitens due to dense coverage with foam; visibility reduced due to wind-blown spray; winds at whole gale force, 48–55 knots.	6.1–9.1 m 20–30 ft
SS8	Mountainous seas; very high-rolling breaking waves; sea surface foam-covered; very poor visibility; winds at storm level, 56–63 knots.	9.1–13.7 m 30–45 ft
SS9	Air filled with foam; sea surface white with spray; winds 64 knots and above.	13.7 m and above 45 ft and above

(112 ft) high. The period of the wave was measured at 14.8 seconds, and the wave speed was calculated at 27 meters (90 ft)/s.

From early times, those who went to sea recognized that there was a relationship between wind speed and wave height. From this knowledge grew the Universal Sea State Code used by the U.S. Navy today, but originally suggested by British Admiral Sir Francis Beaufort in 1806 (table 8.2). Thus it is also called the Beaufort scale of sea state.

Unrelated to local conditions, large waves or **episodic waves** can suddenly appear at sea. Episodic waves have a height equal to a seven- or eight-story building (20–30 meters) and may move at a speed of 25 meters per seconds (60 mph or 50 knots), with a wavelength approaching a kilometer (one-half mile). They occur most frequently near the edge of the continental shelf, in water about 200 meters (660 ft) deep, and in certain geographic areas with strong wind, wave, and current

patterns. The area where the Agulhas Current sweeps down the east coast of South Africa and meets the storm waves from the Southern Ocean is noted for these waves (fig. 8.8). In the North Atlantic, strong northeasterly gales send large storm waves into the edge of the northward-moving Gulf Stream near the edge of the continental shelf, producing large waves. The shallow North Sea also appears to provide suitable conditions for extremely high episodic waves during severe winter storms.

Why does an 8-foot-high wave do more damage than a 4-foot wave?

Where in the world's oceans would you expect to find consistently high waves?

FIGURE 8.8
A giant wave breaking over the bow of the *ESSO Nederland* southbound in the Agulhas Current. The bow of the supertanker is about 25 meters (82 ft) above the water.

8.7 WAVE STEEPNESS

There is a maximum possible wave height for any given wavelength. This maximum value, determined by the ratio of the wave's height to the wave's length, is the measure of the **steepness** of the wave:

$$\text{Steepness} = \frac{\text{Height}}{\text{Length}}$$

or

$$S = H/L$$

If the ratio of the height to the length exceeds 1:7, the wave becomes too steep and breaks. Under these conditions, the angle formed at the wave crest approaches 120°; the wave becomes unstable, cannot maintain its shape, collapses, and breaks (fig. 8.9).

Small unstable, breaking waves are quite common. When wind speeds reach 8 to 9 meters per second (16 to 18 knots), waves known as "whitecaps" can be observed. These waves have short wavelengths (about 1 m), and as each wave reaches its critical steepness it breaks and is replaced by another wave produced by the rising wind. There is rarely sufficient wind at sea to force long waves to their maximum steepness. More frequently, the tops of high waves are torn off by the wind and cascade down the wave faces.

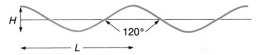

FIGURE 8.9
Wave steepness. When H/L approaches 1:7, the wave's crest angle approaches 120° and the wave breaks.

Relate the steepness of a wave to the wave's breaking.

Why do waves at sea rarely break?

8.8 SHALLOW-WATER WAVES

As a deep-water wave moves into shallow water, the water particle orbits become flattened circles or ellipses (fig. 8.10). The waves begin to "feel the bottom," the orbits are compressed, and the forward speed of the wave is reduced due to friction with the bottom.

Remember that (1) the speed of any wave (deep, intermediate or shallow) is equal to the wavelength divided by its

Waves and Tides

FIGURE 8.10

Deep-water waves become intermediate waves and then shallow-water waves as depth decreases and wave motion interacts with the sea floor.

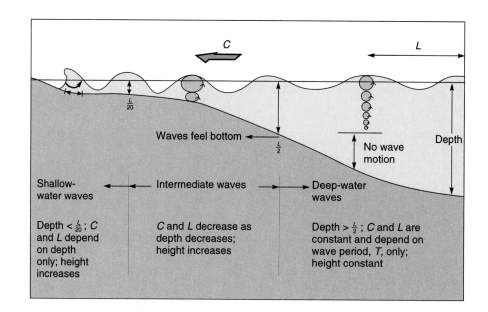

FIGURE 8.11

Shallow-water wave particles move in elliptical orbits. The orbits flatten with depth due to interference from the sea floor.

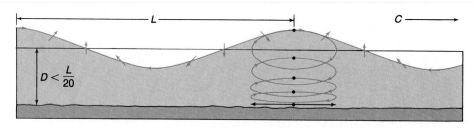

FIGURE 8.12

Waves moving inshore at an oblique angle to the depth contours are refracted. One end of the wave reaches a depth of $\frac{L}{2}$ or less and slows while the other end of the wave maintains its speed in deeper water. Wave rays drawn perpendicular to the crests show the direction of wave travel and the bending of the wave crests.

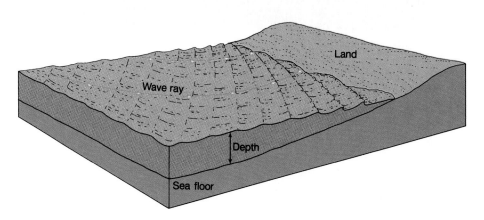

period, and (2) the period of a wave does not change. Therefore, when the intermediate or shallow wave feels bottom, it slows, and the accompanying reduction in the wavelength and speed results in increased height and steepness as the wave's energy is condensed in a smaller water volume. Figure 8.10 shows the transition of wave properties as the water depth changes.

When the wave enters water with a depth of less than one-twentieth the wavelength, the wave becomes a **shallow-water wave.** The shallow-water wave's length and speed are controlled only by the water depth. Here the speed and wavelength are determined by the square root of the product of the earth's acceleration due to gravity (g, 9.81 m/s²) and depth (D):

$$C = \sqrt{gD} \text{ or } 3.13 \sqrt{D};$$
$$L = \sqrt{gD} \, T \text{ or } 3.13 \sqrt{D} \, T.$$

In shallow-water waves, the elliptical orbits of the water particles become flatter with depth until at the sea floor only a back-and-forth motion remains (fig. 8.11). The horizontal dimension of the orbit remains unchanged in shallow water. The group speed, V, of shallow-water waves is equal to the speed of the wave, C. Shallow-water waves are formed when deep-water waves enter coastal waters; they are also present in the open ocean as tsunamis (sec. 8.10) and tides (sec. 8.12).

Waves are refracted, or bent, as they begin to feel the bottom and change wavelength and speed. When waves approach the beach at an angle, one end of the wave crest comes into shallow water and begins to feel bottom while the other end is in deeper water. The shallow-water end moves more slowly than the part in deeper water. The result is that the wave crests bend, or **refract,** and tend to become parallel to the shore (fig. 8.12).

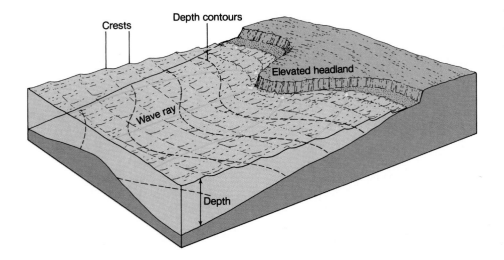

FIGURE 8.13
Waves refracted over a shallow submerged ridge focus their energy on the headland. The converging wave rays show the wave energy being crowded into a smaller volume of water, increasing the energy per unit length of wave crest as the height of the wave increases.

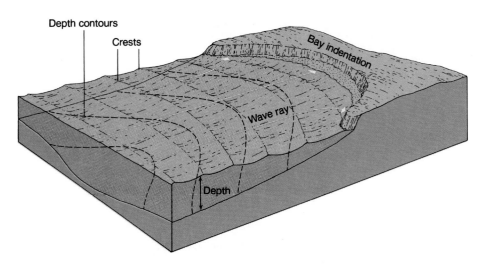

FIGURE 8.14
Waves refracted by the shallow depths on each side of the bay deliver lower levels of energy inside the bay. The diverging wave rays show the energy being spread over a larger volume of water, decreasing the energy per unit length of wave crest as the wave height decreases.

Along an irregular coastline, there is often a submerged ridge seaward from a headland and a depression in front of a bay. When shallow-water waves approach this coastline, the part over the ridge slows down more than the part on either side. The central area at the mouth of the bay is usually deeper than the areas to each side, and so the advancing waves slow down more on the side than in the center. This pattern is shown in figures 8.13 and 8.14. More energy is expended on a unit length of shore at the point of the headland because of increased wave height, while the lower-energy environment of the bay provides sheltered water.

A straight, smooth vertical barrier in water deep enough to prevent waves from breaking reflects the waves. The barrier may be a cliff, steep beach, breakwater, bulkhead, or other structure. The reflected waves pass through the incoming waves to produce interference patterns, and steep, choppy seas often result. If the waves reflect directly back on themselves, the resulting waves appear to stand still, rising and falling in place.

Another phenomenon associated with waves as they approach the shore is **diffraction.** Diffraction is caused by the spread of wave energy sideways to the direction of wave travel. If waves move toward a barrier with a small opening (two landmasses separated by a channel or an opening in a breakwater),

some wave energy passes through the small opening. Once through the opening, the wave crests decrease in height, radiating outward from the gap. This effect is shown in figure 8.15. If the waves approach a barrier without an opening and move past its end, diffraction can still occur, because energy is transported at right angles to the wave crests as the waves pass the end of the barrier.

In areas of the world where the winds blow steadily and from one direction, as in the trade wind belts, waves at sea are very regular in their direction of motion, allowing a vessel to maintain a constant course relative to the waves. Because waves change speed, shape, and height with water depth, and because waves change direction and pattern due to refraction, diffraction, and reflection, it is possible to deduce the presence of shoals, bars, islands, and coasts from the changes in wave pattern. Although these changing patterns are subtle, they can be detected far from land by those who live with the sea and become sensitive to its variations. The Polynesians of the past made long canoe voyages using their knowledge of wave patterns. This understanding plus their knowledge of star positions and their observations of cloud forms over land and sea allowed them to sail many hundreds, even thousands, of kilometers across the open ocean to reach their destinations.

Waves and Tides

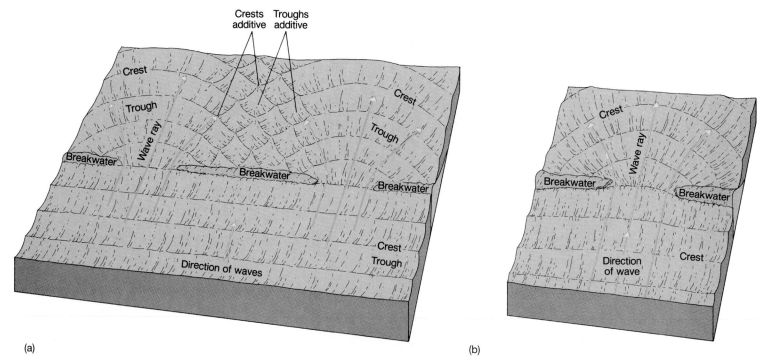

(a)   (b)

FIGURE 8.15

(a) Diffraction patterns produced by waves passing through narrow openings; note the interference pattern. (b) Diffraction occurs behind the breakwater.

How does a deep-water wave change when it enters water with a depth of less than one-twentieth the wavelength?

Which property of the waves does not change?

Why do waves change direction in shallow water?

Compare reflection and diffraction of waves.

Why are waves higher around a point of land extending into the sea?

8.9 THE SURF ZONE

The surf zone is the shallow area along the coast in which the waves slow rapidly, steepen, break, and disappear in the turbulence and spray of expended energy. The width of this zone is variable and is related to both the length and height of the arriving waves and the changing depth pattern. Longer, higher waves, which feel the bottom before shorter waves, become unstable and break farther offshore in deeper water. If shallow depths extend offshore for some distance, the surf zone is wider than it is over a sharply sloping shore.

Breakers form in the surf zone because the wave motion at depth is affected by the bottom. Orbital motion is slowed and compressed vertically; as the wavelength shortens, the wave steepens and breaks. The two most common types of breakers are plungers and spillers (fig. 8.16).

Plunging breakers form on narrow, steep beach slopes. The curling crest outruns the rest of the wave, curves over the air below it, and breaks with a sudden loss of energy and a splash.

(a)

(b)

FIGURE 8.16

Breaking waves. (a) A plunger loses energy more quickly than (b) a spiller.

(a)

FIGURE 8.17
A surfer rides the tube of a large curling wave.

The more common spilling breaker is found over wider, flatter beaches, where the energy is extracted more gradually as the wave moves over the shallow bottom. This action results in the less dramatic wave form, consisting of turbulent water and bubbles flowing down the collapsing wave face. On some breakers there is a slow curling-over of the crest beginning at a point on the steepest part of the crest and moving along the crest as the wave approaches shore at an angle. The result is the "tube" so sought after by surfers (fig. 8.17).

(b)

> Why does a gently sloping beach protect the shore from heavy storm waves?
>
> Which wave maintains its wave form longer, a spiller or a plunger?

8.10 TSUNAMIS

Tsunamis are produced by sudden movements of the earth's crust, or earthquakes. These seismic sea waves are often incorrectly called tidal waves (fig. 8.18).

If an underwater area of the earth's crust is suddenly displaced, it may cause a sudden rise or fall in the sea surface above it. This results in waves with extremely long wavelengths 100 to 200 kilometers (54 to 108 nmi) and long periods as well (10–20 min). Since the average depth of the oceans is about 4000 meters (13,000 ft), tsunamis are shallow-water waves. They radiate out from the point of the seismic disturbance at a speed determined by the ocean's depth and move across the oceans at about 200 meters per second (400 mph). As shallow-water waves, tsunamis may be refracted, diffracted, or reflected in midocean.

When a tsunami leaves its point of origin, it may have a height of 1 to 2 meters (3 to 6 ft), but this height is distributed over its extremely long wavelength. A vessel in the open ocean may not see the wave and is in little or no danger if a tsunami

(c)

FIGURE 8.18
(a,b,c) Sequential photos of the major wave of a tsunami at Laie Point, Oahu, Hawaii, from an earthquake in the Aleutian Islands, March 9, 1957. The highest wave in Hawaii was 3.6 meters above sea level (11.8 ft).

passes. When its path is blocked by a coast or island, the wave behaves like any other shallow-water wave, and its height builds rapidly. The loss of energy is also rapid, and a tremendous amount of moving water races up over the land. The large volume of water surging onto the coast causes large water level changes and destructive currents.

The leading edge of the tsunami wave group may be either a crest or a trough. If a trough arrives at the shore before the first crest, sea level drops rapidly, exposing the sea floor with its plants and animals. People have drowned after following the receding water to inspect the marine life, for they find, too late, that there is a wall of advancing water that they cannot outrun. The Pacific Ocean, ringed by crustal faults and volcanic activity, is the birthplace of many tsunamis. The Aleutian Trench produced the 1946 tsunami that heavily damaged Hilo, Hawaii, killing more than 150 people. In 1957 Hawaii was hit again, but due to early warning and evacuation, no lives were lost. The 1964 Alaska earthquake produced tsunamis that caused severe damage in Alaska and areas on the west coast of Vancouver Island and the northern coast of California.

In September 1992 a coastal submarine earthquake created a tsunami that struck Nicaragua causing extensive damage and killing 170 people, mostly children. In December of this same year an earthquake in the Flores Island region of Indonesia created a 25 meter (82 ft) tsunami that killed 1000 people in the town of Maumere and swept away two-thirds of the population of the island of Babi. In July of 1993 a near-shore earthquake of 7.8 magnitude on the Richter scale occurred near Okushiri Island, off the southwest coast of Hokkaido, Japan causing a tsunami that ran 15 to 30 meters (50 to 100 ft) up the beach over a 20 kilometer (12 mi) length of shoreline. This event caused about $600 million in damages and claimed more than 185 lives.

Tsunamis also appear in the Caribbean Sea, which is bounded by an active island arc system, and in the Mediterranean Sea and along the west coast of South America.

Why are tsunamis not seen in the open ocean?

Why is the tsunami a shallow-water wave in the deep sea?

8.11 STANDING WAVES

Standing waves are waves that reflect back on themselves. They occur in ocean basins, partly enclosed bays and seas, and estuaries. A standing wave can be demonstrated by slowly lifting one end of a container partially filled with water and then rapidly but gently returning it to a level position. If this is done, the surface alternately rises at one end and falls at the other end of the container. The surface oscillates about a point at the center of the container, the **node,** and the alternations of low and high water (troughs and crests) at each end are the **antinodes** (fig. 8.19a). The wavelength is twice the length of the basin.

By tilting the basin back and forth at a faster rate, it is possible to produce a wave with more than one node, as shown in figure 8.19b. In the case of two nodes, there is a crest at either end of the container and a trough in the center. This arrangement alternates with a trough at each end and the crest in the center; the wavelength is equal to the basin length.

Standing waves in bays or inlets with an open end behave somewhat differently than standing waves in closed basins. A node is usually located near the entrance to the open-ended bay, so that one-quarter of a wavelength is present in the bay. There is little or no rise and fall of the water surface at the entrance, but a large rise and fall occurs at the closed end of the bay. Multiple nodes may also be present in open-ended basins. The dimensions of each individual basin determine the wavelength and period of its standing wave.

Standing waves can be triggered by seismic events that suddenly tilt or oscillate the water in natural and artificial basins, even swimming pools. If storm winds create a change in surface level at one side of a basin, the surface may continue to oscillate when the wind ceases until friction causes the oscillation to die out. The water in the basin oscillates at a period governed by the shape and size of the basin. A standing wave occurring in a natural basin is called a **seiche,** and the process of oscillation is called seiching. Severe storms moving rapidly across the Great Lakes often produce a tilted water surface that oscillates as the storm moves on; water level fluctuations of about 0.5 meters (1.5 ft) occur at the shore due to seiching. Tides can also produce standing waves or seiches in oceanic and coastal basins if the oscillation period of the basin is a multiple of the tidal period, for example the Bay of Fundy in southeastern Canada; see section 8.12.

How is a standing wave related to a wind wave?

Compare standing waves in open and closed basins.

What is a seiche?

8.12 THE TIDES

The tides are best known as the rise and fall of the sea around the edges of the land. In some coastal areas there is a regular pattern of one high tide and one low tide each day; this is a **diurnal tide.** In other areas there is a cyclic high water–low water sequence that is repeated twice in one day; this is a **semidiurnal tide.** In a semidiurnal tidal pattern, both high tides reach about the same height, and both low tides drop to about the same level with each cycle. A tide in which the high tides regularly reach different heights and the low tides drop regularly to different levels is called a **mixed tide.** Curves for typical tides at some U.S. coastal cities are shown in figure 8.20.

In a uniform diurnal or semidiurnal tidal system, the greatest height to which the tide rises on any day is known as **high water,** and the lowest point to which it drops is called **low**

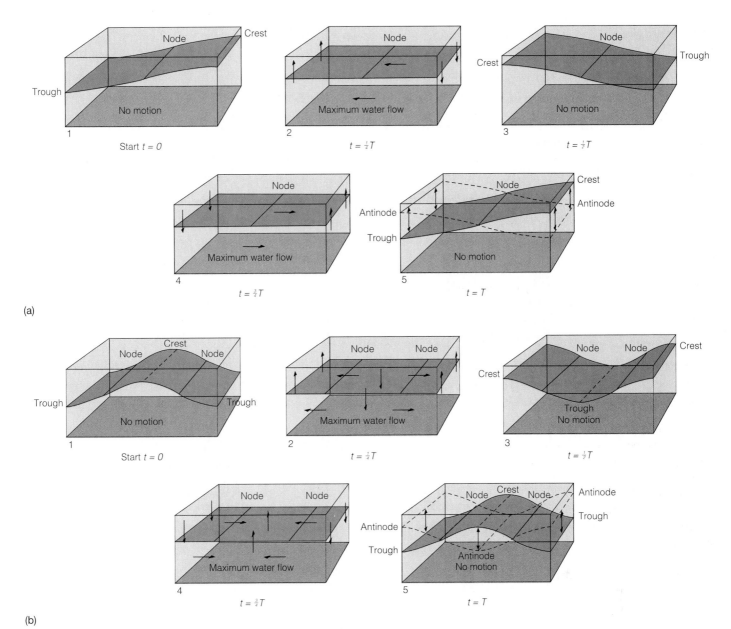

FIGURE 8.19

(a) A standing wave in a basin oscillating about a single node. (b) A standing wave oscillating about two nodes. The time for one oscillation is the period of the wave, T.

water. In a mixed tide it is necessary to refer to **higher high water, higher low water, lower high water,** and **lower low water.** Tidal measurements taken over many years are used to calculate the average or mean tide levels, also mean high waters and low waters.

In areas of uniform diurnal or semidiurnal tide patterns, the zero depth reference on navigation charts is usually equal to mean low water; in regions with mixed tides, mean lower low water is used. Occasionally the low tide falls below the mean value, producing a **minus tide.** A minus tide exposes shoreline usually covered by seawater; it can be a hazard to boaters but is cherished by clam diggers and students of marine biology.

As the water level along the shore increases in height, the tide is said to be rising or flooding; a rising tide is a **flood tide.** When the water level drops, the tide is falling or ebbing; a falling tide is an **ebb tide.**

When tides are studied mathematically in response to the laws of physics, they are known as **equilibrium tides.** To simplify the study of relationships between the oceans and the moon and the sun, this method uses tides on an earth model covered with a uniform layer of water. The tides are also studied as they occur naturally; these tides are called **dynamic tides.** Oceanographers study dynamic tides modified by the landmasses, the geometry of the ocean basins, and the earth's rotation.

Waves and Tides

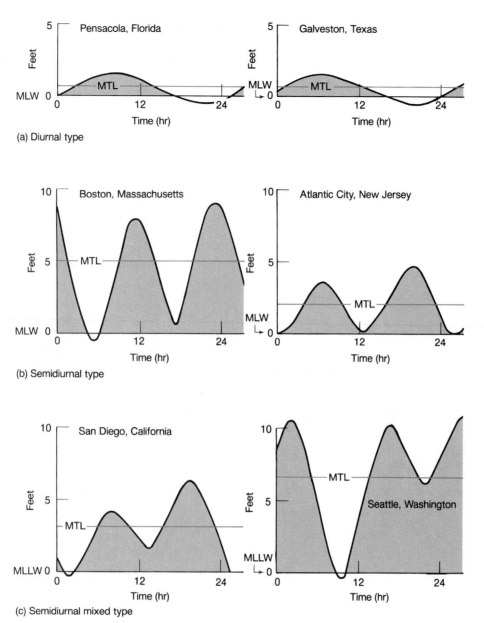

FIGURE 8.20

Tide types and tidal ranges vary from one coastal area to another. The zero tide level equals mean low water (MLW) or mean lower low water (MLLW), as appropriate. MTL equals mean tide level. All tide curves are for the same date.

The effect of the sun's and moon's gravity and the rotation of the earth on tides is most easily explained by studying equilibrium tides. In this discussion consider the earth and the moon as a single unit, the earth-moon system that orbits the sun; refer to figure 8.21a. The moon orbits the earth, held by the earth's gravitational force acting on the moon (B). There is also a force acting to pull the moon away from the earth and send it spinning out into space (B'); this is considered a **centrifugal force** in our discussion. Forces (B) and (B') must be equal and opposite to keep the moon in its orbit. Likewise, the moon's gravitational force acting on the earth (C) must be balanced by the centrifugal force (C'). B' and C' are caused by the earth-moon system rotating about an axis at the center of the system's mass (fig. 8.21b), which is a point 4640 kilometers

(2500 mi) from the earth's center along a line between the earth and moon. The earth-moon system is held in orbit about the sun by the sun's gravitational attraction (A). A centrifugal force again acts to pull the earth-moon system away from the sun (A'). To remain in this orbit, the earth-moon system requires that the gravitational forces equal the centrifugal forces (fig. 8.21c).

The moon's gravitational force is stronger on particles on the side of the earth closest to the moon, while the centrifugal force is stronger on particles on the side of the earth farthest from the center of mass of the earth-moon system. This distribution of forces tends to pull surface particles away from the center of the earth and creates the tide-raising force field on the earth. Because the water covering the earth is liquid and deformable, the moon's

Chapter 8

174

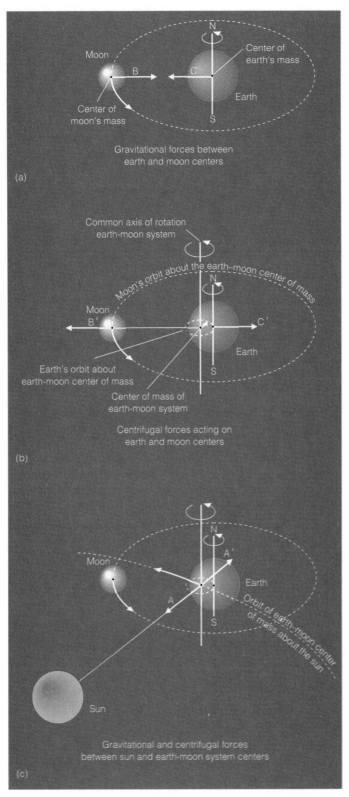

FIGURE 8.21
Gravitational and centrifugal forces act to keep the earth-moon system in balance.

with two depressions in between the bulges, or two crests and two troughs, or two high tide levels and two low tide levels (fig. 8.22). The wavelength is equal to half the circumference of the earth.

The earth makes one rotation in 24 hours, and the bulges or crests in the water covering tend to stay under the moon as the earth turns. A point on earth that is initially centered under a bulge or high tide passes to a trough or low tide, to another high tide, to another low tide, and back to the original high tide position during one revolution of the earth.

While the earth turns, the moon is also moving eastward along its orbit about the earth. After 24 hours the earth point that began directly under the moon is no longer directly under the moon. The earth must turn another 12°, requiring an additional 50 minutes, to bring the starting point on earth back in line with the moon. Therefore, a lunar **tidal day** is not 24 hours long, but 24 hours and 50 minutes. This also explains why corresponding tides arrive at any location about one hour later each day (fig. 8.23).

The sun also acts to produce a tide wave. Because the sun is far away, the gravitational and centrifugal forces that cause tides are only 46 percent of those associated with the moon. The length of the tidal day of the sun tide is 24 hours, but it does not greatly affect the time of the average tidal day, 24 hours 50 minutes.

It takes the moon 29½ days to orbit the earth. Using figure 8.24, follow the motions of the earth, moon, and sun during this period. During the period of the new moon, the moon and sun are on the same side of the earth so that the high tides or bulges produced independently by each reinforce each other. Tides of maximum height and depression produced during this period are known as **spring tides.** In a week's time, the moon is in its first quarter and is located approximately at right angles to the earth and sun. At this time, the crests of the moon tide coincide with the troughs of the sun tide to produce low-amplitude tides known as **neap tides.** The tides follow a four-week cycle with spring tides every two weeks and a period of neap tides occurring in between the spring tides. This spring-neap pattern is shown in figure 8.25.

If the moon or sun stands north or south of the earth's equator, one bulge or crest is found in the Northern Hemisphere and the other in the Southern Hemisphere. Under these conditions, point A in figure 8.26 passes through only one crest or high tide and one trough or low tide each tidal day; a diurnal tide is formed. Compare the diurnal tide at point A with the semidiurnal tide at point B in this figure.

The sun moves from 23½°N (the Tropic of Cancer) to 23½°S (the Tropic of Capricorn) and returns each year. Therefore, tides are more diurnal during the summer and winter solstices. The moon moves from 28½°N to 28½°S and returns in a period of 18.6 years. Occasionally the sun and the moon both reach their most northern or southern points in these cycles at the same time; when this happens, tides become more strongly diurnal at midlatitudes. Because of the earth's elliptical orbit about the sun, the earth is closest to the sun during the Northern Hemisphere's winter, and tides tend to be more extreme (higher highs, lower lows) at that time due to the sun's greater gravitational effect. When the moon's orbit brings it closer to the earth, the tides are also more extreme.

gravity moves water toward a point directly under the moon, producing a bulge in the water covering. At the same time, the centrifugal force acting on the surface opposite the moon creates a similar bulge of water. The earth model develops two bulges

FIGURE 8.22

Distribution of tide-raising forces on the earth. Excess lunar gravitational and centrifugal forces distort the earth model's water envelope to produce bulges and depressions.

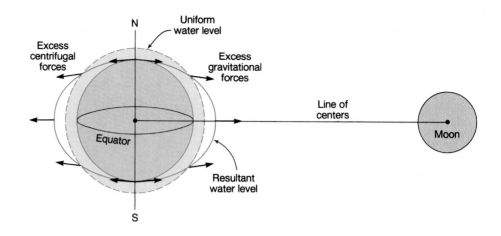

FIGURE 8.23

Point A requires 24 hours to complete one earth rotation. During this time the moon moves 12° east along its orbit, carrying with it the tide crest. To move from A to A' requires an additional 50 minutes to complete a tidal day.

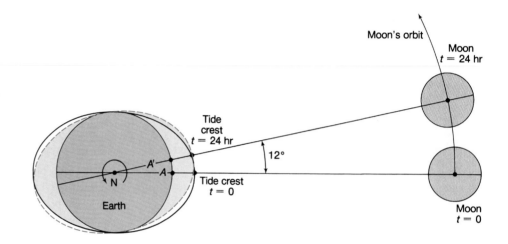

FIGURE 8.24

Spring tides result from the alignment of the earth, sun, and moon during the full moon and the new moon. During the moon's first and last quarters, neap tides are produced.

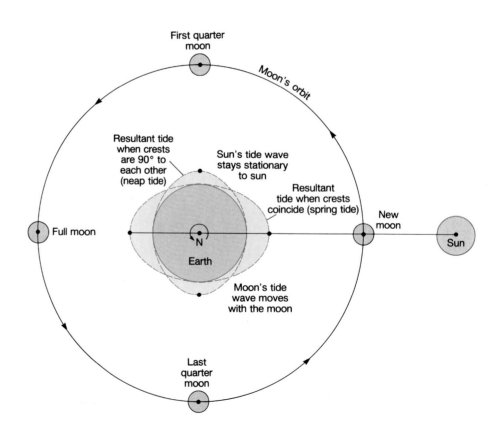

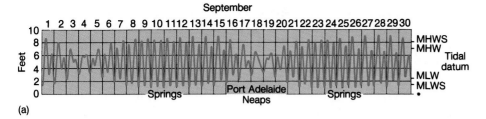

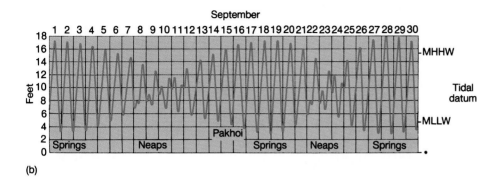

(a)

(b)

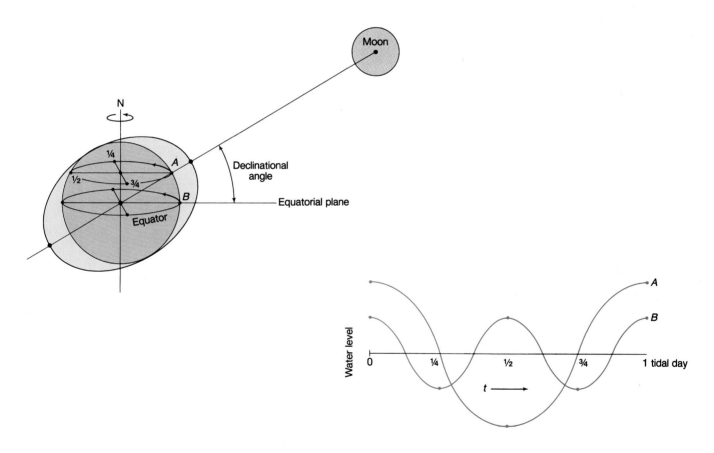

FIGURE 8.26
The declination of the moon produces a diurnal tide at latitude A and a semidiurnal tide at latitude B. Fractions indicate portions of the tidal day.

Although equilibrium tides allow us to understand the basic concepts associated with the oceans' tides, investigating the actual tides that occur in the discontinuous ocean basins requires the dynamic approach. Because continents separate the oceans from each other, the tide wave is discontinuous except in the Southern Ocean around Antarctica. The tide wave has a long wavelength as compared to the oceans' depths; therefore tides behave as shallow-water waves. The tide wave is also reflected from the edges of landmasses and mid-ocean ridges, refracted by the change in water depth, and diffracted as it passes through gaps between continents. In some basins, the tide wave is reflected from the edge of the continents and becomes a standing wave. Because the scale and duration of tidal motion are great, the Coriolis effect on this standing wave deflects the moving water as it flows from crest toward trough. This causes standing-wave type tides to rotate in their ocean basins. Together these factors produce the dynamic tides. In addition, different kinds of tides interact with each other along their boundaries, and the results are exceedingly complex.

Along the coasts in narrow bays, a standing wave with the node at the entrance to the bay and the antinode at the head of the bay may produce an extremely high tidal range at the head of the bay, if the water in the bay naturally oscillates at a period close to the tidal period. For example, the Bay of Fundy in southeast Canada has a tidal range near the entrance node of about 2 meters (6 ft) and a range of 10.7 meters (35 ft) at the head of the bay (fig. 8.27).

When water moves into a coastal region on the rising tide and out of it on the falling tide, **tidal currents** form. These currents may be extremely swift and dangerous as they move through narrow channels into large bays and harbors. When the tidal current changes from an ebb to a flood or vice versa, there is a period of **slack water,** during which the tidal currents slow, stop, and then reverse. Slack water may be the only time that a vessel can safely navigate a narrow channel with swiftly moving tidal currents, sometimes in excess of 20 kilometers per hour (10 knots).

> Draw and label the tide stages of a diurnal, semidiurnal, and mixed semidiurnal tide.
>
> What is the difference between an ebb tide and a flood tide?
>
> Explain how gravity and centrifugal force cause the ocean tides.
>
> Why is a tidal day longer than a solar day?
>
> Explain spring tides and neap tides.
>
> Why is a tide sometimes more diurnal or semidiurnal than at other times?
>
> List reasons for the differences in behavior between equilibrium tides and natural tides.
>
> How does a tidal current differ from an open-ocean surface current?

8.13 PREDICTING TIDES AND TIDAL CURRENTS

Because of all the natural combinations of tides and the factors that affect them, it is not possible to predict the earth's tides from knowledge of the sun and moon forces alone. Accurate, dependable, daily tidal predictions are made by combining local measurements with astronomical data. Water-level recorders are installed at coastal sites, and the rise and fall of the tides are measured over a period of years. Primary tide stations make these water-level measurements for at least 19 years to allow for

the 18.6-year period of the moon. From these data, mean tide levels are calculated. Oceanographers separate the tide record into components with magnitudes and periods that match the tide-raising forces of the sun and moon. They are then able to isolate the local effect from the astronomical data. This local effect is used with astronomical data to predict future tides in the area.

The National Oceanic and Atmospheric Administration (NOAA) of the U.S. Department of Commerce has the responsibility for determining and publishing annual predicted tide tables for North and South America, Alaska, Hawaii, and the coast of Asia. These tables give the dates, times, and water levels for high and low water at primary tide stations and give the correction values required to convert these data for use at auxiliary stations. NOAA also publishes tidal current data that are derived from predicted tide data.

> Why are astronomical data not sufficient to predict tides for a specific area?

8.14 TIDAL BORES

In some areas of the world, large-amplitude tides cause large and rapid changes in water volume along shallow coasts, bays, or river mouths. Under these conditions, the tide level increases more rapidly than the tide wave can normally move. The tide wave forms a spilling, breaking wave front that moves rapidly into the shallow water or up the river. This broken wave front appears as a wall of turbulent water and is called a **tidal bore,** which produces an abrupt change in water level as it passes. The bores are usually less than a meter in height but can be as much as 8 meters (26 ft) high, as in the case of spring tides on the Qiantang River of China. The Amazon, Trent, and Seine Rivers have bores. Fast-rising tides also send bores across the sand flats surrounding Mont-Saint-Michel in France. These bores can be hazardous, because they suddenly flood areas that have been open stretches of beach only minutes before.

> Explain the hazards of tidal bores.

8.15 ENERGY FROM WAVES AND TIDES

A tremendous amount of energy exists in ocean waves. The power of all waves is estimated at 2.7×10^{12} watts, which is about equal to 3000 times the power-generating capacity of Hoover Dam. Unfortunately for human needs, this energy is widely dispersed and is not constant at any given location or time.

(a)

(b)

FIGURE 8.27

(a) Low tide and (b) high tide at The Rocks Provincial Park, Hopewell Cape. The tidal range at the head of the bay exceeds 10 meters.

Waves and Tides

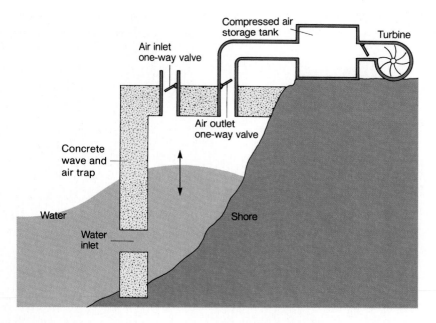

FIGURE 8.28

Each rise and fall of the waves pumps pulses of compressed air into a storage tank. A smooth flow of compressed air from the storage tank turns a turbine that generates electricity.

Wave energy can be harnessed in three basic ways: (1) using the changing level of the water to lift an object; (2) using the orbital motion of the water particles or the changing tilt of the sea surface to rock an object to and fro; and (3) using the rising water to compress air in a chamber. In each case, the wave motion may be used either directly or indirectly to turn a generator and generate electricity.

The 1991 estimate for total world energy produced by all operating wave-energy systems is less than 5.0×10^5 watts. One energy system allows waves to rush into a tapered channel pushing the water to an elevation of 2 to 3 meters (6 to 10 ft). The water spills over the channel and down through a turbine, which produces electricity. This system is used to generate power by a 7.5×10^4-watt plant on Scotland's Isle of Islay, a 3.50×10^5-watt plant at Toftestalen, Norway, and two 1.5×10^5-watt plants under construction, one in Java and the other in Australia.

Along a wave-exposed coast, air traps can be installed so that the crest of a wave moving into the trap compresses a large volume of air, forcing it through a one-way valve. This compressed air powers a turbine to produce energy. The trough of the wave allows more air to enter the trap, readying it for compression by the following wave crest (fig. 8.28). This system is in use in northern Norway and western Ireland.

Average wave power along the coast of Great Britain is calculated at about 5.5×10^4 watts per meter of coastline. If the wave energy could be completely harnessed along 1000 kilometers (620 mi) of coast, it would generate enough power to supply 50 percent of Great Britain's present power needs. Along the northern California coast, waves are estimated to expend 2.3×10^{10} watts annually. It is thought that 20 percent of this power could be harvested, and a northern California utility is considering installing a power-generating device in a breakwater planned for the town of Fort Bragg.

However, for wave-energy systems to help us in our quest for new energy sources requires the thoughtful consideration of certain questions. If all the wave energy were extracted from the waves in a coastal area, what effect would this action have on the shore area? If the nearshore areas are covered with wave energy absorbers, what will the effect be on other ocean uses? Since the individual units collect energy at a slow rate, can they collect enough energy over their projected life span to exceed the energy used to fabricate and maintain them? Harvesting wave energy is not without an effect on the environment; it may be neither cost nor energy effective, and its location may present enormous problems for installation, maintenance, and transport of energy from sites of energy generation to sites of energy use.

The possibility of obtaining energy from the tides exists in coastal areas with large tidal ranges. Commercial power is generated by a tidal power station on the Rance River estuary in France. The province of Nova Scotia in Canada began its power station project in the tidal estuary of the Annapolis River in 1984 (fig. 8.29). Tidal ranges at the Annapolis site vary from 8.7 meters (29 ft) during spring tides to 4.4 meters (14 ft) during neap tides. Annual production from the unit is between 30,000 and 40,000 megawatt-hours. Plants of this kind require building a dam across a bay or estuary so that seawater can be held in the bay at high tide and released through turbines as the outside tide level drops. These systems generate power either on the ebb tide or both the ebb and flood tides (fig. 8.30).

FIGURE 8.29
The Annapolis River tidal power project is the first tidal power plant in North America. It was completed in 1984.

However, there are few places in the world where the tidal range is sufficient, 7 meters (23 ft) or more, and where natural bays or estuaries can be dammed at their entrances at reasonable cost and effort. Neither are these places necessarily located near the population centers that need the power. The costs of the installations and the power distribution, in addition to periodic low power production because of the changing tidal amplitude over the tide's monthly cycle, make this type of power expensive in comparison to other sources.

Explain the sources of power in waves and tides.

Where is commercial electric power being generated by waves and tides at the present time?

What are the reasons for and against the production of electrical power from waves and tides?

Table 8.3 is a summary of the characteristics of the various waves discussed in this chapter.

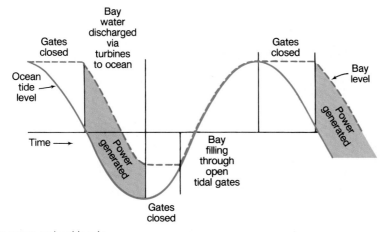

(a) Single-action power cycle; ebb only

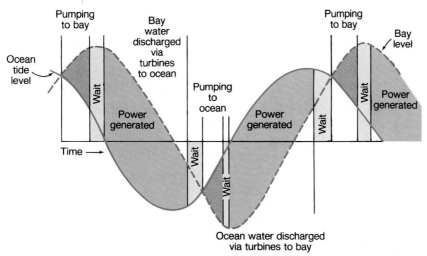

(b) Double-action power cycle; ebb and flood

FIGURE 8.30
(a) A single-action tidal power system. Power is generated on the ebb tide. (b) A double-action tidal power system. Power is generated on the ebb and flood tides.

Waves and Tides

Table 8.3 Wave Summary

Type	Cause	Period	Class of Wave	Open-Ocean Height	Coastal Height	Wave Speed	Group Speed	Orbit Type
Wind Wave	Wind	<0.1–20 s	Deep-water $D>L/2$	Up to 30 m (100 ft)	Up to 6 m (20 ft)	$C = 1.56T$ $C^2 = 1.56L$ 0–111 km/hr (0–60 kts)	$V = C/2$	Circle
			Shallow-water $D<L/20$	—	Up to 6 m (20 ft)	$C = \sqrt{gD}$	$V = C$	Ellipse
Tsunami	Seismic upheaval	10–30 min	Shallow-water $D<L/20$	Small	10–18 m (30–60 ft)	$C = \sqrt{gD}$ 740 km/hr (400 kts)	$V = C$	Ellipse
Standing	Wind, tsunamis, tides, pressure changes	10 min– 30 hrs	Normally shallow-water $D<L/20$	0–2.5 m (0–8 ft)	0–15 m (0–50 ft)	None	None	None
Tide	Gravity of sun and moon	12 hrs 25 min and 24 hrs 50 min	Shallow-water $D<L/20$	0–1.5 m (0–4 ft)	0–15 m (0–50 ft)	$C = \sqrt{gD}$ >740 km/hr (>400 kts)	None	Ellipse

SUMMARY

When the water's surface is disturbed, a wave is formed. The highest point of a wave is the crest; the lowest point is the trough. The wavelength is the distance between two successive crests or troughs. The wave height is the vertical distance between the crest and the trough. Wave period measures the time required for two successive crests or troughs to pass a location. The moving wave form causes water particles to move in orbits. A wave's speed is related to wavelength and period.

Deep-water waves occur in water deeper than one-half the wavelength. Wind waves are formed in storm centers as deep-water waves with a period that never changes. As waves move out from the storm center, the longer, faster waves move through the shorter, slower waves and form groups of waves. This process is known as sorting or dispersion. Waves of uniform characteristics are known as swell. The speed of a group of waves is half the speed of the individual waves in deep water. Swells from different storms cross, cancel, and combine with each other as they move across the ocean.

Wave height depends on wind speed, wind duration, and fetch. Single large waves unrelated to local conditions are called episodic waves. The energy of a wave is related to its height. When the ratio of the height to the length of a wave, or its steepness, exceeds 1:7, the wave breaks.

Shallow-water waves occur when the depth is less than one-twentieth the wavelength. The speed of a shallow-water wave depends on the depth of the water. As the wave moves toward shore and decreasing depth, it slows, shortens, and increases in height. Waves coming into shore are refracted, reflected, and diffracted. The patterns produced by these processes helped peoples in ancient times to navigate from island to island.

In the surf zone, breaking waves produce a water movement toward the shore. Breaking waves are classified as plungers or spillers.

Tsunamis are seismic sea waves. They behave as shallow-water waves and can produce severe coastal flooding and destruction. Standing waves or seiches form when displaced water oscillates in a basin.

The three tidal patterns are diurnal, semidiurnal, and mixed. Rising tides along coasts create flood currents; falling tides create ebb currents.

Equilibrium tides are explained as a balance between gravitational and centrifugal forces. A semidiurnal tide is a tide wave with two crests and two troughs per tidal day. A diurnal tide has one crest and one trough per tidal day. The tidal day is 24 hours 50 minutes long. Spring tides have the greatest range between high and low water; they occur at the new and full moons. Neap tides have the least tidal range; they occur at the moon's first and last quarters. When the moon or sun stands far to the north or south of the equator, the tides become more strongly diurnal.

Dynamic tides are the actual tides as they occur in the ocean basins. The tide wave is a discontinuous shallow-water wave that oscillates in some ocean basins as a standing wave with motions that persist long enough to be acted on by the

Coriolis effect. The tide wave is reflected, refracted, and diffracted along its route. In narrow basins, tides may oscillate about a node at the entrance to the bay and the antinode at the head of the bay.

A rapidly moving tidal bore is caused by a large-amplitude tide wave moving into a shallow bay or river.

Tidal heights and currents are predicted from astronomical data and actual local measurements. NOAA determines and publishes annual tide and tidal current tables.

It is possible to harness the energy of the waves by using water-level changes or the changing surface angle. Difficulties include cost, location, environmental effects, and lack of wave regularity. Tidal power plants are in use in France and Canada. Few places have large enough tidal ranges and suitable locations for tidal dams.

KEY TERMS

surface tension	swell	node	minus tide
crest	group speed	antinode	flood tide
trough	constructive interference	seiche	ebb tide
wavelength	destructive interference	diurnal tide	equilibrium tide
wave height	fetch	semidiurnal tide	dynamic tide
amplitude	episodic wave	mixed tide	centrifugal force
wave period	wave steepness	high water	tidal day
orbit	shallow-water wave	low water	spring tide
deep-water wave	refract	higher high water	neap tide
storm center	diffraction	higher low water	tidal current
dispersion/sorting	tsunami	lower high water	slack water
wave train	standing wave	lower low water	tidal bore

SUGGESTED READINGS

Alper, J. 1993. A Wave is Born. *Sea Frontiers* 39 (1):22–27.

Bascom, W. 1980. *Waves and Beaches: The Dynamics of the Ocean Survey*, rev. ed. Doubleday, Garden City, N.J., 366 pp.

Booda, L. 1985. River Rance Tidal Power Plant Nears Twenty Years in Operation. *Sea Technology.* September: 22–26.

Changery, M. J., and R. G. Quayle. 1987. Coastal Wave Energy. *Sea Frontiers* 33 (4):259–62.

Dutton, G. 1992. Catch a Wave for Clean Electricity. *The World & I.* July:290–97.

Fisher, A. 1989. The Model Makers. *Oceanus* 32 (2):16–21.

Folger, T. 1994. Waves of Destruction. *Discover* 15 (5):66–73.

Garrett, C., and L. R. M. Maas. 1993. Tides and Their Effects. *Oceanus* 36 (1):27–37.

Greenberg, D. A. 1987. Modeling Tidal Power. *Scientific American* 257 (5):128–31.

Lockridge, P. A. 1989. Tsunami: Trouble for Mariners. *Sea Technology* 30 (4):53–57.

Lynch, D. K. 1982. Tidal Bores. *Scientific American* 247 (4):146–56.

McCormick, M. E. 1986. Ocean Wave Energy Conversion. *Sea Technology.* June:32–34.

McCreedie, S. 1994. Tsunamis: The Wrath of Poseidon. *Smithsonian* 24 (12):28–39.

Vadus, J. R. 1989. Technology Need for Ocean Resources Utilization. *Sea Technology* 30 (8):14–25.

CHAPTER

9

Coasts, Estuaries, and Environmental Issues

Outline

Learning Objectives

After reading this chapter, you should be able to

■ Understand the differences between primary and secondary coasts and explain their features.

■ Describe the principal features of a beach.

■ Describe different types of beaches.

■ Understand the processes that form a sand beach.

■ Relate sediment transport in a coastal circulation cell to beach formation.

■ Relate shoreline changes to changes in sea level.

■ Define an estuary and describe the different types.

■ Compare a bay in a high-evaporation environment with an estuary.

■ Understand estuary circulation and how water is exchanged with the ocean.

■ Appreciate the importance of coastal wetlands.

■ Describe the effects of people and their structures on beaches and beach processes.

■ Discuss degradation of marine coastal environments and give examples.

◀ Sea stacks. Big Sur, California.

We visit the coast to see the ocean, to play along the beach, to enjoy the constant interaction between moving water and what appears to be the stable land. On any visit the tide may be high or low, the logs and drift may have changed position, and the dunes and sandbars may have shifted—coasts and beaches are dynamic, not static. No two regions are exactly the same, nor do they change in the same way. Along these coasts are indentations where tidal water from the oceans and freshwater drainage from the land meet and mix. The processes that regulate this mixing are not constant: river flow and tides vary; severe storms and river flooding occur from time to time. Coastal areas share certain characteristics, but each is also unique.

People play an important role in coastal areas, because people and their structures change these areas more than they realize. As the world's population grows, and as more and more people gather in these areas to live, work, and play, they affect them to a greater and greater degree.

9.1   The Coast

The **coasts** of the world's continents are the areas where the land meets the sea. The coast includes cliffs, dunes, beaches, and sometimes the hills and plains that form the edge of the land. The distance to which the coast extends inland varies and is determined by local geography, climate, and vegetation as well as the perception of the limits of marine influence in social customs and culture. The coast is most generally described as the land area that is or has been affected by marine processes such as tides, winds, currents, and waves. The seaward limit of the coast usually coincides with the beginning of the beach or shore, but sometimes includes nearby offshore islands.

The term **coastal zone** includes the open coast as well as the bays and estuaries found in coastal indentations. This phrase incorporates both land and water areas, and has become the standard term used in legal and legislative documents affecting U.S. coastal areas. Under these circumstances the coastal zone's landward boundary is defined by a distance from some chosen reference such as high water; the seaward boundary's limit is defined by state and federal laws.

What are the boundaries of a coast?

9.2   Types of Coasts

Coasts are not static features of the earth environment. Coasts are considered **emergent** if they are rising relative to sea level. This occurs if sea level falls, if the coast is uplifted by tectonic

processes, or if there is a decrease in the weight on the coastal crust (for example, receding glaciers). Coasts are considered **submergent** if they sink relative to sea level. This happens if sea level rises, if the land is lowered by tectonic processes, or if the weight on the crust increases (for example, glacier deposits, river sediment deposits). An emergent coast shows old beaches and shore features above present sea level (fig. 9.2d), while along a low-lying, submergent coast the water moves inland and reclaims the land (fig. 9.1a). Changes in sea level in excess of 100 meters (300 ft) have occurred as the volume of the ocean has changed in response to changes in the amounts of water stored in the continental ice sheets.

To distinguish between the many types of coasts, table 9.1 and figures 9.1 and 9.2 use the system of the late Francis P. Shepard of the Scripps Institution of Oceanography. Classification depends on large-scale features created by tectonic, depositional, erosional, volcanic, or biological processes. All coastal areas belong in one of two major categories: (1) primary coasts that owe their character and appearance to processes that occur at the land-air boundary, and (2) secondary coasts that owe their character and appearance to processes that are primarily of marine origin. Primary coasts have not yet been substantially altered or modified by marine processes. Features of secondary coasts may have been originally land-derived, but are now distinctly the result of ocean processes.

Under what circumstances do emergent and submergent coasts form?

How do primary and secondary coasts differ?

Give examples of primary coasts; explain their features.

Give examples of secondary coasts; explain their features.

9.3   Anatomy of a Beach

The **beach** is an accumulation of sediment (sand or gravel) that occupies a portion of the coast. The beach is not static, but is ever changing and dynamic, because the beach sediments are constantly being moved seaward, landward, and along the shore by nearshore wave and current action. Above the high tide mark, there may be dunes or grass flats spotted with occasional drift logs left behind by an exceptionally high tide or severe storm (fig. 9.3).

A beach profile (fig. 9.4) represents a cut through a beach perpendicular to the coast, and in this case shows the major features that might appear on any beach. Use figure 9.4 while reading this section, remembering that a given beach may not have all these features, for each beach is unique. The **backshore** is the dry region of a beach that is submerged only during the highest tides and severest storms; the **foreshore** extends out past the low tide level, and the **offshore** extends from the shallow-water areas seaward of the low tide level to the outer limit of wave action on the bottom.

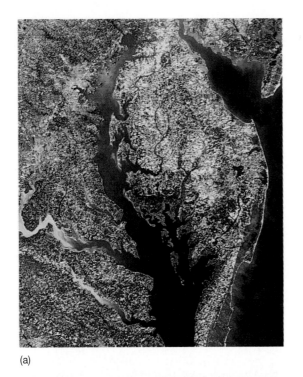

(a)

(b)

(c)

(d)

(e)

(f)

FIGURE 9.1

Primary coasts:
(a) Delaware Bay (upper right) and Chesapeake Bay (center) are examples of drowned river valleys.
(b) The narrow channel of a fjord, Milford Sound, New Zealand. (c) The Mississippi River delta. Because the river is being kept in its channel by levees, river-borne sediments are continuing to the river's mouth and being deposited in deep water instead of spreading out and replenishing the delta.
(d) Coastal sand dunes near Florence, Oregon. (e) A volcanic crater coast. Hanauma Bay, Hawaii, is a crater that has lost the seaward portion of its rim.
(f) The San Andreas Fault runs along the California coast to the west of San Francisco, separating Point Reyes from the Mainland. Bolinas Bay and Tomales Bay are fault bays produced by displacement along the fault line.

Coasts, Estuaries, and Environmental Issues

Table 9.1   Coastal Types

Primary Coasts	Features	Examples
Erosional Examples		
Drowned river valleys (ria coasts) Figure 9.1a	Multibranched shallow areas; sea level has risen or the land has sunk.	Chesapeake or Delaware Bay systems, Rio Plata, Argentina.
Fjord coasts Figure 9.1b	Glacially carved drowned channels. Deep and elongate.	Mountainous coasts of previously glaciated temperate and subpolar regions; Alaska, Greenland, New Zealand, Scandinavia.
Depositional Examples		
Deltaic coasts Figure 9.1c	Heavy river deposition in shallow coastal regions.	Mouths of rivers such as the Mississippi, Nile, Ganges, and Amazon.
Glacial moraine coasts	Deposits left by glaciers.	Long Island and Cape Cod, Baltic Sea, Sea of Okhotsk.
Dune coasts Figure 9.1d	Sand deposits on coasts by wind from the beach or wind from the land.	Central and northern Oregon coast. West coast of Sahara Desert and west coast of Kalahari Desert, Africa.
Volcanic Examples		
Lava coasts	Areas of active volcanoes with lava that reaches the sea.	East coast island of Hawaii.
Cratered coasts Figure 9.1e	Craters of coastal volcanoes at sea level filled with seawater.	Hawaiian islands, and islands and coastal areas of the Aegean Sea.
Tectonic Examples		
Fault coasts Figure 9.1f	Coastal faults filled with seawater.	Tomales Bay, California, coast of Scotland, Gulf of California, Red Sea.

Secondary Coasts	Features	Examples
Erosional Examples		
Regular cliffed coasts Figure 9.2a	Irregular coast made more uniform by wave erosion.	Hawaii, southern California, southern Australia, Cliffs of Dover.
Irregular cliffed coasts Figure 9.2b	Coast made more irregular by wave erosion.	Sea stack coasts of northern Oregon and Washington, New Zealand.
Depositional Examples		
Barrier coasts Figure 9.2c	Shallow coast with abundant sediments bordered by barrier islands and sand spits.	East coast of the United States, Gulf of Mexico, Netherlands.
Beach plains Figure 9.2d	Flat areas of elevated coastal seashore. May be terraced by wave erosion during uplift.	Northern California, Sweden, New Guinea.
Mudflats and salt marshes Figure 9.2e	Sediments deposited in wave-protected environments.	Common in many areas.
Built by Marine Organisms		
Coral reef coasts	Shallow tropical waters; corals produce massive coastal structures.	Tropical regions of the Pacific, Indian, and Atlantic oceans.
Mangrove coasts Figure 9.2f	Tropical and subtropical areas with freshwater drainage.	Florida, northern Australia, Bay of Bengal, other tropical areas.
Marsh grass coasts Figure 9.2g	Shallow, protected areas of fine sediments where plants root in intertidal land.	Eastern United States; coast of Europe; temperate climates.

(a)

(b)

(c)

(d)

(e)

(f)

(g)

FIGURE 9.2

Secondary coasts: (a) A cliffed coast made more regular by erosion, New Zealand. (b) Sea stacks are common features along the wave-eroded cliff coasts of northern California, Oregon, and Washington, occurring in a wide variety of shapes and sizes. (c) Sea Island, Georgia, is a barrier island that has been extensively developed. The shallow water between the island and the coast is visible on the left. (d) A series of pocket beaches cut an elevated marine terrace along the California coast. The flat elevated land is an old beach plain. (e) A protected mudflat shore along Puget Sound, Washington. (f) Mangrove trees growing along the shore of Florida Bay. (g) Coastal saltwater marsh, Gaspé, Québec, Canada.

FIGURE 9.3
Driftwood accumulates at the high tide line in areas where timber is plentiful.

Coasts, Estuaries, and Environmental Issues

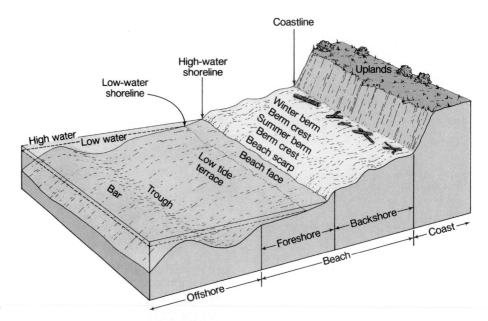

FIGURE 9.4
A typical beach profile with associated features.

FIGURE 9.5
Berms on a gravel beach. The winter storm berm with drift logs is located high on the beach. The summer berm is between the winter berm and the beach face in the foreground.

FIGURE 9.6
A wave-cut scarp on a gravel beach.

Berms appear in the area where foreshore and backshore meet. A berm is formed by wave-deposited material and has a relatively flat top like a terrace, or it may have a **berm crest** that runs parallel to the beach. When two berms are found on a beach (fig. 9.5), the berm higher up the beach is the winter berm, formed during severe storms when the waves reach high up the beach and pile up material along the backshore. The berm closer to the water is the summer berm, produced by the gentle waves of spring, summer, or fall that do not reach as far up the beach. Once the winter berm is formed, it is not disturbed by the less intense storms during the year, but the waves that produce a winter berm erase the summer berm.

Between the berm and the water level there may be a wave-cut **scarp** at the high-water level. The scarp is an abrupt change in the beach slope that is caused by the cutting action of waves at normal high tide (fig. 9.6). Berms and scarps do not form between normal high and low tide levels because of the wave action and continual rise and fall of the water over the foreshore. The portion of the foreshore below the low tide level is often flat, forming a **low tide terrace;** the upper portion between high and low tide levels has a steeper slope and is known as the **beach face.** The slope of a beach face is related to the particle size of the loose beach material and the wave energy.

Seaward of the low tide level, in the offshore region, there may be **troughs** and **bars** running parallel to the beach. Troughs and bars change seasonally as beach sediment moves seaward, enlarging the bars during winter storms, and then shoreward in summer, diminishing the same bars. When a bar accumulates enough sediment to break the surface and is stabilized by vegetation, it becomes an island of the barrier island coast type (fig. 9.2c).

Beaches do not occur along all shores. Along rocky cliffs, no beach area may be exposed between low and high tide levels. However, there may be small pocket beaches between headlands, each separated from the next by rock headlands or cliffs (fig. 9.2d).

Distinguish between backshore, foreshore, and offshore.

What is a berm? Distinguish between a summer berm and a winter berm.

Why is a beach face without features?

What is the function of bars along a beach?

9.4 BEACH TYPES

Beaches are also described in terms of (1) shape and structure, (2) composition of beach material, (3) size of beach materials, and (4) color. In the first category, beaches are described as wide, or narrow, steep or flat, and long or discontinuous (pocket beaches). A beach area that extends outward from the main beach and then turns and parallels the shore is called a **spit** (fig. 9.7). If a spit extends to an offshore island, it is called a **tombolo.** A spit extending offshore in a wide, sweeping arc is often called a hook. See figure 9.8 for an example of a tombolo and figure 9.9 for an example of a hook.

Along some coasts bars form parallel to the shore in shallow water where there are abundant sediments. When these bars accumulate sufficient material to break the sea surface they become **barrier islands.** Once a barrier island is formed, vegetation helps to stabilize the sand and increase the island's elevation by trapping sediments and accumulating organic matter. Elongate barrier islands along the East and Gulf Coasts of the United States protect low-lying coasts from direct wave damage. Here towns and recreation sites have been built on the larger islands; see figure 9.2c.

FIGURE 9.7
A spit has formed across the entrance to Sequim Bay, Washington.

FIGURE 9.8
A spit connected to an offshore island forms a tombolo.

Coasts, Estuaries, and Environmental Issues

FIGURE 9.9

Ediz Hook is a long spit forming a natural breakwater at Port Angeles, Washington (infrared false color photo).

Materials that form beaches include shell, coral, rock, and lava. Shingles are flat, circular, smooth stones formed from layered rock when wave action slides the stones back and forth with the water movement. When stones roll rather than slide, round stones or cobbles are formed. The size of the beach particles is described by the terms sand, mud, pebble, cobble, and boulder; refer to Chapter 4.

Beaches are formed from whatever loose material is present in the environment. Land materials are brought to the coast by the rivers or come from local cliffs by wave erosion. Coral and shell particles broken by the pounding action of the waves are carried onshore by the moving water. In Hawaii there are white sand beaches derived from coral, and black sand beaches derived from lava. Green sands can be found in areas where a specific mineral (olivine or glauconite) is available in large enough quantities, and pink sands occur in regions of sufficient shell material.

Distinguish between a spit, a hook, and a tombolo.

What forms a beach?

9.5 BEACH PROCESSES

A sand beach exists because there is a balance between the supply and the removal of the material that forms it. A sand beach that does not appear to change with time is not necessarily static, but is more likely to represent a dynamic equilibrium, with equal supply and removal of beach material.

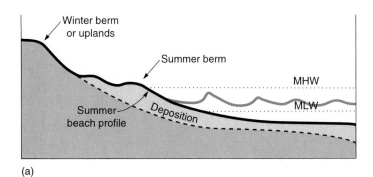

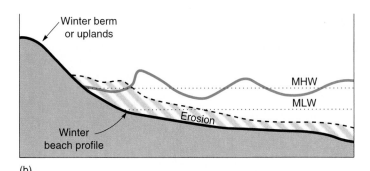

FIGURE 9.10

Seasonal beach changes. (a) The small waves and average tides of summer move sand from offshore to the beach and form a summer berm. (b) The high waves and storm tides of winter erode sand from the beach and store it in offshore bars. Winter conditions remove the summer berm, leaving only the winter berm on the beach.

The gentle waves of summer move sand up the beach and deposit it, where it remains until the winter season. The large storm waves of winter remove the sand from the beach and transport it back offshore to the sandbars. This process lowers the beach each winter and covers it with sand during the summer (fig. 9.10).

Waves moving toward a beach produce a current in the surf zone that moves water onshore and along the beach. The landward motion of water is called the **onshore current,** and this current transports suspended sediment toward the shore. Waves usually do not approach a shore with their crests completely parallel to the beach but instead strike the shore at a slight angle. This pattern sets up a current in the surf zone that moves down the beach; this is the **longshore current.** Both onshore and longshore flow in the surf zone are illustrated in figure 9.11. The turbulence that occurs as the waves break in the surf zone tumbles the beach material into suspension in the water. The wave-produced longshore current displaces this wave-suspended sediment down the shore in the surf zone, producing a **longshore transport.** In addition the uprush and backwash, or swash, of water from each breaking wave moves sediment in a zigzag pattern down the beach face as a part of the longshore transport (fig. 9.11). Along both coasts of North America, the predominant longshore transport steadily moves the sediments in a southerly direction.

How can sand beaches change between summer and winter?

How is the longshore current formed, and how does it function to keep a beach in a dynamic equilibrium?

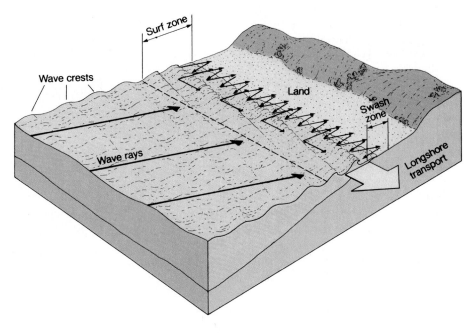

FIGURE 9.11

Waves in the surf zone produce a longshore current that transports sediments down the beach. Arrows indicate onshore and longshore water movement.

Coasts, Estuaries, and Environmental Issues

9.6 COASTAL CIRCULATION CELLS

The onshore current accumulates water as well as sand against the beach. This water must return seaward one way or another. A headland may deflect the longshore current seaward, or the water flowing along the beach can return seaward in areas of quieter water with smaller waves and less onshore flow.

Regions of seaward return are frequently narrow and some distance apart, located above troughs or depressions in the sea floor. The returning water flow must be swift in order to carry enough water beyond the surf zone to balance the flow toward the beach. These narrow areas of rapid seaward flow are called **rip currents** (fig. 9.12). Rip currents can be a major hazard to surf swimmers; in the spring of 1994, five swimmers were drowned and six others hurt in a single rip current system at American

Beach, Florida. To escape a rip current one must swim parallel to the beach, or across the rip current, and then return to shore.

Rip currents are visible as streaks of discolored turbid water extending seaward through the clearer water of the outer surf zone. The discoloration of the water is from the particles stirred up by waves and carried along the beach until they reach the rip current.

Just seaward of the surf, the rip current dissipates into an eddy. Some of the sediment is deposited in deeper water and is lost from the down-beach flow. However, some is returned to shore with the onshore transport on either side of the rip current. The path followed by sediment from its source to its place of deposit is called a **coastal circulation cell** (fig. 9.13). In a major coastal circulation cell, the seaward transport of sediment results in deposits far enough offshore to prevent any recycling. At the end point of

FIGURE 9.12

Rip currents form along a beach in areas of low surf and reduced onshore flow.

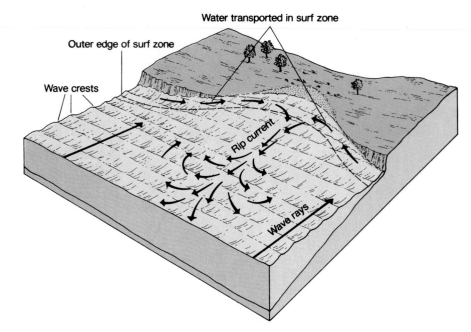

FIGURE 9.13

A coastal circulation cell extends down the coast from the sediment source to the area of deposit.

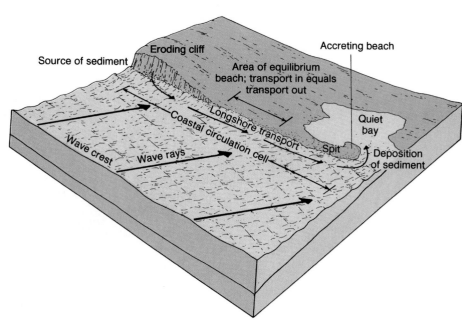

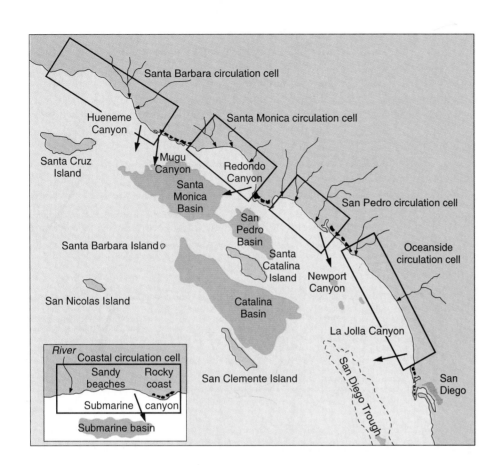

FIGURE 9.14

Major coastal circulation cells along the southern California coast. Each cell starts with a sediment source and ends where beach material is transported into a submarine canyon.

such a circulation cell, the longshore current is deflected away from the beach, and the sand is carried and deposited offshore, often moving through submarine canyons into the offshore basins. This action removes the sand from its journey along the coast, and the sand beaches disappear below this point until a new source contributes sand to form the next major coastal circulation cell.

South of Point Conception in southern California, oceanographers recognize four distinct major coastal circulation cells (fig. 9.14). Each cell begins and ends in a region of rocky headlands where submarine canyons are found offshore. (Refer to Chapter 4 for a review of submarine canyons.)

> Why are rip currents hazardous?
>
> What are the components of a coastal circulation cell?

9.7 RISING SEA LEVEL

Sea level has risen about 15 centimeters (6 in) during the past century, but the rate of rise has recently increased, and it is thought that sea level may gain another 30 centimeters (1 ft) in the next fifty years because of melting land ice and warming ocean water. If such an increase happens, the edge of the coast will move inland from hundreds to thousands of meters along low-lying coasts, and coastal erosion will be accelerated.

The increasing rate of sea-level rise has been primarily blamed on the greenhouse effect discussed in Chapter 6. If the amount of CO_2 in the atmosphere doubles during the next 100 years, there are estimates that the ocean surface water will warm an additional 2° to 4°C. Under these circumstances it is thought that the greatest warming will occur at the high latitudes, where continental ice masses will melt and increase the volume of the oceans. The rise in sea level caused by land ice melting is estimated at between 31 and 110 centimeters (1–3.5 ft) by the year 2100. The expansion of the warming water will also affect sea level by raising it an additional 60 centimeters (24 in) for each degree of increase in the average ocean temperature.

During intense storms, barrier islands and any dwellings, towns, or recreational facilities built on these islands sustain the damage that the continental coast is spared. The islands may be flooded and breached in some places, and old channels may be filled in other places. Because of their susceptibility to frequent and severe storm damage as well as rising sea level, the 1982 Coastal Barrier Resources Act prohibits federal subsidies for repairing roads, bridges, piers, and flood insurance coverage for homes built after 1983 in certain designated undeveloped barrier-island areas. States from Maine to North Carolina to Washington have regulated construction of seawalls and other permanent structures in the coastal zones.

> What are some possible causes of the projected rise in sea level?
>
> Why are barrier islands poor places for major development projects?

Coasts, Estuaries, and Environmental Issues

9.8 ESTUARIES

An **estuary** is a portion of the ocean that is semi-isolated by land and diluted by freshwater drainage. An estuary's circulation, water exchange with the ocean, and current patterns are determined by the tides, the river flow, and the geometry of the inlet. The geometry of an inlet may be used to classify an estuary; however, this method does not give a clear indication of the dynamic behavior of an estuary as a body of water. Therefore, circulation and distribution of salinity are used to distinguish between the four basic types of estuaries, listed in table 9.2 and shown in figure 9.15a–d.

Salt wedge estuaries at the mouths of rivers are strongly stratified, as measured by salinity-density changes over depth. These estuaries are never well mixed vertically, except above the boundary between the river's surface and the upper surface of the salt wedge. Other estuary types have various degrees of vertical stratification: (1) the **fjord,** highly stratified or poorly mixed, (2) the weakly stratified or well-mixed, and (3) the partially stratified and mixed. The degree of stratification is governed by the speed of tidal currents and associated turbulence, the rate of freshwater addition, the roughness of the bottom topography, and the average depth of the estuary.

Some parts of a complex estuary system may function as one type of estuary, and other parts may function as other estuary types.

The net circulation of an estuary, out (or seaward) at the top and in (or landward) at depth, exchanges water between the estuary and the ocean. Tidal flow that oscillates with the rise and fall of the tide produces the energy for mixing and should not be confused with the net flow in estuaries. The net circulation carries wastes and accumulated debris seaward and disperses them in the larger oceanic system. It is also through this circulation that organic materials and juvenile organisms produced in the estuaries and their marsh borderlands are moved seaward, while nutrient-rich salt water is brought inward at depth to replenish the estuary's water and salt content.

What is an estuary?

Describe the circulation in each of the four types of estuaries.

Why is it important to understand estuary circulation?

Table 9.2 Types of Estuaries

Type	Characteristics	Examples
Salt Wedge Estuary In mouths of rivers flowing strongly and directly into seawater. Figure 9.15a	Has tilted, sharply defined boundary between salt and fresh water. Boundary moves up and down stream with changes in tide and river flow. Salt water mixes upward into river flowing seaward above the boundary. Salt distribution and salt transfer governed by river flow. Strong density stratification over saltwater wedge.	Columbia, Mississippi, and Hudson Rivers. In the mouth of any reasonable size river entering an ocean or a saltwater bay.
Well-Mixed Estuary In shallow bays with large intertidal volumes, strong tidal mixing, and weaker river flows. Figure 9.15b	Salt content varies in the horizontal. Little or no vertical change in salt, and no strong density stratification. Weak net flow of water seaward at all depths. Salt moves toward river by mixing, not by flow. Also in constricted channels with strong turbulent tidal flow.	Chesapeake Bay, Delaware Bay, Humboldt Bay, California.
Partially-Mixed Estuary Deeper than shallow bays with both strong tidal flows and relatively high rates of freshwater discharge. Figure 9.15c	Salt content varies both vertically and horizontally. Moderate density stratification. Salt moves toward the river by strong net seawater flow at depth and some horizontal mixing. Vertical mixing and flow combine and form well-developed net seaward surface flow. Strong density stratification near surface when freshwater input high. Net flows assure good exchange of water with the sea.	San Francisco Bay, Puget Sound, Strait of Georgia, British Columbia.
Fjord-Type Estuary Long, narrow, deep, glacially carved channel, often with shallow entrance sill. Weak tidal currents except at the entrance. Figure 9.15d	Water at depth uniform in salt content, isolated from sea by sill, and exchanged very slowly. Surface layer very dilute with strong changes in salt content and density at sill depth. Strong stratification at sill depth acts as false bottom to seaward-moving surface layer. Net seaward flow strong in surface layer. Water at depth may stagnate.	Southeast Alaska, British Columbia, Norway, Greenland, New Zealand, Chile, and Argentina.

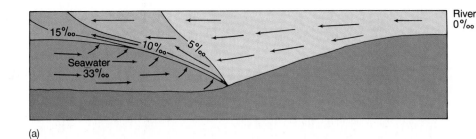

(a)

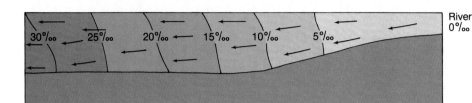

(b)

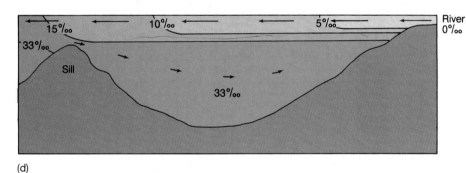

(c)

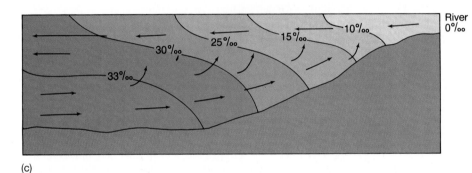

(d)

FIGURE 9.15
(a) The salt wedge estuary. The high flow rate of the river holds back the lesser flow of salt water. The salt water is drawn upward into the fast-moving river flow. Salinity is given in parts per thousand (‰).

(b) The well-mixed estuary. Strong tidal currents distribute and mix the seawater throughout the shallow estuary. The net flow is weak and seaward at all depths. Salinity is given in parts per thousand (‰).

(c) The partially-mixed estuary. Seawater enters below the mixed water that is flowing seaward at the surface. Seaward surface net flow is larger than river flow alone. Salinity is given in parts per thousand (‰).

(d) The fjord-type estuary. River water flows seaward over the surface of the deeper seawater and gains salt slowly. The deeper layers may become stagnant due to the slow inflow rate of salt water. Salinity is given in parts per thousand (‰).

9.9 AREAS OF HIGH EVAPORATION

Bays and seas located near latitudes 30°N and S have low precipitation and high evaporation rates. Although they may have rivers, these areas are not estuaries if they do not receive sufficient fresh water to become diluted. Instead, evaporation increases the surface water salinity and density, causing the surface water to sink and accumulate at depth, where it flows seaward to exit into the ocean. The ocean water is less dense than the outflowing high-salinity water, and ocean water flows into the sea or bay at its surface (fig. 9.16). Inflow and outflow from these seas and bays are the reverse of an estuary's circulation, and they are sometimes called "inverse estuaries." The Red Sea and the Mediterranean Sea are examples of such areas. These evaporative seas with their continuous overturn often

have a better exchange with the ocean than do true estuaries. Because density stratification is diminished, the entire water column circulates and exchanges.

Compare the circulation of a bay in a high-evaporation environment to a partially-mixed estuary.

9.10 WATER EXCHANGE

The time required for an estuary to exchange its water with the ocean is its **flushing time.** If the net circulation is rapid and the total volume of an estuary is small, flushing is rapid. A

Coasts, Estuaries, and Environmental Issues

FIGURE 9.16

The evaporative sea. Evaporation at the surface removes water, and river flow is negligible. Seawater flows inward at the surface, and the seaward flow is at depth.

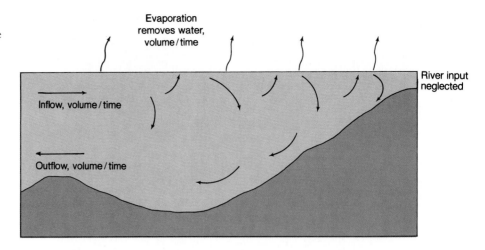

rapidly flushing estuary has a high carrying capacity for wastes, because the dissolved or suspended wastes are moved rapidly out to sea and diluted. A slowly flushing estuary risks accumulating wastes and building up high concentrations of land-derived pollutants. The flushing time of estuaries is determined by dividing their volume by the rate of net seaward flow. Understanding the circulation and flushing time of our estuary systems is essential to maintaining them as healthy, productive, and useful bodies of water.

Some bays and harbors that do not qualify as estuaries because they have no freshwater inflow may be flushed by tidal action. On each change of the tide, a volume of water equal to the area of the bay times the water level change between high and low waters exits and enters the bay. When this volume leaves the bay on the ebb tide, it may be carried down the coast by a prevailing coastal current. With the next tidal rise, an equivalent volume of different ocean water enters with the rising tide. Here flushing time is measured in the number of tidal cycles required to remove and replace the average volume of the bay. However, the flushing is often not complete because if coastal currents are weak, part of the ebbing water may recycle back into the bay on the next flood tide. In addition, the seawater entering the bay may not mix completely with the bay water on the rising tide and may return to the sea on the next ebb tide without displacing bay water.

Recycling may also occur in partially-mixed estuaries with two-layer flow. If tidal current turbulence is strong at the estuary entrance, some of the seaward moving surface layer may be mixed down into the deeper inflowing seawater. This action increases the flushing time by decreasing the net rate of seaward-moving water, and flushing time is determined by the portion of the seaward flow that does not recycle. This recycling of surface water also has the advantage of aerating the deeper parts of the estuary by mixing oxygen-rich surface water downward.

Define flushing time.

Explain the role of recycling in estuary circulation.

9.11 WETLANDS

The value of estuarine areas as centers of biological productivity and nursery areas for the coastal marine environment is well known to biologists. Freshwater and saltwater marshes and swamps, known as **wetlands,** border the estuaries and provide nutrients, food, shelter, and spawning areas for marine species including such commercially important organisms as crabs, shrimp, oysters, clams, and many kinds of fish.

In the past, coastal wetlands were routinely filled to produce usable ground for industry and port facilities, and channels were dredged to permit the entry of deep-draft vessels into ports and harbors (fig. 9.17). The San Francisco Bay estuary is considered the most modified major estuary in the United States. It has a surface area of 1240 square kilometers (480 mi²), and its river systems drain 40 percent of the surface area of California. The marshlands of the bay and its river deltas were diked, first to increase agricultural land and then for homes and industries. The conversion of wet land to dry land has reduced the tidal marshes from an area of about 2200 square kilometers (860 mi²) to 125 square kilometers (49 mi²). The increased cropland required increased irrigation, and the state built a series of dams, reservoirs, and canals. Today, 40 percent of the Sacramento and San Joaquin Rivers' flow is removed for irrigation, and another 24 percent is sent by aqueduct to central and southern California. By the year 2000, it is expected that San Francisco Bay will receive only 30 percent of its 1850 freshwater input. Proposals to prevent further degradation of the bay include restrictions on freshwater diversion to agriculture and limits on urban discharge and development.

Each year 100 to 150 square kilometers (40 to 60 mi²) of Mississippi River wetland marsh disappear under the Gulf of Mexico. The wetlands of the Louisiana coast are a great natural resource, 4 million acres, providing habitat for fish, shellfish, birds, and other animals that are important to the economy, biology, and history of the region. The delta of a river, such as the Mississippi, loses a portion of its sediments to erosion and dispersal by coastal waves and currents, but if the supply of sediments exceeds the rate of erosion, the delta grows and wetlands increase. If the delta does not receive its normal supply of sediment because of levees, dams, and other flood-control efforts, the delta land is lost rapidly.

FIGURE 9.17
The industrialized estuary of the Duwamish River at Seattle, Washington.

Industrial growth and diversification as well as increasing populations in coastal areas create enormous pressures for coastal wetlands. Although U.S. federal and state environmental legislation regulates the development and modification of many of these areas, over half of the coastal wetlands in the conterminous states have been destroyed, and the remainder continues under pressure for residential, resort, and marina use (fig. 9.18).

> Why are wetlands important?
>
> Why have the majority of U.S. wetlands been altered and eliminated?

9.12 ENVIRONMENTAL ISSUES

Historically, people have depended on access to salt water for trade and transport and access to fresh water for their living requirements; this has resulted in the establishment of major population centers along the shores of estuaries and coastal seas. In the United States today, over one-half of the population lives within 80 kilometers (50 mi) of the coasts (including the Great Lakes). This population, with its necessary industries, energy-generating facilities, and waste-treatment plants has created a tremendous burden on coastal zones.

In the United States, the 1960s and 1970s produced the environmental laws and agencies that presently monitor and control many practices in these oceanographically complex, biologically sensitive, and economically important areas. In the 1980s a policy based on environmental principles coincided with a series of environmental disasters, and the legacy of the 1990s is now being written. There appears to be a greater concern for acting with care and judgment in coastal areas, but as the economic burden of protecting the environment increases, these rules may be relaxed or changed. There is also a legacy from past practices, for in many cases the industries responsible for environmental problems no longer exist. In these cases the often difficult and always expensive cleanup becomes a burden to the taxpayers rather than to the perpetrators.

Beaches

People like beaches; their desire for property along this sensitive fringe of land is great. People change beaches, and one way in which they do so is by damming rivers to control floods and

FIGURE 9.18

The wetlands of Barnegat Bay, New Jersey, were replaced by a housing and recreational complex.

FIGURE 9.19

These groins have been placed along the beach at Hastings in southern England to trap sand moving along the coast. Beach height difference at this groin is 4 meters (12 ft).

generate power. The sand that once moved down these rivers to supply coastal beaches is now deposited in the lakes behind the dams. An important source of material to the continued balance of the beaches has been removed. Although the sediment supply to the beaches is reduced, the longshore current continues to carry away suspended beach material, and the beaches are eroded.

To protect harbors and coastal areas from the force of waves, people build breakwaters and jetties. Breakwaters are usually built parallel to the shore, while jetties extend seaward to protect or partially enclose a water area. The protected areas behind jetties and breakwaters are quiet; materials carried by the longshore current settle here, for they are no longer held in suspension by wave action. The result is less material available for a beach further down the coast. Along some coasts, people trying to maintain their beaches place groins of rock or timber perpendicular to the beach, figure 9.19. These groins trap sand moving in the longshore transport, and the result is a buildup of sand on the up-current side of the groins and loss of sand on the down-current side.

Harbors are often constructed without taking into consideration all the factors governing longshore transport. They fill with the sand that falls out in their quiet waters and require constant dredging to keep them open. Santa Barbara harbor, shown in figure 9.20 is one example. In this case the jetty extending seaward on the harbor's north side acts as a groin trapping sediment. When this area behind the groin is filled, the sediment sweeps southward along the jetty and settles out in the quiet water behind the jetty's south end until it is dredged and pumped back into the longshore current further down the coast.

In their efforts to maintain their homes and beaches, many people have constructed bulkheads or seawalls to prevent the loss of their property. These expensive structures are made of timbers, poured concrete, or large boulders. They armor the land in hopes of preventing erosion from storm waves, high tides and boat wakes. Many times these structures fail, and

FIGURE 9.20

Santa Barbara harbor as seen from the south. In the foreground, a dredge removes the sediment that has built up in the protected, quiet water behind the jetty. At the upper left, sand accumulates against the jetty and widens the beach.

often they make the erosion problem worse. If only a portion of the coastline is protected, wave energy may be concentrated at the ends of the seawall and the breaking waves move behind the wall and erode the shore. Because the seawall causes the waves to expend their energy over a very narrow portion of the beach, severe erosion of the beach materials at the foot of the seawall may occur, and the seawall may collapse. Seawalls may reflect waves that combine with incoming waves to cause higher more damaging waves at the seawall or at some other place along the shore.

Eroding beaches are often artificially maintained by supplying the beach with sand and gravel mined from upland areas or dredged from offshore bars. Such programs are expensive but may be necessary to compensate for the erosion of beach sediment. If these programs are to be successful, the particle sizes of the material must be carefully chosen to maintain the beach under its own particular wave-energy conditions. If too fine a material is used, it is eroded swiftly and transported to some other area where it piles up, producing unwanted shoaling or even smothering bottom-dwelling organisms.

The natural shore allows waves to expend their energy over a wide area, eroding the land but maintaining the beaches, all a part of nature's processes of give and take. Time and time again, coastal facilities have been built in response to the demands of people, only to trigger a chain of events that results in newly created problems of either erosion or deposition. Forty-three percent of the shoreline of our national and territorial areas, excluding Alaska, is now losing more sediment than it receives. Our beaches are eroding and disappearing into the sea.

Water Quality

Dumping solid waste and pouring liquid pollutants into estuaries and near-shore waters are now recognized as the wrong solutions to industrial and domestic waste problems. The costs of cleaning up and of building plants to process these wastes are high, requiring increased taxes and increased product prices. While indiscriminate industrial dumping of materials in estuaries and coastal waters has substantially halted in response to

Coasts, Estuaries, and Environmental Issues

recent regulations, pesticides such as DDT (dichloro-diphenyl-trichloroethane), long-lived toxic organic compounds such as polychlorinated biphenyls (PCBs), polycyclic aromatic hydrocarbons (PAHs), and heavy-metal ions such as lead, mercury, zinc, and chromium are still making their way through the environment, even though many of these materials are closely controlled or no longer manufactured. For example, U.S. manufacture, sale, and use of PCBs stopped in 1978 and 1979, but electrical transformers filled with PCBs are still being found, often leaking into the marine environment.

Surface runoff from agriculture carries pesticides and nutrients, which can poison or overfertilize the waters. By producing an excess of plant material, nutrients can be as destructive as pesticides, because the plants eventually die and decay, removing large quantities of oxygen. Lack of oxygen then kills other organisms, which in turn decay and continue to remove oxygen from the water.

Urban surface runoff via storm sewers adds a wide mix of materials, including silt, hydrocarbons from oil, lead from gasoline, residues from industry, pesticides and fertilizers from residential areas, and coliform bacteria from animal wastes. Even the chlorine added to drinking water and used to treat sewage effluent as a bactericide may form a complex with organic compounds in the water to produce chlorinated hydrocarbons that can be toxic to marine organisms.

Considering all the pathways, sources, and types of materials that can be classed as pollutants, their management and eventual exclusion from the coastal zone is an extremely difficult and complex problem. However, efforts to reduce the discharge of toxic materials into the marine environment are showing some signs of success. Improved sewage treatment decreases the discharge of nutrients and toxicants, and the decrease in use of leaded gasoline appears to be reducing the amount of lead being deposited in the marine environment. Lead carried by the Mississippi River to the Gulf of Mexico has been reduced about 40 percent in the last decade.

Other countries may not be able to match the U.S. efforts to reduce the discharge of toxic materials because of economic and political difficulties. Eighty-five percent of the pollutants flowing into the Mediterranean come from the land, and 80 percent of the sewage flowing into the Mediterranean is untreated. A plan to improve this situation was adopted in 1976, and by 1992 the number of contaminated swimming beaches dropped by 30 percent, but few nation participants have established adequate guidelines to meet the plan's requirements.

Toxic Sediments

Many toxicants reaching the estuaries do not remain in solution in the water but become adsorbed onto the small particles of clay and organic matter suspended in the water. These particles clump together, then settle out due to their increased size; this increases the concentration of toxins in the sediments. Once the sediments have been contaminated they become a source of continual leaching of toxicants to the water and to organisms living in contact with the bottom. Correcting this kind of contamination requires either the removal of the contaminated sediments, which must be stored in specialized landfills or, when water depth is sufficient, the covering of the contaminated sediments with a thick layer of clean material, isolating the contaminated sediments from both water and organisms. Both of these methods are very expensive and should not be considered unless the original source of the contamination has been removed.

Some of the particulate matter that adsorbs toxicants has a high organic content and forms a food source for marine organisms. In this way, heavy metals and organic toxicants find their way into the body tissues of organisms, where they may accumulate and be passed on to predators and human harvesters. Polycyclic aromatic hydrocarbons (PAHs) from combustion of fossil fuels and other organic materials are known to cause lesions in bottom fish living in contact with contaminated sediments. Shellfish concentrate heavy metals at levels many thousands of times higher than the levels found in the surrounding waters. Between 1953 and 1960 in Minamata, Japan, mercury from a local industrial source was released into a bay from which shellfish were harvested by a local village. The contaminated shellfish caused severe mercury poisoning and death among the villagers who ate the shellfish. The physical and mental degenerative effects of the poisoning were especially severe on children whose mothers had eaten the shellfish during pregnancy. The condition produced by this tragic episode has been named Minamata disease.

Since 1986 the Mussel-Watch program has used tissue analysis from mussel and oyster populations to monitor certain chemicals as indicators of human activity and environmental quality. More than two hundred U.S. coast and estuary sites are used in this ongoing effort. The study has shown more decreases than increases in chemical concentrations between 1986 and 1990 and appears to support the limits being set on discharges of chemical contaminants and the laws supporting these limits.

Plastics

It is estimated that every year more than 77 tons of plastic trash are routinely dumped or lost at sea by naval and merchant ships (fig. 9.21a). The National Academy of Sciences estimates that the commercial fishing industry yearly loses or discards about 135 million kilograms (298 million lbs) of fishing gear (nets, ropes, traps, and buoys) made mainly of plastic and dumps another 24 million kilograms (52 million lbs) of plastic packaging materials. Recreational vessels, passenger vessels, and oil and gas drilling platforms all add their share.

Nobody knows how long plastic remains in the marine environment, but an ordinary plastic six-pack ring could last 450 years. So-called biodegradable plastic available in the United States shows no evidence of solving plastic problems at sea. The light-absorbing molecules that break down the plastic after a few months' exposure to sunlight are unlikely to be very helpful in an environment that keeps the plastic cool and coated with a thin film of organisms that shade it from the light.

FIGURE 9.21
The impact of plastics on the marine environment: (a) Persistent litter includes plastic nets, floats, and containers. (b) A common murre entangled in a six-pack yoke. (c) A young gray seal caught in a trawl net. (d) A sea turtle with a partially ingested plastic bag. Photos (a), (b), and (c) are from Sable Island in the North Atlantic Ocean, approximately 240 kilometers (150 mi) east of Nova Scotia, Canada.

Thousands of marine animals are crippled and killed each year by plastics (fig. 9.21b,c,d). As many as 30,000 fur seals a year are estimated to become entangled in lost or discarded plastic fishing nets and choked to death in plastic cargo straps. Seabirds and marine mammals swallow plastics. Porpoises and whales have been suffocated by plastic bags and sheeting. Fish are trapped in discarded netting. Sea turtles eat plastic bags and die. Discarded plastics have become as great a source of mortality to marine organisms worldwide as oil spills, toxic wastes, and heavy metals.

The Marine Plastic Pollution Research and Control Act of 1987 is a U.S. law that prohibits dumping plastic debris everywhere in the oceans; other types of trash may be dumped at specific distances from shore. Ports and terminals are required to provide waste facilities for debris, and enforcement is delegated to the U.S. Coast Guard. There is no international enforcement. It may be that only people educated to act in a responsible manner will reduce the tide of plastics rising around the world.

Oil Spills

The transport of oil and oil products on the high seas and the production of oil from offshore wells create the potential for accidents that release large volumes of oil. The average discharge of oil into the world's oceans and navigable waters during each of the last ten years has been 117×10^4 metric tons (340×10^6 gal), not including the occasional megaspill of 3.8×10^4 metric tons (11×10^6 gal) or more. Spills far at sea are difficult to assess, because direct visual and economic impacts on coastal areas do not occur, and damage to marine life cannot be accurately evaluated. Spills due to the grounding of vessels or due to accidents during transfer or storage occur in coastal areas, where the environmental degradation and loss of marine life can be readily observed. Three oil spills have occurred in the last 25 years that have become ecological landmarks; these are the sinking of the *Amoco Cadiz*, the grounding of the *Exxon Valdez*, and the wartime discharge of crude oil into the Persian Gulf.

Coasts, Estuaries, and Environmental Issues

FIGURE 9.22

(a) The tanker *Amoco Cadiz* aground and broken in two off the coast of France, March 1978. (b) A bay along the French coast was fouled with oil. (c) Hot water under high pressure is used to clean Prince William Sound beaches after the 1989 *Exxon Valdez* oil spill. (d) Thick oil collects among the rocks on a beach in Prince William Sound. (e) Oil along the Persian Gulf coast after the 1991 Gulf War. (f) Cleaning up a 1993 oil spill along the beaches of Tampa Bay, one more time in 1994.

For example, in March 1978 the supertanker *Amoco Cadiz* lost its steering in the English Channel and broke up on the rocks of the Brittany coast of France (fig. 9.22a and b). Gale winds and high tides spread the oil over more than 300 kilometers (180 mi) of the French coast. Of the approximately 210,000 metric tons of oil spilled, it is considered that 30 percent evaporated and was carried in the air over the French countryside, 20 percent was cleaned from the beaches by the army and volunteers, another 20 percent soaked into the sand of the beaches to stay until winter storms washed it away, 10 percent (the lighter, more toxic fraction) dissolved in the seawater, and still another 20 percent sank to the sea floor in deeper water, where it will continue to contaminate the area for years to come.

Eleven years later and 40 kilometers (25 mi) out of Valdez, Alaska, the *Exxon Valdez*, loaded with 170,000 metric tons of crude oil, spilled 32,600 tons, or 10 million gallons, of oil into Prince William Sound (fig. 9.22c and d). Local contingency plans for cleanup were not prepared to handle a spill of this size; delays as well as lack of equipment and personnel, the rugged coastline, the weather, and the tidal currents of the enclosed area combined to intensify the problem. The oil spread quickly, distributing itself unevenly over more than 2300 square kilometers (900 mi^2) of water, and took a large toll of seabirds, marine mammals, fish, and other marine organisms. The oil moving out of Prince William Sound did not wash out to the open sea but was captured by the nearshore currents and repeatedly oiled the rocky wilderness beaches in the weeks that followed. In the cold subarctic conditions of Prince William Sound, the degradation of the oil by sunlight and microbes proceeds more slowly than in warmer regions. Plant and animal populations also recover more slowly in the cold water, where organisms tend to live longer and reproduce more slowly. Researchers inspecting the area three years after the disaster, in 1992, reported that the area was recovering and repopulating with organisms as the oil aged and degraded. They also noted that the areas most intensively cleaned to remove oil are recovering more slowly and with less-balanced populations than the areas in which the oil has been allowed to degrade naturally.

The average depth of the Persian Gulf is 40 meters (130 ft) and the circulation of its high-salinity water is sluggish. During the eight-year Iran-Iraq war the bombing of Persian Gulf oil facilities produced major spills, including one from an oil rig that poured out 172 metric tons (50×10^3 gal) per day for nearly three months. The Gulf endured an even greater catastrophe during the 1991 Gulf War as an estimated 0.86 to 1×10^6 metric tons (250–300×10^6 gal) of crude oil gushed into Gulf waters, the world's greatest oil spill to date. Some of this oil was deliberately released, some came from a refinery at a Gulf battle site, and bombing contributed additional quantities. Changes in the wind in the weeks after this spill stopped the oil from drifting the entire length of the Gulf, but some 570 kilometers (350 mi) of Saudi Arabia's shoreline was oiled (fig. 9.22e). International oil-spill experts rushed to help and were able to protect the water-intake pipes of desalination plants and refineries. About half of the oil evaporated and about 3×10^5 metric tons have been recovered. Some shore areas were bulldozed clean; others became test plots to assess natural recovery. Much has sunk to the bottom of the Gulf, and it is expected to take decades for the damaged areas to recover.

The tragedy of these spills continues to demonstrate that there is no adequate technology to cope with large oil spills, particularly under difficult weather and sea conditions along rugged, remote coastlines. On the average, no more than 15 percent of the oil spilled under these conditions has been recovered from any oil spill, and this figure may be an inflated number, because the recovered oil mixture has a high water content.

The technology for oil cleanup at sea includes oil booms and oil skimmers (fig. 9.23). These devices are useful in confining and recovering small spills in quiet protected waters, but are not very effective in open and exposed areas. The onshore cleanup is often as destructive as the spill, especially in wilderness situations where the numbers of people, their equipment, and their wastes further burden the environment. The toxicity of weathered crude oil is low, and many oil-spill experts believe that cleanup efforts should be concerned not with removing oil from the beaches but with moving oil seaward to prevent it

FIGURE 9.23

A catamaran oil skimmer. The vessel cruises at about 3 knots, guiding surface oil between the twin hulls. The oil adheres to a moving belt and is lifted into onboard storage tanks.

Coasts, Estuaries, and Environmental Issues

from reaching the beaches. They believe that many of the cleanup methods used along shorelines cause more immediate and long-term ecological damage than leaving the oil to degrade naturally. Habitats treated with hot water take longer to recover than those left untreated; backhoeing and high-pressure washing (100 lbs/in^2) destabilize gravel and sand beaches, killing live animals and working the oil into the sediments; see figure 9.22c. To prevent oil from reaching the beaches, some cleanup experts support newly developed, low-toxicity dispersants; the dispersed oil moves into deeper water where it is diluted and its toxicity lessened.

More data are needed to assess the effectiveness of attempts to speed up natural degradation by adding nutrients to the beaches in order to increase populations of biodegrading microorganisms. Releasing other strains of bacteria has been tested with limited success; the new organisms may not compete effectively with natural populations.

Once oil is removed from a beach, the first cleaning operation may not be the last. On the east coast of Florida the beaches of Tampa Bay are lined with resorts and condominiums, and the economy depends heavily on these beaches for tourism. A freighter and two oil barges collided here in August of 1993. One barge carried diesel fuel and gasoline that caught fire and burned. The second barge carried heavy fuel oil; this barge sank and released the fuel oil to mix with the sediments on the bottom of the bay. Each time strong winds stir the bay, contaminated sediments are moved onshore to create a new oil contamination problem. Figure 9.22f shows the end of a costly, January 1994, cleanup episode along St. Petersburg beach.

The immediate damage from a large spill is obvious and dramatic; by contrast, the effects of the small but continuous additions of oil that occur in every port and harbor are much more difficult to assess, because they produce a chronic condition from which the environment has no chance to recover. Refined products such as gasoline and diesel fuel are more toxic to marine life than crude oil, but they evaporate rapidly and disperse quickly. Crude oil is slowly broken down by the action of water, sunlight, and bacteria, but the portion that settles on the sea floor moves down into the sediments, where it continues to contaminate for years.

What activities contribute to the degradation of the marine coastal environment?

What is the effect of each structure when built along a beach: (a) breakwater, (b) jetty, (c) seawall?

What are the sources and pathways for toxicants entering the coastal zone?

Why do coastal sediments have higher toxicant concentrations than the overlying water?

Why are plastics so harmful to marine animals?

Discuss the international effort to reduce plastics in the marine environment. How can it help the situation? How may it not help it?

What happens to spilled oil in the marine environment?

Discuss the technology used for oil-spill cleanup. Why is it not fully effective?

How does constant addition of small amounts of oil compare to the effects of a single large oil spill?

SUMMARY

The coast is the land area that is affected by the ocean; the coastal zone includes coastal bays, estuaries, and inlets. Primary coasts are formed by nonmarine processes and include drowned river valleys, fjords, deltaic coasts, glacial moraines, dune coasts, lava and cratered coasts, and fault coasts. Secondary coasts are modified by ocean processes and include regular and irregular cliffed coasts, barrier coasts, beach plains, mudflats, coral reefs, mangrove, and marsh grass coasts.

Typical beach features include an offshore trough and bar and a beach area with a low tide terrace, beach face, beach scarp, and berms. Winter and summer berms are produced by seasonal changes in wave action. Beaches are described by their shape and the size, color, and composition of beach material.

Beaches exist in a dynamic equilibrium, where supply balances removal of beach material. Gentle summer waves move the sand toward the shore; high-energy winter waves scoop the sand off the beach and deposit it in an offshore sandbar. The breaking waves of the surf zone produce a longshore current that moves the sediment down the shore in longshore transport. Coastal circulation zones are defined by the path followed by beach sediments from source to region of deposit. Rip currents are narrow, fast seaward movements of water and sediment through the surf zone. Rising sea level along low-lying coasts may move the coast a significant distance inland in the next fifty years. Barrier islands are particularly susceptible to coastal flooding and storm damage.

Estuaries are semi-isolated portions of the ocean that are diluted by fresh water. There are four basic estuary types based on circulation patterns and salt distribution. In bays and seas with high evaporation rates, there is continuous overturn, and water at depth flows seaward while ocean water enters at the surface.

Estuaries that flush rapidly have a greater capacity for dissipating wastes than those that flush slowly. In bays and harbors that are flushed only by tidal action, recycling of water and wastes can occur. Partial recycling is also possible in estuaries with two-layered flow.

Wetlands border coasts and estuaries; they are important areas that often have been destroyed by filling, dredging, and development. Artificial structures change beaches; they are often responsible for unwanted beach erosion. Human interference with the natural processes of the coastal zone also results in harbors filled with sand, like that of Santa Barbara, California.

Coastal and estuary water quality is affected by dumping solid waste and liquid pollutants. Pesticides and long-lived toxicants move through the environment; runoff from agricultural and urban areas does much damage. Toxicants become concentrated in the coastal sediment, and organisms concentrate toxicants and pass them along the marine food chains. Plastic trash is an increasing problem to marine animal life. Oil spills are a special problem in inshore waters, for which there is no adequate cleanup technology.

KEY TERMS

coast	offshore	bar	rip current
coastal zone	berm	spit	coastal circulation cell
emergent coast	berm crest	tombolo	estuary
submergent coast	scarp	barrier islands	salt wedge
beach	low tide terrace	onshore current	fjord
backshore	beach face	longshore current	flushing time
foreshore	trough	longshore transport	wetlands

SUGGESTED READINGS

Atlas, R. M. 1993. Bacteria and Bioremediation of Marine Oil Spills. *Oceanus* 36 (2):71.

Boicourt, W. C. 1993. Estuaries, Where the River Meets the Sea. *Oceanus* 36 (2):29–37.

Canby, T. Y. 1991. After the Storm. *National Geographic* 180 (2):2–32. Oil spill cleanup in the Persian Gulf.

Capuzzo, J. E. M. 1990. Effects of Wastes on the Ocean: The Coastal Example. *Oceanus* 33 (2):39–44.

Hass, P. M., and J. Zuckmann. 1990. The Med Is Cleaner. *Oceanus* 33 (1):38–42.

Hawley, T. M. 1990. Herculean Labors to Clean Wastewater. *Oceanus* 33 (2):72–75. Technology of sewage treatment.

Hodgson, B. 1990. Alaska's Big Spill. *National Geographic* 177 (1):5–43. The *Exxon Valdez* oil spill.

Holloway, M. 1991. Soiled Shores. *Scientific American* 265 (4):102–16. Cleaning up oil spills.

Horgan, J. 1991. Muddled Cleanup in the Persian Gulf. *Scientific American* 265 (4):99–110.

Horton, T. 1993. Chesapeake Bay. *National Geographic*. 183 (6):2–35.

Kusler, J. A., W. J. Mitsch, and J. S. Larson. 1994. Wetlands. *Scientific American*. 270 (1):64–71.

Land, T. 1990. European Scientists Cooperate to Save the North Sea. *Sea Frontiers* 36 (1):52–55.

Lowenstein, F. 1985. Beaches or Bedrooms—The Choice as Sea Level Rises. *Oceanus* 28 (3):20–29.

Milliman, J. 1989. Sea Levels: Past, Present and Future. *Oceanus* 32 (2):40–43.

NOAA. 1990. *Coastal Environmental Quality in the United States, 1990*. National Oceanic and Atmospheric Administration, Rockville, Maryland. 34 pp.

Parker, P. A. 1990. Clearing the Oceans of Plastics. *Sea Frontiers* 36 (2):18–27.

Pilkey, O. H. 1990. Barrier Islands. *Sea Frontiers* 36 (6):30–36.

Wacker, R. 1991. The Bay Killers. *Sea Frontiers* 37 (6):44–51. Environmental degradation's effect on fisheries.

Wanless, H. R. 1989. The Inundation of Our Coastlines. *Sea Frontiers* 35 (5):264–71. Emphasis on South Florida.

Weber, P. 1994. It Comes Down to the Coasts. *World Watch* 7 (2):20–29.

Wilber, R. J. 1987. Plastic in the North Atlantic. *Oceanus* 30 (3):61–68.

Zedler, J. B., and A. N. Powell. 1993. Managing Coastal Wetlands. *Oceanus* 36 (2):19–28.

Biological Invaders

When settlers and traders came to the North American continent more than three and a half centuries ago they brought with them many organisms, plant and animal, intentional and unintentional. Among the unintentional were large communities of widely varied organisms that attached to and bored into the wooden hulls of their ships in the harbor or bay where their journeys began. Many of the organisms transported across the oceans in this way would have been swept from the ships' sides by the waves and currents, but a few invaders would have survived when the ships anchored at their journey's ends. Hundreds of years later there is no way of knowing when the first European barnacles or periwinkle snails began to colonize the east coasts of the United States and Canada. Today these two organisms are considered typical of this region.

The Ballast Water Problem

Today's ships do not carry such communities, for their steel hulls are protected by antifouling paints, and they move through the water at speeds sufficient to sweep away many of the organisms that wooden hulls carried with them. However, today's ships do carry ballast water that is loaded and unloaded to preserve the stability of the ship as it unloads and loads cargo. Tens of thousands of vessels with ballast tanks ranging in capacity from hundreds to thousands of gallons of water move across our oceans. These vessels rapidly transport this ballast water and its population of organisms across natural oceanic barriers; the release of ballast water and organisms may occur days or weeks later, thousands of kilometers from their point of origin.

Biologists sampled ballast water from 159 cargo ships from various Japanese ports that docked in Coos Bay, Oregon.* The ballast water contained members of all the major groups of floating or planktonic organisms: copepods, marine worms, barnacles and barnacle larvae, flatworms, jellyfish, and shellfish. The copepod density was estimated to be greater than 1500 per cubic meter; the larvae of marine worms, barnacles, and shellfish were greater than 200 organisms per cubic meter.

Some Examples

Whether organisms arrive in ballast water, in the packing of commercially harvested fish and shellfish, attached to the floats of seaplanes, or are released from home aquariums, they all leave behind the natural controls of predators and disease found in their native environments. If these invaders are introduced into a hospitable new environment, they may flourish and severely disrupt its biological relationships, forcing out some species and destroying others. The results may be ecological catastrophes.

In 1985 an Asian clam (*Potamocorbula amurensis*) (fig. 1) was discovered in a northern arm of San Francisco Bay. This clam was previously unknown in the area and

FIGURE 1
Asian clam (*Potamocorbula amurensis*).

* Source: J. T. Carlton and J. B. Geller, "Ecological Roulette: The Global Transport of Nonindigenous Marine Organisms" in *Science* 261(5117): 78–82,1993.

FIGURE 2
Green crab (*Carcinus maenas*).

FIGURE 3
Colonies of *Spartina* proliferate over the mudflats of Willapa Bay, Washington.

appears to have arrived in its juvenile or larval form in the ballast water of a cargo ship from China. During the next six years the clam spread southward into the bay and formed dense colonies, as many as 10,000 clams per square meter. The Asian clam feeds on microscopic plants and the larvae of crustaceans, and the food requirements of these huge populations reduce amounts available for native species, stressing the system's species balance and food chains.

In 1990, a second intruder, the European green crab (*Carcinus maenas*) (fig. 2), was recognized in the southern part of San Francisco Bay. This crab moved from Europe to the East Coast of the United States in the 1820s and to Australia in the 1950s. It may have hitchhiked to the bay in ballast water or perhaps in seaweed wrapped around Maine lobsters. This crab is less than 6.5 centimeters broad (3 in), voracious and belligerent. It feeds on clams and mussels and may spawn several times from a single mating. It has populated most of the bay in the last two to three years and feeds enthusiastically on the Asian clam. The green crab may clean the bay of the Asian clam, but it can also feed on native species and outcompete them for food.

In 1982 the ballast water of a ship from America carried a jellyfish-like organism known as a comb jelly (*Mnemiopsis leidyi*) into the Black Sea. From the Black Sea it spread to the Azov Sea and has recently moved into the Mediterranean. It has no predator in these areas and devours huge quantities of animal plankton, small crustaceans, fish eggs, and fish larvae. Russian scientists claim this small organism now dominates the Black Sea, devastating the fish populations and catches of the last five years. The cost to the Black Sea fisheries is an estimated $250 million and the fisheries of the Azov Sea shut down after catches dropped an estimated 200,000 tons.

Ballast water carrying the resting cells of Japanese red-tide-causing organisms was responsible for shutting down the natural and aquaculture shellfish harvests of Tasmania and southern Australia in the 1980s. Fish can also be transported this way; for example, Japanese sea bass were introduced into the Sydney harbor region in 1982–83.

Cordgrass (*Spartina alterniflora*) (fig. 3) is a deciduous, perennial, flowering plant native to the highly productive salt marshes of the U.S. Atlantic and Gulf coasts. It is a very

continued . . .

Coasts, Estuaries, and Environmental Issues

competitive plant spreading by underground stems, seeds, and colonies that form when pieces of the root system or whole plants float into an area and take root. Once established it begins to trap sediment, eventually producing a high marsh system. In its native estuaries cordgrass prevents erosion and marshland deterioration and is used for wetland restoration projects. It is valued as a stabilizer, sediment trap, and nursery area for fish and shellfish.

During the last two centuries cordgrass has been introduced, intentionally and unintentionally, along the coasts of England, France, New Zealand, China, and the western United States. Free of its native insect predators, it has spread slowly and steadily, crowding out the native plants and trapping sediment. It turns tidal mudflats into high marshes inhospitable to the many fish and waterfowl that depend on the mudflats. It hampers shellfish harvesting and interferes with recreational use of beaches and waterfronts. Efforts to control cordgrass outside its natural environment have included burning, flooding, shading plants with black plastic, smothering plants with mud or clay, applying herbicides, and mowing repeatedly. Little success has been reported; mowing small accessible patches and biological controls for long-term regulation may be the most realistic approach.

Solutions

In the case of ballast water, the introduction of zebra mussels into the Great Lakes and their subsequent invasion of the rivers and streams of the central United States sounded the alarm for the freshwater environment. In 1990 the Nonindigenous Aquatic Nuisance Prevention and Control Act (NANPCA) was enacted in the United States. Under this law the United States adopted voluntary regulations for exchanging ballast water on the high seas for vessels bound for the Great Lakes; this provision became law in 1993. Congress included major studies to determine the extent of ballast management to be required along U.S. coasts: should ballast control be extended to all U.S. coastlines, and to what extent do nonindigenous species threaten the environment and economy of U.S. waters? Exchange of ballast water at sea can be attempted if there is no danger to ship or crew, but not all the organisms may be flushed out, and the sediment in the bottom of the ballast tanks may not be moved. Proposals have included treating the ballast water by adding poisonous chemicals, heating the water, filtering the water, and exposing it to ultraviolet radiation. However, none of the world's cargo vessels are designed for ballast management, and all these proposals require some redesign or refit of vessels.

Controlling ballast water will not close all doors to invading marine species, and it will be costly, but it will lead to fewer foreign invasions. However, it is well to keep in mind that according to James T. Carlton of the Maritime Studies Program at Williams College, "No introduced marine organism, once established, has ever been successfully removed or contained."

Readings

Carlton, J. T. 1993. Exotic Invaders. *The World & I* 8 (12):220–25.

Hedgpeth, J. W. 1993. Foreign Invaders. *Science* 261 (5117):34–35.

Travis, J. 1993. Invader Threatens Black, Azov Seas. *Science* 262 (5138):1366–67.

10

Productivity of the Oceans

Outline

Learning Objectives

After reading this chapter, you should be able to

- Relate photosynthesis and respiration to primary production.

- Discuss the factors that control primary production and explain its significance to all forms of marine life.

- Explain how primary production varies over the world's ocean regions.

- Trace the cycling of nitrogen and phosphorous in the marine environment.

- Relate flows of energy and nutrients through the levels of a trophic pyramid to food chains and food webs.

- Understand chemosynthesis and give examples of marine chemosynthetic communities.

- Discuss the efficiency of harvesting at various trophic levels.

◀ Diatoms, unicellular plants of the sea.

In all the environments of all the oceans, organisms prey upon each other in the series of prey-predator relationships called food chains. Understanding these chains, from the minute, microscopic organisms with which they start to the more familiar large carnivores such as tuna, salmon, and seals with which they end, requires a knowledge of the ocean's productivity, or the rate at which organic material is produced in the sea. The ocean, like the land, requires plant life to harness the energy of the sun. However, global ocean productivity is based not on the large seaweeds and sea grasses found along the shore but on the microscopic members of the plant kingdom floating at the surface of all the world's oceans. To understand the productivity of the oceans one must understand the variations in the abundance of these tiny plant cells.

10.1 PRIMARY PRODUCTION

Plants and animals that float or drift with the movements of the water are the **plankton.** Plant plankton are called **phytoplankton;** they are mainly unicellular (single-celled) plants. Like all plants, the phytoplankton require sunlight, nutrients or fertilizers, carbon dioxide gas, and water for growth. Phytoplankton cells contain the pigment **chlorophyll** that traps the sun's energy for use in **photosynthesis.** The photosynthetic process converts the carbon dioxide and water to sugars or high-energy organic compounds from which the cell forms new plant material. The production of new plant material from photosynthesis and nutrients by using the sun's energy is termed **primary production.** The total amount, or mass, of organic material produced by photosynthesis per volume or surface area per time is the gross primary production.

Photosynthesis is represented by the equation:

$$6\,CO_2 + 6\,H_2O \xrightarrow{\text{sunlight/chlorophyll}} C_6H_{12}O_6 + 6\,O_2$$

$$6 + 6 \xrightarrow{\text{sunlight/chlorophyll}} 1 + 6$$

| molecules | molecules | | molecule | molecules |
| carbon dioxide | water | | sugar | oxygen |

The sugars produced by photosynthesis are broken down by the plant cells with the addition of oxygen to yield energy, carbon dioxide, and water in the process known as **respiration.** Respiration provides both plants and animals with energy for their life processes.

$$C_6H_{12}O_6 + 6\,O_2 \rightarrow 6\,CO_2 + 6\,H_2O + \text{life-support energy}$$

$$1 + 6 \rightarrow 6 + 6 + \text{life-support energy}$$

| molecule | molecules | | molecules | molecules |
| sugar | oxygen | | carbon dioxide | water |

After deducting the part of gross primary production that is broken down by respiration to yield energy for life processes, the remaining gain in plant mass per volume or surface area per time is the net primary production. The net primary production is available to the food chains for consumption by animals and decomposition by bacteria.

Net and gross primary production are usually reported for a unit volume of water, or the volume of water under a fixed area of the sea surface, and over a given period of time. The organic matter produced can be expressed as the number of organisms or as their weight, known as **biomass.** Primary production is most often given as the mass or dry weight of organic carbon in grams, produced under a square meter of sea surface over a time, or g C/m^2/time; it is a rate of change of biomass.

The total plant biomass under any area of sea surface at any instant in time is the **standing crop.** It is the result of growth, reproduction, death, and grazing (fig. 10.1). When a population of herbivores is grazing the phytoplankton population at the same rate as the net primary production increases it, a nearly constant standing crop results. See figure 10.7 for a satellite view of the global phytoplankton standing crop.

Because of their size and abundance in coastal areas, seaweeds are the most conspicuous plants of the oceans. However, they represent only 5 to 10 percent of the total plant material produced by the oceans. Seaweeds are discussed in more detail in Chapter 12.

Distinguish between photosynthesis and respiration.

Relate primary production and photosynthesis.

Distinguish between gross primary production and net primary production.

Define standing crop and relate it to primary production.

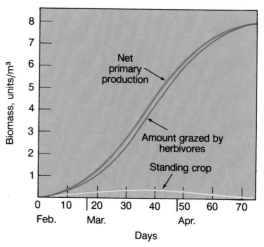

FIGURE 10.1

Net cumulative primary production is balanced by the grazing of herbivores. Both populations increase during the spring; the standing crop remains nearly constant throughout the year.

10.2 LIGHT AND PRIMARY PRODUCTION

Phytoplankton growth is controlled in part by the light at the ocean surface. At polar latitudes, the summer days of nearly continuous but low-intensity daylight and the little sunlight available during the rest of the year result in a short period of rapid growth during the midsummer (fig. 10.2a).

At the middle latitudes, the sunlight's intensity and duration vary with the seasons. The spring increase in solar radiation warms the surface and increases the density stability of the water column, helping the phytoplankton to remain at the surface. Figures 10.2b and 10.3 show an initial increase in phytoplankton biomass with the lengthening of daylight hours in the spring. A second increase in the population occurs during the late summer.

Compare figures 10.2a and 10.2b with figure 10.2c, showing the lower level but extended growth of phytoplankton in the tropics. The population increase associated with the seasonal change in tropic-latitude sunlight is slight; the year-round sunlight provides abundant high-intensity solar energy at all times.

The phytoplankton must remain in the ocean's surface layers to take advantage of the available sunlight. The warm surface layer of low-density water in the tropics exists all year long, and the phytoplankton are not displaced easily into the underlying more dense water. At midlatitudes a low-density, warm surface layer is found only in summer; if the warm weather is late or broken by cool periods, both the formation of the low-density layer and the active growth of the phytoplankton are delayed. During the winter, surface cooling and storms mix the surface layers into deeper water. Refer to figure 7.4. At polar latitudes the growing season is short; the low level of sunlight warms the sea surface to form only a weakly stable density layer. The addition of low-salinity water from melting sea ice helps to maintain a stable water column at high latitudes.

If light were the sole factor influencing plant growth, the level of phytoplankton growth in the tropics would be expected to maintain year-round levels similar to the summer values shown for the polar and middle latitudes. But nutrients as well as light are required if the phytoplankton population is to increase.

How does light control primary production in polar seas and in equatorial seas? Relate the phytoplankton biomass to sunlight in each of the four seasons at midlatitudes.

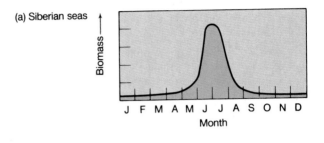

(a) Siberian seas

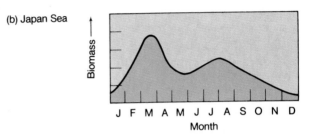

(b) Japan Sea

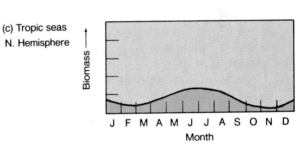

(c) Tropic seas
N. Hemisphere

FIGURE 10.2

Phytoplankton growth cycles vary with the light available at different latitudes. (a) At high latitudes, growth is limited to a brief period in midsummer. (b) Seasonal light changes at the middle latitudes increase growth in early spring and continue it through the summer. (c) Sunlight levels vary little at tropic latitudes, where phytoplankton growth is nearly uniform throughout the year.

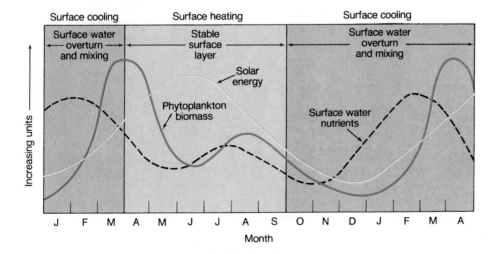

FIGURE 10.3

Phytoplankton biomass, nutrient supply, and surface water stability respond to solar energy changes at the middle latitudes in the Northern Hemisphere.

Productivity of the Oceans

10.3 NUTRIENTS AND PRIMARY PRODUCTION

At polar latitudes, winter overturn resupplies the surface waters with nutrients. The growing season is short, and the nutrients are more than sufficient. At these latitudes the availability of light controls phytoplankton growth.

In middle latitudes, the surface nutrients are replaced each year by winter storms and the cooling of the surface water to produce a shallow overturn. Nutrients are available to the phytoplankton cells when the increase in spring sunlight warms the surface and decreases the surface water density. Together the sunlight and nutrient supply trigger the first period of phytoplankton growth. Grazing organisms decrease the phytoplankton and release nutrients for the second phytoplankton increase during the late summer. The phytoplankton bloom continues, although controlled by the rate of nutrient resupply, and the population declines with the reduction in light levels as winter begins. Note the interaction between the sunlight, the stability of the surface water, the nutrients in the surface water, and the phytoplankton biomass in figure 10.3. In the tropics, lack of surface mixing and overturn mean low levels of nutrients and a low plant biomass despite the constant high level of solar energy.

When a plant or an animal dies, organisms known as **decomposers** (bacteria and fungi) release the energy within the organism's tissues to the environment and break down the organism's complex organic molecules to basic molecules, such as carbon dioxide, nutrients, and water. Since no significant amount of new matter comes to the earth from space, living systems must recycle inorganic molecules to form the organic compounds required for their systems and life processes.

Nitrogen is essential in the formation of proteins, and phosphorus is required in energy reactions, cell membranes, and nucleic acids. Nitrogen in the form of nitrates, and phosphorous in the form of phosphates are two soluble nutrients that are removed from the water by the phytoplankton as their populations grow and reproduce. Nitrogen and phosphorus are cycled into the animal populations as the animals feed on the plants and are returned to the water as organisms die and decay. Excretory products from the animals are also added to the seawater, broken down, and used again by a new generation of plants and animals. The cyclic pathways for nitrogen and phosphorus are illustrated in figures 10.4 and 10.5.

Nitrogen gas is useless to plants; it must be converted to nitrate to be incorporated into plant tissue. This is done by a few species of microorganisms that convert nitrogen gas to ammonia, other species that convert ammonia to nitrite, and still other species that change nitrite to nitrate, the form most easily absorbed by the plants.

It is possible to follow the cyclic path of any molecule required for life. Water is a requirement for life; it is part of all life processes and is cycled over and over through life-forms, time, and space; refer to the hydrologic cycle in Chapter 2.

What role does the nutrient supply play in the primary productivity of polar, midlatitude, and equatorial oceans?

Explain the concept of nutrient cycles.

What part do decomposers play in nutrient cycles and primary productivity levels?

FIGURE 10.4

The nitrogen cycle. Although nitrogen makes up 78 percent of the earth's atmosphere, only a few microorganisms are able to change nitrogen gas to the nitrate that is used by land and sea plants, which are in turn eaten and recycled by animals.

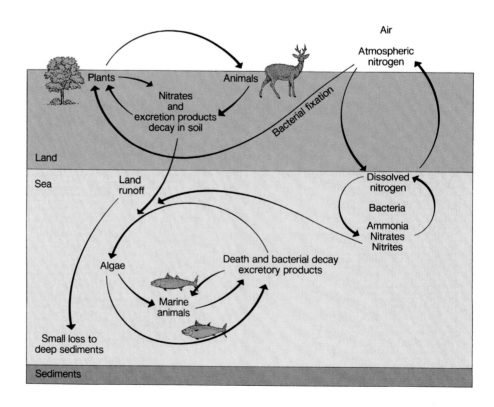

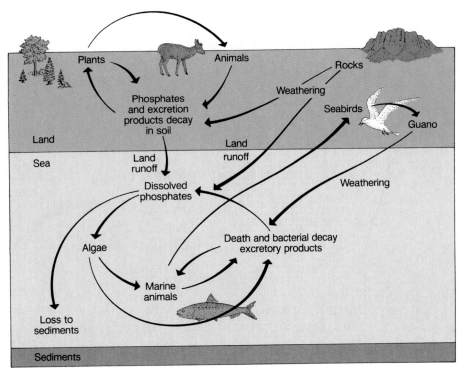

FIGURE 10.5

The phosphorous cycle. Dissolved phosphorus is carried to the sea by runoff from the land, where it is taken up by plants and recycled through animals until it is released from dead tissue by bacterial action.

10.4 MEASURING PRIMARY PRODUCTION

To determine the amount of plant material present in a water sample one can filter out the phytoplankton, count the cells, and multiply by the average mass per individual cell. Other less direct but less tedious methods are also available. Biomass may be estimated by extracting the chlorophyll from a sample and determining the concentration of pigment present. Another method exposes the chlorophyll in the phytoplankton cells to certain wavelengths of light that cause the chlorophyll pigment to fluoresce. The intensity of the fluorescence gives a measure of the chlorophyll, and therefore the phytoplankton biomass, present in a given volume of water.

A simple technique used extensively during the first part of this century collects seawater samples with their natural phytoplankton population at selected depths and light levels. By measuring the amount of oxygen present at the time of collection and then remeasuring it after several hours, during which subsamples are incubated at the light of the sampled depth and in the dark, it is possible to calculate gross primary production, net primary production, and respiration. Similar experiments add radioactive isotope carbon-14 to each sample and, after a period of time, the phytoplankton are filtered from the water and the amount of carbon-14 incorporated into the plant cells is measured to determine primary production.

Direct measurement of primary production is expensive and limited to the sampling areas of oceanographic and fisheries research vessels. The map in figure 10.6 was compiled from many years of such sampling; however satellites have been recording sea-surface chlorophyll measurements for more than fifteen years. Figures 10.7a–d are images of global distribution of chlorophyll concentrations or phytoplankton biomass prepared from satellite data collected between 1978 and 1986. Each image shows chlorophyll concentration averaged over three months. Changes in chlorophyll concentration from one image to the next, document standing-crop changes which are governed by the combined effects of seasonal cycles of light intensity, depth of light penetration, nutrients, water temperature, currents, grazing by herbivores, and population die-off. In these satellite images phytoplankton abundance is shown to be low in the central ocean areas (magenta to deep blue) and higher along coastal and upwelling areas (yellow, orange, and red); areas of moderate phytoplankton population are green-blue. Black ocean areas indicate insufficient data.

10.5 GLOBAL PRIMARY PRODUCTION

Figures 10.6 and 10.7 show the global geographic distribution of primary production and biomass. Coastal areas are generally more productive than open ocean, because along the coasts and in the estuaries rivers and land runoff supply nutrients. Also, currents cause mixing and turbulence that supply nutrients from shallow depths to surface layers. The fresh water helps to create a stable, low-density surface layer, which keeps the plant cells up where sunlight is plentiful.

Productivity of the Oceans

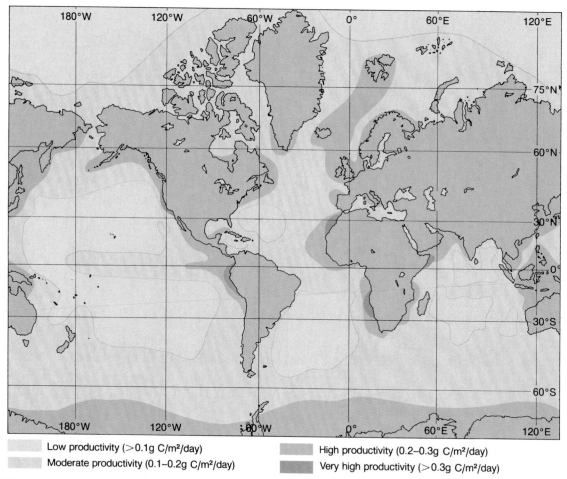

Low productivity (>0.1g C/m²/day)

Moderate productivity (0.1–0.2g C/m²/day)

High productivity (0.2–0.3g C/m²/day)

Very high productivity (>0.3g C/m²/day)

FIGURE 10.6

The distribution of primary production in the world's oceans.

FIGURE 10.7

Satellite observations of global chlorophyll concentration. (a) January, February, March. (b) April, May, June. (c) July, August, September. (d) October, November, December. Color is equivalent to milligrams of chlorophyll pigment per cubic meter of surface water; see color bar in (a).

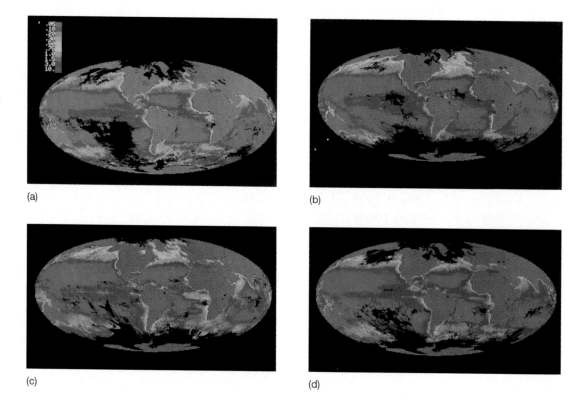

(a)

(b)

(c)

(d)

Narrow areas of very high productivity are found against the west coasts of North and South America, the west coast of Africa, and along the west side of the Indian Ocean. These are major upwelling zones on the sheltered sides of landmasses in the trade wind belts and in areas of abundant sunlight. These are also the areas of the world's greatest fish catches. In these nutrient-rich zones, the populations of phytoplankton form the first step in the feeding relationships that have produced great schools of commercially valuable fish.

The equatorial Pacific demonstrates the influence of upwelling on primary production caused by the equatorial divergence, as does the surface divergence surrounding Antarctica. In contrast, the downwelling centers of the large ocean gyres, where surface convergences occur, are areas of low nutrient availability that depresses primary production. Return to Chapter 7, section 7.3 and figure 7.6. Compare areas of surface divergence and convergence, upwelling and downwelling, with the distribution of primary production in figure 10.6.

On the average, upwelling areas are three times more productive than coastal areas and six times more productive than the open ocean for the same units of area and time. This relationship is shown in table 10.1. Notice that there is about one hundred times more coastal area than upwelling area, and nearly ten times more open-ocean area than coastal area. As a result, most of the organic carbon plant-compounds are made by the oceans' plants is produced at low rates over the large areas of the open sea. To compare primary production on land with that at sea, study table

10.2. Primary production in the open sea is about the same as that of the deserts on land, while upwelling areas are comparable to pastureland and lush forests, and certain estuaries approach the most heavily cultivated land. However, this generally accepted concept of low productivity in ocean areas away from coasts and upwelling zones may be revised. New techniques provide data that show two to three times as much organic matter per surface area of open ocean than has been reported in the past.

Shallow coastal areas and estuaries in which the sea bottom is exposed to sunlight support masses of attached seaweeds and single-celled plants. These provide a large amount of organic material to the animal population. Primary production occurs at all depths in shallow estuaries, and organic material is also produced in the bordering marshlands. The result is an extraordinary rate of primary productivity, supporting the animals within the estuary and also contributing organic matter that is exported to coastal waters.

Where are the areas of highest and lowest marine primary production? What are their equivalents in land vegetation?

What phenomena increase the primary production of an area and what phenomena decrease it?

Why is the total primary production of the open ocean higher than that of upwellings and coastal areas?

Table 10.1 World Ocean Primary Production

Area	Primary Production (g C/m²/yr)	World Ocean Area (km²)	World Ocean Area (%)	Total Primary Production (metric tons of carbon/year)
Upwellings	300	0.36×10^6	0.1	0.1×10^9
Coasts	100	36×10^6	9.9	3.6×10^9
Open oceans	50	325×10^6	90.0	16.3×10^9

Source: Data from J. H. Rhyther, "Photosynthesis and Fish Production in the Sea" in *Science* 166 (3901):72–76, 1969.

Table 10.2 Gross Primary Production: Land and Ocean

Ocean Area	Amount (g C/m²/yr)	Land Area	Amount (g C/m²/yr)
Open ocean	50	Deserts Grasslands	50
Coastal ocean	25–150	Forests Common crops Pastures	25–150
Upwelling zones deep estuaries	150–500	Rain forests Moist crops Intensive agriculture	150–500
Shallow estuaries	500–1250	(Sugarcane and sorghum)	500–1250

Source: After E. P. Odum, *Fundamentals of Ecology,* 3d edition, Saunders, Philadelphia, 1971.

Productivity of the Oceans

10.6 FOOD CHAINS, FOOD WEBS, AND TROPHIC LEVELS

Primary production forms the first link in the **food chain** that connects plants and animals. Where there is high primary production by phytoplankton, seaweeds, or other marine plants, there are large populations of animals. Animals are consumers; they feed on primary producers or other consumers. The **herbivores** eat plants directly; the **carnivores** feed on the herbivores or other carnivores. The most numerous (and the greatest biomass) of herbivores are the plant-eating animal plankton, or **zooplankton.** These are the primary consumers that convert plant tissue to animal tissue, and then become the food for other zooplankton, the flesh-eating carnivores or secondary consumers. Food chains may be long or short, but they are rarely simple and linear. Food chains are more likely to show complex interrelationships among organisms, in which case it is more appropriate to call the pattern a **food web.** Examples are the herring food web diagram, shown in figure 10.8, and the overall ocean food web, shown in figure 10.9.

Food chains and food webs represent the pathways followed by nutrients and food energy as they move through the succession of plants, grazing herbivores, and carnivorous predators. These relationships are often demonstrated in the form of a pyramid made up of **trophic levels** that represent links in a food chain (fig. 10.10). Trophic levels are numbered from the bottom of the pyramid to the top; the primary producers are always the first trophic level. The herbivorous zooplankton are the second trophic level, and the carnivores form the upper levels up to the top carnivore, a predator on which no other marine organism preys (for example, sharks and killer whales).

In general, moving upward from the first trophic level, the size of the organisms increases and the numbers and biomass of organisms decrease. The larger numbers of small organisms at the lower trophic levels collectively have a much larger biomass than the smaller numbers of large organisms at the upper levels. Figure 10.10 relates trophic levels, the primary energy source of the sun, decomposition, recycling of nutrients, and energy loss.

The overall efficiency of energy transfer up each layer of an open-ocean trophic pyramid is estimated at about 10 percent. If, in order to add 1 kilogram (2.2 lbs) of weight, a person ate 10 kilograms of salmon, to attain that weight the salmon had to consume 100 kilograms of small fish, and the fish needed to consume 1000 kilograms of carnivorous zooplankton, which in turn required 10,000 kilograms of herbivorous zooplankton,

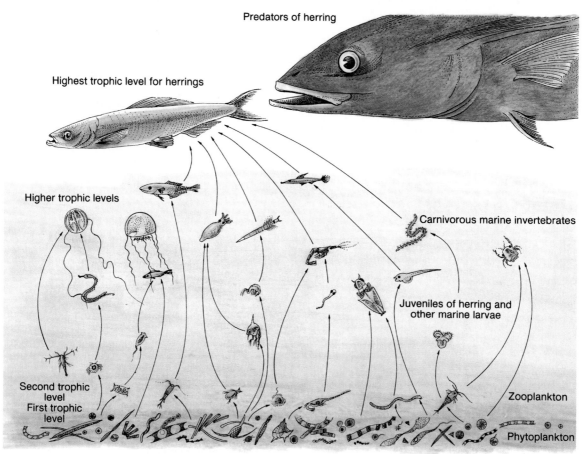

FIGURE 10.8
The food web of the herring at various stages in the herring's life.

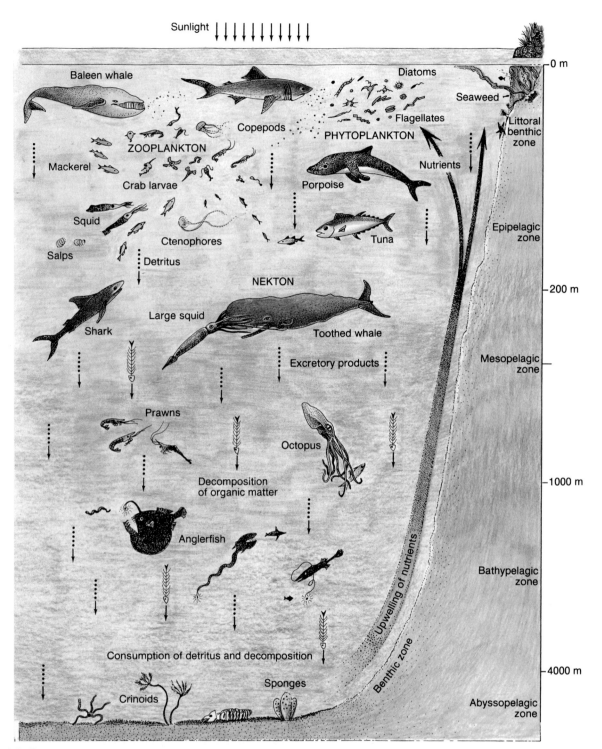

FIGURE 10.9

The ocean food web. Plant, animal, and bacterial populations are dependent on the flow of energy and the recycling of nutrients through the food web. The initial energy source is the sun, which fuels the primary production in the surface layers. Herbivores graze the phytoplankton and benthic algae and are in turn consumed by the carnivores. Animals at lower depths depend on organic matter from above. Upwelling recycles nutrients to the surface, where they are used in photosynthesis.

needing 100,000 kilograms of phytoplankton to supply the eventual 1 kilogram gain at the top of the pyramid. The 90-percent energy loss at each trophic level goes to the metabolic needs of the organisms, feeding, breathing, moving, reproduc-

ing, and heat loss. An organism that consumes 100 units from the level directly below uses 90 units for its own metabolic needs and converts only 10 units to body tissue available to predators in the level above.

Productivity of the Oceans

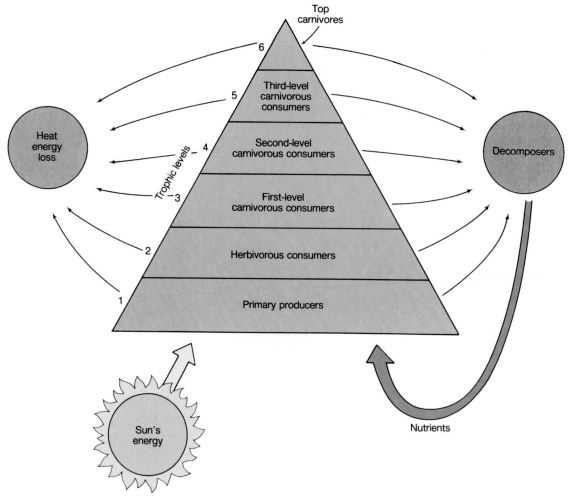

FIGURE 10.10

A trophic pyramid. Trophic levels are numbered from base to top. The first trophic level requires nutrients and energy. Nutrients are recycled at each level; energy is lost as heat at each level.

Compare a food chain and a food web.

Sketch a trophic pyramid of four levels; label primary producers, herbivores, and carnivores. How are energy and nutrients related to these trophic levels?

Why does energy lessen when moving upward from one trophic level to the next?

10.7 CHEMOSYNTHETIC COMMUNITIES

In March 1977, an expedition from Woods Hole Oceanographic Institute using the research submersible *Alvin* discovered communities of animals living around hot-water, hydrothermal vents of the Galápagos Rift (a branch from the East Pacific Rise) at depths of 2500–2600 meters (8000–8500 ft). Since then, other communities have been found on the Juan de Fuca Ridge at 1600 meters (5000 ft), off the west coast of Florida at

3300 meters (11,000 ft), in the central Gulf of Mexico at 700–800 meters (2300–2600 ft), and along equatorial sections of the Mid-Atlantic Ridge. Animals in these areas include filter-feeding clams and mussels, in addition to anemones, worms, barnacles, limpets, crabs, and fish. The clams are very large and show the fastest growth rate of any known deep-sea animal, up to 4 centimeters (2 in) per year. The tube worms are startling in size, up to 3 meters (10 ft) long. These crowded communities have a biomass five hundred to one thousand times greater than the biomass of the normal deep sea.

Dense clouds of bacteria are the base of the food pyramid for these communities. These bacteria use dissolved chemicals in the seawater to obtain energy by a process known as **chemosynthesis.** Instead of being dependent on sunlight for energy to produce organic matter by photosynthesis, these ocean-bottom populations rely on dissolved hydrogen sulfide and particles of sulfur in the hot vent water. The bacteria use chemosynthesis to fix carbon from carbon dioxide into organic molecules such as simple sugars. The other vent animals feed on the bacteria; no sunlight is necessary, and no food is needed from the surface. These self-contained vent communities are among the most productive in the world. In areas where the vents have become inactive, the animal communities have died when their energy source has been removed.

Small animals, perhaps small shrimplike organisms, may graze the bacteria directly. Soft-bodied organisms may absorb dissolved organic molecules released by dead bacteria. The tube worms (fig. 10.11) have no mouth and no digestive system; the soft tissue filling their internal body cavity is filled with bacteria. The mussels have only a rudimentary gut, and the clams and mussels have large numbers of bacteria in their gills. Both the clams and the tube worms have red flesh and red blood based on the oxygen-binding molecule hemoglobin.

During studies of the large shrimp populations surrounding some vent areas, researchers noticed a reflective spot just behind the head of the shrimp. The reflective spot connects to the shrimp's brain, and light-sensitive pigment related to visual pigment is also present. This system is thought to allow the shrimp to detect radiation associated with the hot vents (refer to fig. 3.27) and may enable the shrimp to orient themselves with the vents and their food supply.

Off the coasts of Florida and Oregon, communities based on chemosynthesis have been found that are associated with cold seepage areas. Along the continental slope of Louisiana and Texas, oil and gas seep up to the surface of the sea floor. In 1985, clams, mussels, and large tube worms (fig. 10.12) were collected from these sites at depths between 500 and 900 meters (1600–3000 ft). The mussels' gills contain bacteria that use methane gas as a carbon source; the mussels are able to use organic carbon compounds produced by the bacteria. The tube worms and clams acquire their carbon from methane, biologically degraded oil, and the decay of other organisms. Carbon compounds from methane have been identified in animals preying on the shellfish, which demonstrates a pathway for chemosynthetic carbon to enter the general deep-sea animal community.

In 1990, salt seeps were reported at 650 meters (2000 ft) on the floor of the Gulf of Mexico. Depressions are filled with salt brine more than 3.5 times the usual salinity of seawater, and the brine also contains methane gas. These extremely dense brine lakes are surrounded by large communities of chemosynthetic mussels.

Ninety-five percent of the nearly three hundred species found at these vents and seeps are new to science, and many are only distantly related to other earth creatures, appearing more closely related to ancient species. The presence of a plentiful food supply allows their rapid growth, despite the surrounding low temperatures and their remoteness from the photosynthetic layer at the sea's surface. These discoveries have changed our view of ocean food webs and the deep sea floor; scientists will never again consider it to be sparsely populated or inhospitable to life.

How is it possible for large animal populations to live without a photosynthetic base?

Draw a three-level trophic pyramid for a chemosynthetic community.

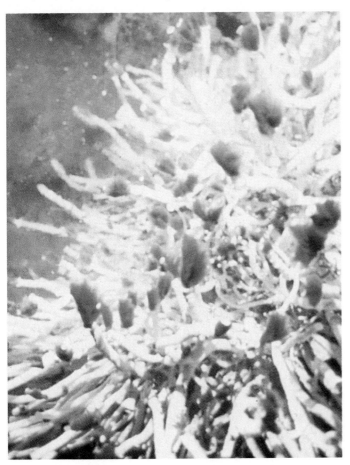

FIGURE 10.11

Tube worms surround a vent area. This photograph was made from the submersible *Alvin*.

FIGURE 10.12

Mussels and tube worms at a Gulf of Mexico oil and gas seep.

10.8 HARVESTING EFFICIENCY

Custom, culture, economics, and availability influence the harvesting of the oceans by humans. Commercial harvests are made from both the higher trophic levels (for example, salmon, tuna, halibut, and swordfish), and the lower levels (for example, herring, shellfish, and anchovy). Harvesting high in the trophic pyramid is less energy efficient than at low trophic levels. Overharvesting any level endangers levels both above and below it by removing food resources from higher levels and preventing recycling of nutrients from higher to lower levels.

The yields of food resources for humans are highest when the harvest is conducted at the lowest possible trophic level. The relationship between total plant production and theoretical production of fish is shown in table 10.3. The third column in this table gives the average efficiency of energy conversion between the trophic levels, and the fourth column

shows the trophic level at which humans usually harvest for food. The well-mixed, nutrient-rich waters of the coasts and estuaries support short, efficient food chains (fig. 10.13a,b). Because of the high rates of plant production in these areas, food is easily obtained and energy is transferred with greater efficiency to the next higher trophic level; consumers obtain food with less effort, and fish production is high. Compare the efficiency of these food chains with the open-ocean food chain in figure 10.13c and the open-ocean fish production in table 10.3.

Why is it more efficient to harvest anchovies than tuna?

Why is the efficiency of energy transfer per trophic level less in the open ocean than in upwelling zones?

Table 10.3 Oceanic Food Production

Area	Plant Production (metric tons of carbon/year)	Efficiency of Energy Transfer per Trophic Level	Trophic Level Harvested	Fish Production (metric tons/year)
Open ocean	16.3×10^9	10%	5	1.6×10^6
Coastal regions	3.6×10^9	15%	4	120×10^6
Upwelling areas	0.1×10^9	20%	2.5	120×10^6

Source: Data from J. H. Rhyther, "Photosynthesis and Fish Production in the Sea" in *Science* 166 (3901):72–76, 1969.

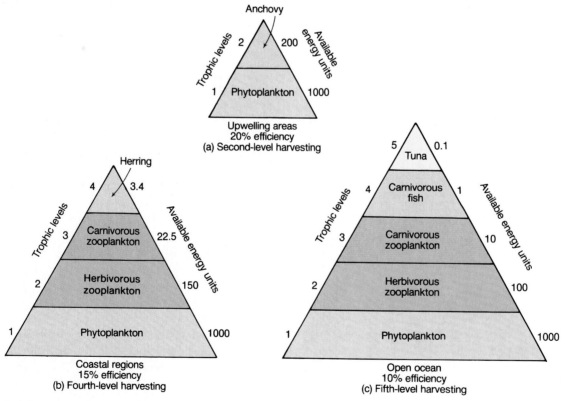

FIGURE 10.13

Trophic level efficiency varies among (a) upwelling areas, (b) coastal regions, and (c) the open ocean. The number of trophic levels and the level at which humans harvest differ with location.

SUMMARY

Plants use photosynthesis to produce organic compounds from carbon dioxide and water in the presence of sunlight and chlorophyll; the new plant material is called primary production. Respiration breaks down organic compounds with the addition of oxygen to yield energy, water, and carbon dioxide.

Gross primary production is the total amount of organic material produced by photosynthesis per volume per unit of time; net primary production is the gain in organic material after deducting the amount used in plant respiration. Biomass is the amount of organic matter produced expressed as number or weight of organisms. The biomass available at a location at a specific time is the standing crop.

Primary production is controlled by the interaction of sunlight, nutrients, and the stability of the surface water in an area. At polar latitudes, the availability of light controls phytoplankton growth; in the tropics, sunlight is available year-round, but the nutrient supply is poor and limits production. At midlatitudes, light, nutrients, and water column stability vary with the seasons. Nutrients cycle through the land and sea, the plants and animals, and are returned to the water by death and bacterial decay.

Primary production is estimated by counting plant cells, measuring chlorophyll concentration, measuring the rate at which oxygen is produced or the rate of carbon-14 incorporation into plant cells. Satellite images provide evidence of standing stock changes over the yearly cycle.

In general, coastal waters are more productive than the open ocean. Good mixing, nutrients in the land runoff, and density-stable water combine to make shallow coastal areas very productive. Upwelling areas are three times more productive than coastal water and six times more productive than the open ocean, but there is ten times more open-ocean area than coastal water area and one hundred times more coastal water area than upwelling area.

The phytoplankton are preyed upon by herbivorous zooplankton that are preyed upon in turn by carnivorous zooplankton. The term "food web" is more appropriate than "food chain" to describe interconnecting feeding patterns. Trophic pyramids present the relationship between primary producers, herbivores, and carnivores in terms of the transfer of biomass and energy.

Self-contained deep-ocean animal communities depend on chemosynthetic bacteria for the first step in their food chains. Communities have been found in the vicinity of hydrothermal vents, around cold seeps of oil and methane gas, and around salt seeps.

In the open ocean, the transfer efficiency between trophic levels is approximately 10 percent. Efficiency is higher in coastal and upwelling areas.

KEY TERMS

plankton	primary production	decomposer	zooplankton
phytoplankton	respiration	food chain	food web
chlorophyll	biomass	herbivore	trophic level
photosynthesis	standing crop	carnivore	chemosynthesis

SUGGESTED READINGS

Alper, J. 1990. The Methane Eaters. *Sea Frontiers* 36 (6):22–29.

Ballard, R. D., and J. F. Grassle. 1979. Return to Oases of the Deep. *National Geographic* 156 (5):689–705.

Grassle, J. F. 1987/88. A Plethora of Unexpected Life. *Oceanus* 31 (4):41–46.

Life in the Sea, Readings from Scientific American. 1982. W. H. Freeman, San Francisco, 248 pp.

Lutz, R. A. 1991/92. The Biology of Deep-Sea Vents. *Oceanus* 34 (4):75–83.

Oceanus. Fall 1984. 27 (3). Issue devoted to hot springs and cold seeps.

Tunnicliffe, V. 1992. Hydrothermal-Vent Communities of the Deep Sea. *American Scientist* 80 (4):336–49.

11

Life in the Water

Outline

Learning Objectives

After reading this chapter, you should be able to

- Separate marine organisms into three categories and give examples of each type.

- Name and describe the major environmental zones of the oceans.

- Describe adaptations made by organisms for life in the ocean.

- Distinguish between phytoplankton and zooplankton and appreciate the diversity of planktonic organisms.

- Explain the phenomenon known as "red tide" and its consequences.

- Understand the role of bacteria in the sea.

- Explain how plankton are sampled at sea.

- Recognize and appreciate the diversity of animals making up the nekton.

- Understand how sharks and rays differ from other fish and how deep-sea bony fish differ from commercial fish species.

- Describe the two categories of whales and discuss the current situation of whale populations.

- Explain the effect of the Marine Mammal Protection Act in U.S. waters.

- Discuss the present threat to sea turtle populations.

- Relate increased world fish catches to the decline of specific fisheries.

- Explain present trends in world fish catches.

- Discuss the role of fish farming as a means of ocean harvest.

◀ Blueline snapper, Hawaii.

The oceans and the coastal seas provide a complex variety of environments for living organisms. At the surface, conditions range from polar to tropical, over depth from light to total and constant darkness, and the sea floor may be rock, sand, or mud. Organisms of the marine environment have much in common with the organisms of the land, but they also have very different survival problems and have developed unique solutions to cope with them. This chapter explains some of the characteristics that allow organisms to live in the sea.

Life in the water ranges from microscopic, single-celled organisms to the largest fish and the greatest whales. This chapter discusses the plankton, the wanderers and drifters of the oceans, existing in great swarms, and moving with the currents. Sharing the vast oceans above the sea floor are the nekton; these are the free swimmers of the oceans, moving through the water independent of the motion of currents and waves. The nekton include marine mammals, seagoing reptiles, and squid, but most of the nekton are fish. The plankton include the unicellular plant life at the beginning of the ocean food chains, and the nekton provide much of the commercial harvest taken from higher levels in the ocean food web.

11.1 CLASSIFICATION OF ORGANISMS

A wide variety of organisms inhabit the environments of the oceans. All organisms are classified and placed in groups to promote identification and to increase understanding of the relationships that exist among them. Examples of the classic biological categories are given in table 11.1. A simple and practical method divides all marine organisms into three groups, based on where and how they live. Plants and animals that float or drift with the movements of the water are the **plankton.** Plant plankton are called **phytoplankton** (also discussed in Chapter 10) and animal plankton are known as **zooplankton.** Organisms that live attached to the bottom or on or in the bottom are the **benthos,** and the animals that swim freely and purposefully in the sea are the **nekton.** See also Appendix B for a more detailed classification of marine organisms.

Name and define the three categories of marine organisms.

11.2 ENVIRONMENTAL ZONES

Because the marine environment is so large and complex, it is divided into subunits called zones. The zone classifications used here are based on the system developed by Joel Hedgpeth in 1957; they are shown in figure 11.1. The water environment is the **pelagic** zone, and the seafloor environment is the **benthic** zone. The pelagic zone is divided into the coastal or **neritic** zone above the continental shelf and the **oceanic** zone or deep water away from the influence of land. The oceanic zone is then subdivided by depth into the subzones shown in figure 11.1. The surface, where there is enough light intensity for plant growth, is known as the **photic** zone, and it extends through both the neritic and oceanic zones. The photic zone has an average depth of 50 to 100 meters (150 to 300 ft). All the other subdivisions of the pelagic zone are without sunlight; they are located in the **aphotic** zone.

The sea floor, or benthic environment, is also subdivided on the basis of depth (fig. 11.1). Tidal fluctuations and waves at the shoreline define the **supralittoral** or splash zone (above high water), the **littoral** or **intertidal** zone (between high and low water), and the **sublittoral** or **subtidal** zone (below low water). These three zones occur in the photic zone, where there is sufficient light to support phytoplankton and benthic plants.

Table 11.1 Taxonomic Categories of Some Marine Organisms

	Killer Whale	Northern Fur Seal	Pacific (Japanese) Oyster	Giant Octopus	Sea Lettuce	Giant Kelp
Kingdom	Animalia	Animalia	Animalia	Animalia	Plantae	Plantae
Phylum	Chordata	Chordata	Mollusca	Mollusca	Chlorophyta	Phaeophyta
Class	Mammalia	Mammalia	Bivalvia (Pelecypoda)	Cephalopoda	Chlorophycae	Phaeophycae
Order	Cetacea	Carnivora (Pinnipedia)	Anisomyaria	Octopoda	Ulvales	Laminariales
Family	Delphinidae	Otariidae	Ostreidae	Octopodidae	Ulvaceae	Lessoniaceae
Genus	*Orcinus*	*Callorhinus*	*Crassostrea*	*Octopus*	*Ulva*	*Macrocystis*
Species	*Orcinus orca*	*Callorhinus ursinus*	*Crassostrea gigas*	*Octopus dofleini*	*Ulva lactuca*	*Macrocystis pyrifera*

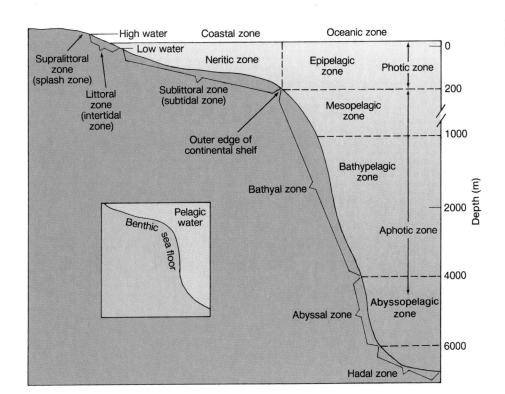

FIGURE 11.1
Zones of the marine environment.

Life in all the zones is influenced by variations in temperature, dissolved gases, substrate, nutrients, and all the factors discussed in previous chapters. The coastal littoral and the pelagic photic zones change greatly with latitude and season, in contrast to regions of greater depth. There are great differences between the shallow zones of polar, middle, and tropic latitudes. The photic zone is deeper in the clear waters of the tropics than in the coastal waters of the midlatitudes. Substrate is an important factor in benthic zones; at shallow depths the bottom may be rock, sand, or mud while in the deeper zones there is less variety. The importance of substrate to the distribution of benthic organisms is further discussed in Chapter 12.

Name and describe the major environmental zones of the oceans.

Name and describe the environmental zones of the intertidal region.

Make a list of factors affecting marine environmental zones.

11.3 LIFE AND THE MARINE ENVIRONMENT

Buoyancy and Flotation

Unlike land organisms that require structural strength to support their bodies in air and against gravity, marine organisms are often exceedingly delicate and fragile. The salt water surrounding marine organisms has a density similar to the bodies of many of the organisms, and since the organisms move with the moving water, they do not require structural strength to withstand its flow. The **buoyancy** of objects in salt water is due to the water's density and helps to keep the floating organisms at the surface. It supports the bodies of the bottom-living creatures and lessens the energy expended by the swimmers.

Many organisms have ingenious adaptations to help them stay afloat. Some jellyfish-type animals secrete gases into a float that enables them to stay at the sea surface. Some plants secrete gas bubbles and form gas-filled floats, which help keep their fronds in the sunlit waters while they are anchored to the sea floor. One floating snail produces and stores intestinal gases; another forms a bubble raft to which it clings. The chambered nautilus (fig. 11.2a), a relative of the squid, lives in its last shell chamber and fills the remaining chambers with gas (mainly nitrogen). Another relative of the squid, the cuttlefish (fig. 11.2b) has a soft, porous, internal shell or cuttlebone. These animals regulate their buoyancy by controlling the relative amounts of gas and liquid within their shells.

Many fish have gas-filled swim bladders that keep them neutrally buoyant. When a fish changes depth, it adjusts the gas pressure in its swim bladder to compensate for the pressure change in the water; this limits its vertical swimming speed. Active, continuously swimming predatory species such as the mackerel, some tuna, and the sharks do not have swim bladders, and bottom fish also lack them.

Planktonic organisms store their food reserves as oil droplets that decrease their density and retard sinking. Many have developed spines, ruffles, and feathery appendages that increase their surface area and decrease their sinking rate, helping

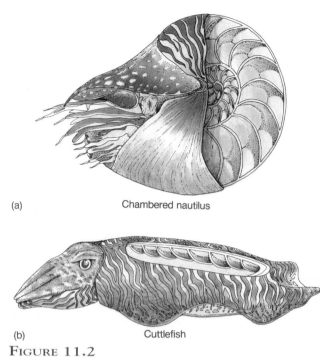

(a)

Chambered nautilus

(b)

Cuttlefish

FIGURE 11.2

(a) Chambered nautilus. (b) Cuttlefish. The chambered shell that provides buoyancy is shown in each organism.

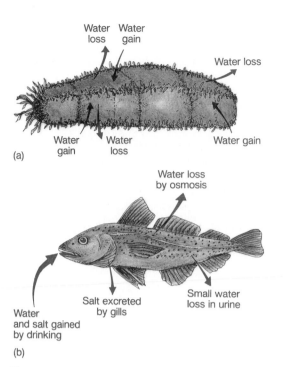

(a)

(b)

FIGURE 11.3

(a) The salt concentration of the seawater is the same as the salt concentration of the sea cucumber's body fluids (35‰). The water moving out of the sea cucumber is balanced by the water moving into it. (b) The salt concentration in the tissues of the fish is much lower (18‰) than that of the seawater (35‰). To balance the water lost, the fish drinks salt water, from which the salt is removed and excreted.

them to float at the sea surface. Large members of the nekton such as whales and seals decrease their density and increase their buoyancy by storing large quantities of blubber, which is mainly low-density fat. Sharks and other varieties of fish store oil in their livers and muscles.

> How are so many marine organisms able to survive without skeletons and rigid structural support?
>
> Give examples of adaptations used by marine organisms to increase their buoyancy.
>
> How does a swim bladder limit the vertical motion of a fish?

Fluid Balance

Special problems are posed for living creatures if the salt content of their body fluids differs from the salinity of the water that surrounds them. Water molecules cross the membranes that separate body fluids from seawater, moving along a gradient from a high concentration of water (low salinity) to a low concentration of water (high salinity); this process is known as **osmosis.** Most fish have body fluids less salty than seawater, that is, their body fluids have a water concentration that is greater than seawater. Their tissues tend to lose moisture, and fish must constantly expend energy to prevent dehydration. Fish stay in fluid balance by drinking seawater nearly continually and excreting its salt across their gills. Sharks and rays do not have this problem, because their body fluids have the same

approximate salt content as seawater. The body fluids of many bottom-dwelling organisms such as sea cucumbers and anemones are also similar to seawater; there is no concentration gradient across their tissues, and the organisms neither gain nor lose water. The fish and the sea cucumber are compared in figure 11.3.

Some organisms can maintain their salt-fluid balance over only limited salinity ranges; others have remarkable abilities to move between high- and low-salinity water. Salmon spawn in fresh water but move down the rivers as juveniles and live their adult lives in the sea. After several years (the time depends on the species), the salmon return to their home streams. The Atlantic common eel reverses this process by migrating downstream to spawn in the Sargasso Sea. The new generation of eels spends one to three years at sea, then returns to fresh water to live for up to ten years before migrating seaward. Many fish and crustaceans use the low-salinity coastal bays and estuaries as breeding grounds and nursery areas for their young, then as adults they migrate farther offshore into higher-salinity waters.

> Why do marine fish lose moisture from their body tissues?
>
> How does the life cycle of a Pacific salmon differ from that of an Atlantic eel?

Bioluminescence

Sunlight illuminates only the surface waters, but there is another source of light in the oceans. On a dark night, when the wake of a boat is seen as a glowing ribbon, or the water flashes with light as oars dip, or a person's hands glow briefly when a net is hauled in, living organisms are producing the light. This phenomenon is **bioluminescence,** produced by the interaction of the compound luciferin and the enzyme luciferase, a biochemical reaction releasing light with 99 percent efficiency.

Bioluminescence is triggered by the agitation of the water that disturbs microscopic organisms, causing them to flash and produce glowing wakes and wave crests. Jellyfish glow if they feed on these organisms, and so do one's hands if they come in contact with crushed tissue. Other bioluminescent organisms in the sea include squid, shrimp, and some fish. Many mid-depth and deep-water fish have light-producing organs, some in patterns on their sides, possibly for identification. Others have light-producing organs on their ventral surfaces, making them difficult to see from below, and still others have glowing bulbs dangling below their jaws or attached to flexible dorsal spines acting as lures for their prey. Flashlight fish living in the reefs of the Pacific and Indian Oceans have specialized organs below each eye which are filled with light-emitting bacteria. These fish use the light to see, communicate, lure prey, and confuse predators.

> What is bioluminescence and what kinds of organisms produce it?
>
> How do deep-water fish use bioluminescence?

Color

Some sea animals are transparent and blend with their background, for example jellyfish and most of the small floating animals in the surface layers. In the clear waters of the tropics, where light penetrates to greater depths, bright colors play their greatest role. Some fish conceal themselves with bright color bands and blotches. These colors disrupt the outline of the fish and may draw the predator's attention away from a vital area to a less important region; for example, a black stripe may hide the eye while a false eye spot appears on a tail or fin. Color is also used to send a warning. Organisms that sting, taste foul, bear sharp spines, or have poisonous flesh are often striped and splashed with color, for example sea slugs and some poisonous shellfish. Among fish that swim near the surface in well-lighted water, for example salmon, rockfish, herring, and tuna, dark backs and light undersides are common. This color pattern allows the fish to blend with the bottom when seen from above and with the surface when seen from below (fig. 11.4).

In the turbid coastal waters of temperate latitudes, there is less light penetration, and drab browns and grays conceal animals seen against the plants and the bottom. Cold-water bottom fish are usually uniform in color to match the bottom, or speckled and mottled with neutral colors. The flatfish have skin

cells that expand and contract to produce color changes, allowing them to conceal themselves by matching the bottom type on which they live (fig. 11.5).

Color is thought to be important in species recognition, courtship, and possibly in keeping schools of fish together. However, we do not know how the animals see the colors, and color may play roles that we do not know or understand.

> What is countershading and what is its benefit?
>
> Discuss several ways in which marine organisms use color.

Barriers and Boundaries

Rapid changes in temperature, density, salinity, and light with depth act as barriers, isolating species and populations from one another. As the water deepens, the effects of such barriers

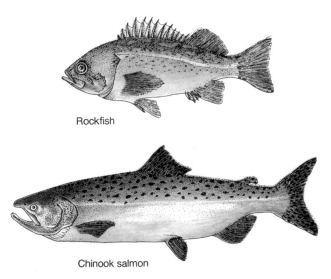

Rockfish

Chinook salmon

FIGURE 11.4
Viewed from above, the dark dorsal surface of the fish blends with the sea floor; viewed from below, the light ventral surface blends with the sea surface. This type of coloration is known as countershading.

FIGURE 11.5
The winter flounder resting on a checkerboard pattern shows its use of camouflage.

Life in the Water

decrease as water properties become more homogeneous. In the horizontal direction similar boundaries exist between zones of surface convergence and divergence in the oceans and between fresh water and seawater in coastal areas. When one type of water is moving through or adjacent to another type, the boundaries are sharp; for example, populations that do well in the warm, salty water of the Gulf Stream may die if they are displaced into the cold, less salty Labrador Current flowing between the North American coast and the Gulf Stream.

Ridges may isolate one deep-ocean basin from another, separating water of differing characteristics with different populations. Isolated seamounts with their peaks in shallow water may support specific isolated communities of sea life. Near-surface and surface species of the tropic regions are prevented from moving between the Atlantic and Pacific Oceans by the land barrier of Central America, and Africa acts as a barrier keeping the tropical species of the Indian Ocean from exchanging freely with the tropical species of the Atlantic. The cold water south of both South America and Africa is a barrier to the survival of the tropical species.

11.4 THE PLANKTON

Although many plankton have a limited ability to move toward and away from the sea surface, they make no purposeful motion against the ocean's currents and are carried from place to place suspended in the seawater. Some plankton are quite large; jellyfish may be the size of a large washtub, trailing 15 meters (50 ft) tentacles. But the phytoplankton and many zooplankton must be observed under a microscope. Bacteria and the smallest phytoplankton cells may be less than 0.005 millimeters in diameter; these are collected using special filtering systems. The zooplankton and phytoplankton between 0.07 and 1 millimeter are known as "net plankton," because they are captured in tow nets made of very fine mesh nylon.

How do plankton move from place to place?

What is the size range of planktonic organisms?

11.5 PHYTOPLANKTON

The phytoplankton (fig. 11.6) are mainly unicellular (single-celled) plants known as **algae.** Each phytoplankton cell is an independent, photosynthesizing individual, and even in species in which the cells attach together in long chains or other aggregations, there is no division of labor between the cells. There is only one large, multicellular planktonic alga, the seaweed *Sargassum* that is found floating in the area of the North Atlantic known as the Sargasso Sea. *Sargassum* reproduces vegetatively by fragmentation to form large mats, which provide shelter and food for a wide variety of organisms, including specialized fish and crabs found nowhere else.

Diatoms and **dinoflagellates** are two major groups of organisms belonging to the phytoplankton. Diatoms are sometimes called golden algae, because their characteristic yellow-brown pigment masks their green chlorophyll. Some diatoms have radial symmetry; they are round and shaped like pillboxes, while others have bilateral symmetry and are elongate. The round diatoms float better than the elongate forms; therefore the elongate diatoms are often found on the shallow sea floor or attached to floating objects, while the round diatoms are more truly planktonic. All are found in areas of cold, nutrient-rich water. Some common diatoms are shown in figure 11.7.

A hard, rigid, transparent cell wall impregnated with silica surrounds each diatom. Pores connect the living portion of the cell inside to its outside environment (fig. 11.8). The buoyancy of the diatoms is increased by the low density of the interior of the cells and the production of oil as a storage product. Their small volume and accompanying relatively large surface area help these cells to stay afloat. Their large surface area also provides them with greater exposure to sunlight and water containing the gases and nutrients for photosynthesis and growth. In addition, some diatoms have spines or other projections that increase their ability to float.

Diatoms reproduce very rapidly by cell division; when their populations discolor the water, it is known as a **bloom.** Diatoms are most important as primary producers. Those that are not consumed by herbivores eventually die and sink to the ocean floor. In shallow areas, the cells reach the sea floor with some organic matter still locked inside, and they may, over a very long period of time, form oil deposits. The cell walls of diatoms that sink to greater depths build up siliceous sediments, the siliceous ooze discussed in Chapter 4.

Dinoflagellates are red to green in color and can exist at lower light levels than diatoms, because they can both photosynthesize like a plant and ingest organic material like an animal. Their external walls do not contain silica; some are smooth and flexible but others are armored with plates of

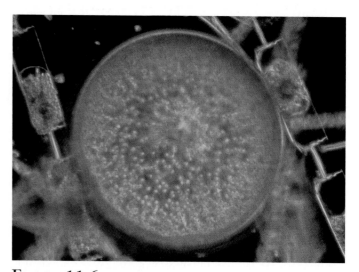

FIGURE 11.6

Phytoplankton. The large diatom, *Coscinodiscus*, between strands of *Ditylum*, another diatom. (×100)

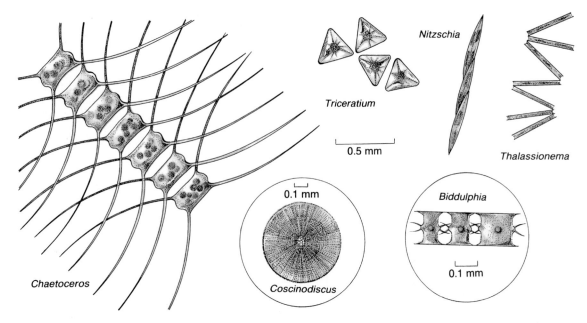

FIGURE 11.7

Diatoms exist as single cells or in chains.

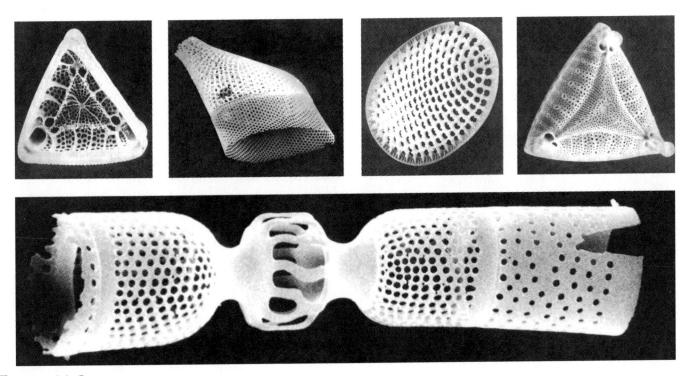

FIGURE 11.8

Stereoscan micrographs of diatom frustules.

cellulose. Dinoflagellates usually have two whiplike appendages or **flagella** that beat within grooves in the cell wall, giving the cells limited motility. Under favorable conditions, they multiply even more rapidly than diatoms to form blooms, but they are not as important as the diatoms as a primary ocean food source. Some dinoflagellates are called fire algae, because they glow with bioluminescence at night. Representative dinoflagellates are shown in figure 11.9.

What is *Sargassum* and where is it found?

Distinguish between diatoms and dinoflagellates.

What allows diatoms to stay in the photic zone?

What is a bloom?

Why are some dinoflagellates called fire algae?

Life in the Water

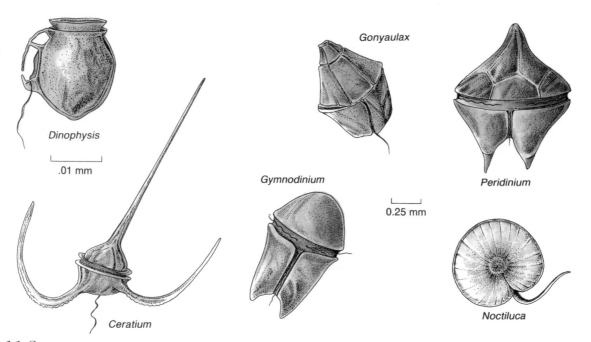

FIGURE 11.9

Dinoflagellates. *Noctiluca, Gymnodinium,* and *Gonyaulax* produce red tides. *Noctiluca* is a bioluminescent, nontoxic dinoflagellate. *Gymnodinium* and *Gonyaulax* produce toxic red tides and paralytic shellfish poisoning.

11.6 RED TIDES AND TOXIC BLOOMS

Certain species of dinoflagellates produce **red tides.** These blooms may or may not be poisonous to fish and other organisms, and may or may not produce symptoms of paralytic shellfish poisoning (PSP), neurotoxic shellfish poisoning (NSP), or diarrhetic shellfish poisoning (DSP) in humans. In North American waters several different dinoflagellates produce red tides: *Gonyaulax, Alexandrium,* and *Gymnodinium* are toxic; *Noctiluca* is not (fig. 11.9). Toxic species are generally not poisonous to shellfish feeding on a bloom, but the toxins are concentrated in the tissues of the shellfish and produce PSP, NSP, or DSP in humans who eat them. *Gymnodinium* and a species of *Gonyaulax* kill fish in the Gulf of Mexico; *Gonyaulax* also kills shrimp and crab.

The toxins that produce PSP are powerful nerve poisons that can cause paralysis and death if the breathing centers are affected. One species of *Gymnodinium* can produce at least five different toxins, and the saxitoxins, which are found in butter clams (*Saxidomus*) but are produced by *Gonyaulax*, are fifty times more lethal than strychnine. The toxin is not affected by heat, so cooking the shellfish does not neutralize the poison. Even after the visible signs of red water due to a dinoflagellate bloom have disappeared, the shellfish can retain the toxin in their tissues for long periods, and so the beaches are kept closed to shellfish harvesting. NSP symptoms are similar to basic food poisoning; unlike PSP, few cases have been reported, and no known deaths have occurred. Only recently have researchers linked DSP outbreaks to the presence of a dinoflagellate, for DSP outbreaks may have been occurring throughout history and been attributed to bacterial contamination. Scientists believe that the toxins are a defense against predators.

No one knows precisely what triggers the sudden blooms of these organisms, but it appears that red tides and toxic blooms are increasing in frequency and severity around the world. They often happen in spring or summer after heavy rains have produced a land runoff of nutrient-rich water, and some scientists believe that the continuous addition of nitrates and phosphates to coastal water from both sewage and agricultural runoff may be partly to blame. High salinities are thought by some to trigger blooms in the Gulf of Mexico waters; along the New England coast temperature and light may play the major roles in activating blooms. Another suggestion is that cysts or thick-walled resting cells, formed by the dinoflagellates, settle to the bottom and may be brought to the surface by upwelling water. Cysts in the shallow waters of bays and harbors may be activated by sudden disturbance of the bottom; in deeper coastal water, the cysts appear to have an annual cycle.

An estimated 10,000–50,000 people eating fish in tropical regions of the world are affected by ciguatera poisoning each year. There is no way to prepare affected fish to make it safe to eat. Symptoms of ciguatera poisoning are extremely variable and may occur in various combinations; they include headache, nausea, vomiting, abdominal cramps, reduced blood pressure, and in severe cases, convulsions, muscular paralysis, hallucinations, and death. Several dinoflagellates are associated with ciguatera, but *Gambierdiscus toxicus* is most often the cause. Researchers in the Gilbert Islands found that dinoflagellates are eaten by herbivores, and the ciguatoxins then move through the food web in a cycle that appears to take at least eight years.

Ciguatera is an international problem; it has delayed development of fisheries in the Red Sea, Sri Lanka, New Guinea, and Puerto Rico; the loss to the Floridean/Caribbean/Hawaiian seafood industry is estimated at $10 million yearly.

Shellfish contaminated with domoic acid, traced to a bloom of the diatom *Pseudonitzschia pungens*, caused three deaths in eastern Canada in 1987. Prior to that time diatoms were not known to produce toxins. Shellfish and crab fisheries along the entire U.S. West Coast were closed in 1991 due to blooms from another species of this diatom. Domoic acid poisoning in humans is known as amnesic shellfish poisoning (ASP) and may cause short-term memory loss as well as nausea, disorientation, and muscle weakness.

How are red tides, dinoflagellates, and paralytic shellfish poisoning related?

What environmental conditions are thought to trigger red tides?

Why are shellfish unsafe to eat during and after red tide episodes?

Distinguish between ciguatera and domoic acid poisoning.

11.7   ZOOPLANKTON

The zooplankton are either herbivores, grazing on the phytoplankton, or carnivores feeding on other members of the zooplankton. Many of the zooplankton have some ability to swim and can even dart rapidly over short distances in pursuit of prey or to flee predators. They may move vertically in the water column, but, like the phytoplankton, they are also transported by the currents. A sample of zooplankton is shown in figure 11.10.

The life histories of zooplankton types are varied and show different strategies for survival in a world where reproduction rates are high and life spans are short. These animals may produce three to five generations a year in warm water, where food supplies are abundant and higher temperatures accelerate life processes. At high latitudes, where the season for phytoplankton growth is brief, the zooplankton may produce only a single generation in a year.

Convergence zones and boundaries between types of water concentrate zooplankton and attract predators. Some zooplankton migrate toward the sea surface each night and return to the depths each day, either in an attempt to maintain their light level or in response to the movement of their food resource. The amount of daily migration varies between 10 meters (30 ft) and 500 meters (1500 ft). Accumulations of these organisms reflect sound waves from depth sounders and are seen on a depth recording as a false bottom or **deep scattering layer,** known as the DSL.

Foraminiferans and **radiolarians** are single-celled microscopic members of the zooplankton; both are shown in figure 11.11. Foraminiferans are encased in compartmented calcareous coverings, while the radiolarians are surrounded by

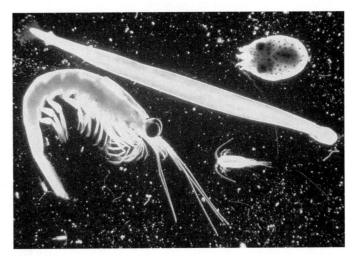

FIGURE 11.10
This zooplankton sample shows an octopus larva (upper right), an arrowworm (center), a euphausiid (lower left), and a copepod (lower right) (×5).

silica walls. The radiolarian coverings are ornately sculptured and covered with delicate spines. Protrusions or pseudopodia ("false feet"), many with skeletal elements, radiate from the cell and catch diatoms and small protozoa. Both foraminiferans and radiolarians are found in the warmer regions of the oceans. After death, their coverings accumulate on the ocean floor contributing to the ocean sediments, as discussed in Chapter 4.

Among the most common and widespread zooplankton types worldwide are the small **crustaceans** (shrimplike animals): **copepods** and **euphausiids** (fig. 11.12). These animals are basically herbivorous and consume more than half their body weight daily. Copepods are smaller than euphausiids; euphausiids move more slowly and live longer than copepods. Both reproduce more slowly than the diatoms, doubling their populations only three to four times a year. They may make up more than 60 percent of the zooplankton in any of the oceans and serve as a food source for small fish. In the Arctic and Antarctic, the euphausiids are the **krill,** occurring in such quantities that they provide the main food for the baleen whales (see section 11.13). Estimates of total krill biomass range from 5 million to 6 billion metric tons, and are probably in excess of 100 million tons. The 1988 commercial krill harvest was 370,000 tons, rising to 396,000 metric tons in 1989. The catch had dropped to 233,000 metric tons in 1991 because of the breakup of the former Soviet Union's fishing fleet.

Arrowworms (fig. 11.12) are common, carnivorous members of the zooplankton. They are abundant in ocean waters from the surface to the great depths. Arrowworms are 2 to 3 centimeters (1 in) long, nearly transparent, and feed on other zooplankton. There are several species found in seawater, and in some cases a particular species is limited to a certain type of water.

Other zooplankton include swimming snails, comb jellies, and true jellyfish. The swimming snails are **pteropods,** modified **mollusks** that may or may not have a small shell, but they

Life in the Water

FIGURE 11.11
Selected members of the
radiolaria (*Acanthonia,
Acanthometron,* and
Aulacantha) and a
foraminifera (*Globigerina*).

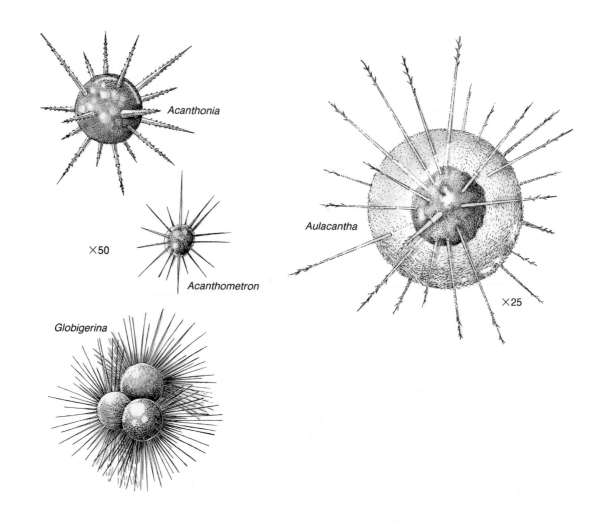

FIGURE 11.12
Crustacean members of the
zooplankton: *Calanus* and
Oithona are copepods;
Euphausia superba is known
as krill. The *chaetognath* is
the arrowworm.

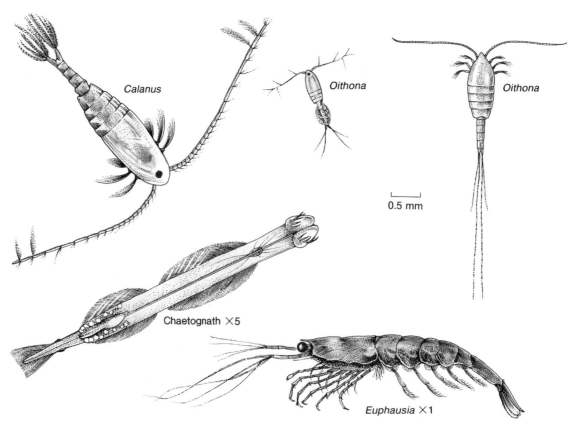

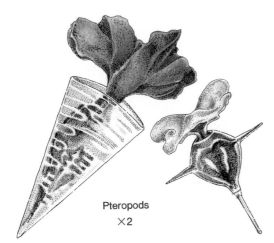

FIGURE 11.13
The pteropods are planktonic snails.

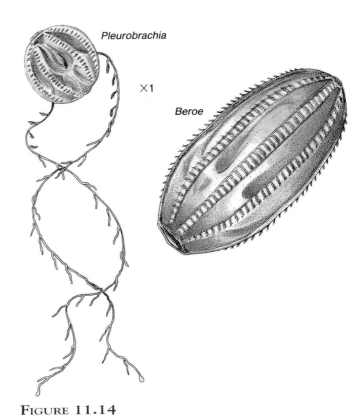

FIGURE 11.14
The comb jellies (ctenophores) *Pleurobrachia* and *Beroe*. *Pleurobrachia* is often called a sea gooseberry.

all have a transparent, gracefully undulating "wing" (fig. 11.13). Some pteropods are herbivores, and some are carnivores. Comb jellies or **ctenophores** (fig. 11.14) float in the surface waters. Some have trailing tentacles; all are propelled slowly by eight rows of beating **cilia** or hairlike cell extensions. The round sea gooseberries or sea walnuts are small, but the beautiful, tropical Venus' girdle may be 30 centimeters (12 in) or more in length. All ctenophores are carnivores, feeding on other zooplankton.

True jellyfish resemble comb jellies, but they come from another unrelated group of animals, the Cnidaria or **coelenterates.** Some jellyfish spend their entire lives as drifters; others are planktonic for only a portion of their lives, as they eventually settle and change to a bottom-dwelling form similar to a sea anemone. Another group of unusual jellyfish include the Portuguese man-of-war, *Physalia*, and the small by-the-wind-sailor, *Velella* (fig. 11.15). Both the Portuguese man-of-war and the by-the-wind-sailor are **colonial** animals; they are collections of individual but specialized organisms, some of which have the task of gathering food, reproducing, or protecting the colony with stinging cells, while some form the float.

All of the zooplankton discussed to this point, with the exception of certain jellyfish, spend their entire lives as plankton and are called **holoplankton.** However, an important portion of the zooplankton spends only part of its life as plankton; these are members of the **meroplankton.** The eggs and **larvae,** or juvenile forms, of oysters, clams, barnacles, crabs, worms, snails, starfish, fish, and many other organisms are a part of the zooplankton for a few weeks. The currents carry these larvae to new locations, where they find food sources and areas to settle. In this way, areas where a species may have died out are repopulated, and overcrowding in the home area is made less likely. Sea animals produce larvae in enormous numbers, and so these meroplankton are an important food source for the zooplankton and other animals. The parent animals may produce millions of spawn, but only small numbers of males and females need survive to adulthood in order to guarantee survival of the stock.

Larvae often look very unlike the adult forms into which they will develop (fig. 11.16). Early scientists who found and described these larvae gave each a name, thinking they had discovered a new type of animal. We keep some of these names today, referring, for example, to the trochophore larvae of worms, the veliger larvae of sea snails, the zoea larvae of crabs, and the nauplius larvae of barnacles. Other members of the meroplankton include fish eggs and juvenile fish. Some large seaweeds release reproductive cells that drift in the plankton until they are consumed or settle out to grow attached to the sea floor.

Where would you expect to find large quantities of zooplankton?

What causes the "deep scattering layer"?

Give an example of a single-celled member of the zooplankton.

What kinds of animals form the majority of the zooplankton?

Why are krill important in ocean food webs?

Describe several larger members of the zooplankton.

What is a colonial animal? Give an example.

Distinguish between holoplankton and meroplankton.

What benefit do planktonic larvae have for attached adult organisms?

237   *Life in the Water*

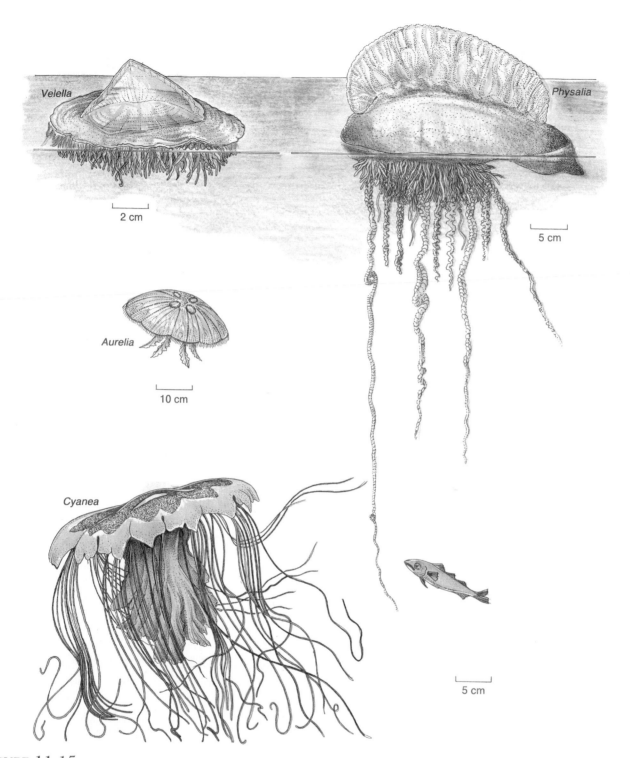

FIGURE 11.15

Jellyfish belong to the Coelenterata or Cnidaria. *Velella*, the by-the-wind-sailor, and *Physalia*, the Portuguese man-of-war, are colonial forms.

11.8 BACTERIA

Bacteria are the smallest living organisms; they are microscopic, single cells and many have the ability to reproduce by cell division every few minutes when they inhabit a favorable environment. Bacteria live free in the seawater and exist on every available surface within it, for example decaying material, the surfaces of organisms, the sea floor, and pieces of floating wood. They produce dense blooms in estuaries and warm-water regions; certain bacteria are associated with the hydrothermal vents of the deep sea floor, where they are the primary producers for the animal communities surrounding the vents (refer to Chapter 12).

Bacteria play an important role in the decay and breakdown of organic matter, returning it to the sea as basic chemicals and compounds to be used again by new generations of plants and animals (refer to section 10.3). Bacteria also serve as a major protein source. A film of bacteria is found on minute particles of floating organic material; these are an ideal

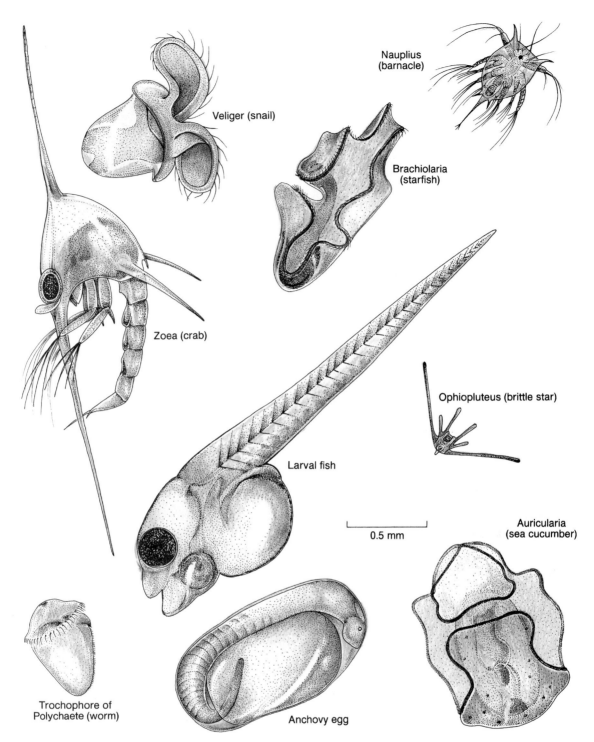

Veliger (snail)

Nauplius (barnacle)

Brachiolaria (starfish)

Zoea (crab)

Ophiopluteus (brittle star)

Larval fish

0.5 mm

Auricularia (sea cucumber)

Trochophore of Polychaete (worm)

Anchovy egg

FIGURE 11.16
Members of the meroplankton. All are larval forms of nonplanktonic adults.

food for planktonic larvae and a variety of single-celled proto-zoans which in turn serve as food for tiny worms, clams, and crustaceans.

What are the roles of bacteria in the sea?

Where are marine bacteria found?

11.9 SAMPLING THE PLANKTON

Traditionally, plankton are sampled by towing conical nets of fine mesh material through the sea behind a vessel, or the nets are dropped over the side of a stationary vessel and pulled straight back up like a bucket (fig. 11.17). After the net is returned to the deck, it is rinsed carefully and the "catch" is collected. To measure the volume of water that has flowed through the net, a flow meter is placed in the mouth of the net.

The speed at which the net is towed is important; the tow must be rapid enough to catch the organisms but slow enough to let the water pass through the net. Fine nets are used for sampling phytoplankton while coarser mesh nets are reserved for the larger zooplankton. Plankton may also be sampled by using a water bottle or a submersible pump to collect water that is then filtered to remove the plankton.

Today's oceanographers sample zooplankton with multiple-net systems mounted on a single frame; the nets are opened and closed on command from the ship. The frame also carries electronic sensors that relay data on salinity, temperature, water flow, light level, net depth, and the angle of the tow to the ship's computer.

Whatever the sampling method, the number and kinds of plankton must be determined. A subsample of the catch is inspected directly under the microscope. If the amount of plankton is of greater interest than the kinds of organisms, an electronic particle counter may be used to find the total count of organisms, or a subsample may be dried and weighed. The biomass of phytoplankton can also be determined by dissolving out the chlorophyll from a sample and measuring the pigment's concentration. New optical instruments measure the natural fluorescent signal coming from chlorophyll pigment in phytoplankton cells. These measurements are made directly and instantaneously at the desired depth and then related to phytoplankton abundance. In order to learn how planktonic organisms behave, researchers at the Woods Hole Oceanographic Institute use a strobe light with four video cameras at four different magnifications to photograph the plankton. Eventually they hope to program the system to recognize different kinds of plankton, allowing the researchers to acquire species and population information.

FIGURE 11.17
A plankton net is retrieved after a near-surface tow.

What different methods are used to sample plankton?

Table 11.2 is a summary of the plankton, their distribution, and their characteristics.

Table 11.2 Plankton

Type	Distribution	Characteristics
Plants		
Phytoplankton		
Diatoms	Cool, nutrient-rich surface waters, inshore and offshore.	Single photosynthetic cells or chains of cells with silica covering, radial or bilateral symmetry.
Dinoflagellates	All surface and near-surface waters, inshore and offshore.	Single cells, two flagella, cellulose covering. Some photosynthetic; some ingest organics. Bioluminescence common; some toxic. Red tides formed by some.
Animals		
Zooplankton		
Holoplankton (permanent plankton)	Worldwide polar to tropic, surface and deeper waters, inshore and offshore.	Many varieties, large range in size, migrate vertically, vary from single cells to complex multicellular and colonial forms. Herbivores and carnivores. Copepods and krill especially important.
Meroplankton (part-time plankton)	Worldwide polar to tropic, surface and deeper waters, inshore and offshore.	Developmental stages (eggs and larvae) of nonplanktonic animals. Important to distribution and food production.

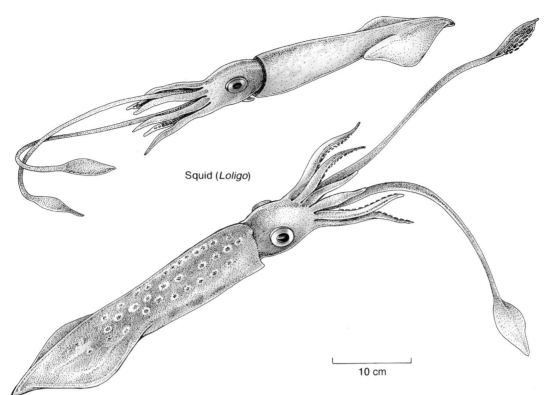

Squid (*Loligo*)

10 cm

11.10 THE NEKTON

Approximately 5000 species of nekton swim freely through the pelagic and neritic regions of the oceans. The only **invertebrate** animals (animals without backbones) among these are the squid and a few species of shrimp. The other members of the nekton are **vertebrates** (animals with backbones); these are the fish, the reptiles, and the mammals.

Squid (fig. 11.18) are abundant in deeper water, travel in large schools, and migrate toward the surface at night. Giant squid (*Architeuthus*), more than 15 meters (50 ft) long, are the food of the sperm whale. Scientists know little about these huge creatures but suspect that there may be large numbers of them in the upper aphotic zone.

The fish dominate the nekton. Fish are found at all depths and in all the oceans, but their distribution is determined directly or indirectly by their dependence on primary producers. Fish are concentrated in upwelling areas, shallow coastal areas, and estuaries. The surface waters support much greater populations per unit of water volume than the deeper zones, where food resources are sparser.

Fish come in a wide variety of shapes related to their environment and behavior. Some are streamlined, designed to move rapidly through the water (tuna and mackerel); others are flattened for life on the sea floor (sole and halibut); while still others are elongate for living in soft sediments and under rocks (some eels). Fins provide the push or thrust for locomotion and are found in a variety of shapes and sizes. Fish use their fins to change direction, turn, balance, and brake. A sample of the variety found among marine fish is discussed in the next two sections.

What invertebrate animals are part of the nekton?

11.11 SHARKS AND RAYS

Sharks differ from other fish by their skeletons of cartilage rather than bone and by their toothlike scales. Shark scales have a covering of dentine similar to vertebrate teeth and are extremely abrasive; sharkskin has been used as a sandpaper and as a polishing material. The shark's teeth are modified scales; they are replaced rapidly if they are lost, and they occur in as many as seven overlapping rows. Sharks are acutely aware of their environment through good eyesight and excellent senses of smell, hearing, mechanical reception, and electrical sense. They are most sensitive to chemicals associated with their feeding and are able to detect them in amounts as dilute as one part per billion. Sharks can detect the movement of water from currents or from injured or distressed animals; they use their electroreception sense to locate prey and recognize food. As a shark swims through the earth's magnetic field, an electric field is produced that varies with direction, giving the shark its own compass.

There are some 300 species of sharks widely spread through the oceans; some are shown in figure 11.19. The whale shark is the world's largest fish, reaching lengths of more than 15 meters (50 ft). This graceful and passive animal feeds on plankton and is harmless to other fish and mammals. The docile basking shark, 5 to 12 meters (15 to 40 ft) long, is another plankton feeder.

Many sharks are swift and active predators, attacking quickly and efficiently using their rows of sharp serrated teeth.

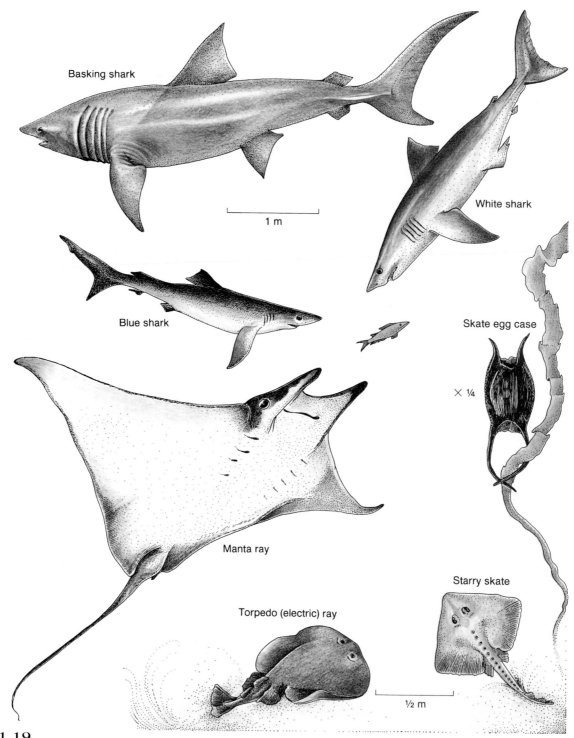

Basking shark

White shark

1 m

Blue shark

Skate egg case

× ¼

Manta ray

Starry skate

Torpedo (electric) ray

½ m

FIGURE 11.19
Representative members of the cartilaginous fish. The leathery egg case, or mermaid's purse, is that of a skate.

They also play an important role as scavengers by eliminating the diseased and aged animals. Sharks can and do attack humans, although the reasons for these attacks and the frenzied feeding sometimes observed in groups of sharks are not understood. A human swimming inefficiently at the surface may look like a struggling, ailing animal and be attacked; a diver swimming completely submerged may appear as a more natural part of the environment and be ignored.

Skates and rays, also shown in figure 11.19, are flattened, sharklike fish that live near the sea floor. They move by undulating their large side fins, which gives them the appearance of flying through the water. The large manta rays are plankton feeders, but most rays and skates are carnivorous, preferring crustaceans, mollusks, and other benthic organisms. Their tails are usually thin and whiplike, and that of the stingray carries a poisonous barb at the base. Some of the skates and a few of the rays have shock-producing electric organs; they are along the sides of the tail in skates and on the wings of the rays. Their purpose appears to be mainly defensive. Like the sharks, most rays bear their young live. Skates enclose the fertilized eggs in a leathery capsule, called a mermaid's purse or a sea purse, that is ejected into the sea and from which the fully formed young emerge in a few months (fig. 11.19).

How do sharks and rays differ from other fish?

At what trophic level do the largest sharks feed?

What roles do sharks play in the ocean food web?

How does a skate or a ray differ from a shark?

11.12   BONY FISH

Most commercially valuable fish are found between the ocean's surface layers and about 200 meters (600 ft) deep, and most of these fish are streamlined, active, predatory, and capable of high-speed, long-distance travel. Among the most important species fished commercially are the enormously abundant small herring-type fish, such as sardine, anchovy, menhaden, and herring. These fish feed directly on plankton and are found in large schools in areas of high primary productivity.

Fish schools may consist of a few fish in a small area or thousands of fish covering several square kilometers. Usually the fish are all of the same species and similar in size. Fish schools have no definite leaders, and the fish change position continually. Most schooling fish have wide-angled eyes and the ability to sense changes in water displacement, allowing them to keep their relationship to one another constant as the school moves or changes direction. Schooling probably developed as a means of protection, since each fish in the school has less chance of being eaten than it does alone. The school may also keep reproductive members of the population together.

Other valuable fish that are harvested commercially include mackerel, pompano, swordfish, and tuna. These fish are caught at sea, out of sight of land. Fish that live on or near the bottom do not swim as rapidly as those that swim freely in the water. The flounder, halibut, turbot, and sole are commercially valuable bottom fish. Perch and snapper tend to congregate along the sea floor in the shallower, near-shore areas. They are often called rock fish because they hide among the rocks and live in cracks of reefs and rocky areas.

Representative bony fish that are fished commercially are shown in figure 11.20. The commercial exploitation and economic significance of food fish are discussed in section 11.15.

The fish of the deep sea are not well known, for the depths at which they live are difficult and expensive to sample. A variety of deep-sea types is shown in figure 11.21.

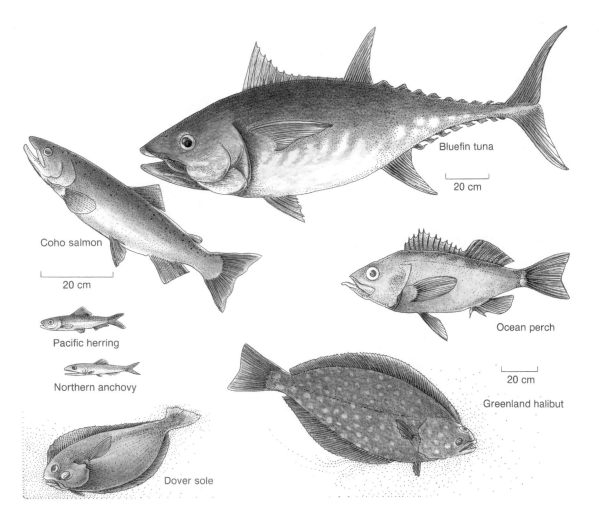

FIGURE 11.20
Commercially harvested members of the bony fish.

Bluefin tuna
20 cm

Coho salmon
20 cm

Ocean perch
20 cm

Pacific herring

Northern anchovy

Greenland halibut

Dover sole

Life in the Water

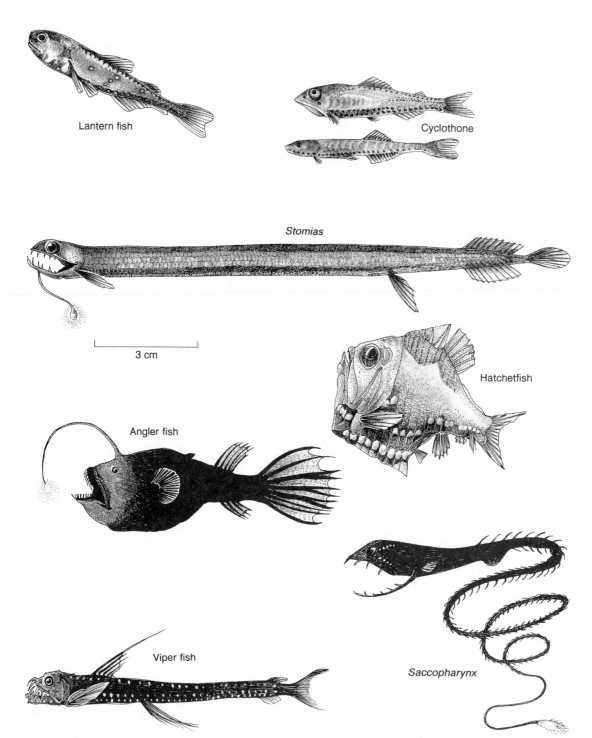

FIGURE 11.21
Fish from the deep sea.

In the dim, transitional layer between 200 and 1000 meters (600–3000 ft), the waters support vast schools of small luminous fish. The many species of *Cyclothone* are believed to be the most common fish in the sea. Deeper-living species are black, while the shallower-living species are silvery, to blend with the dim light. The lantern fish has a worldwide distribution; some 200 species are distinguished by the pattern of light organs along their sides. They are a major item in the diet of tuna, squid, and porpoises. *Stomias*, a fish with a huge mouth, long,

pointed teeth, and light organs along its sides, is also found here, as is the large-eyed hatchetfish. These fish prey on the great clouds of euphausiids and copepods found at this depth.

In the perpetually dark zones are predators with highly specialized equipment for catching their prey. These fish are equipped with light-producing organs used as lures, large teeth that in some species fold backward toward the gullet so their prey cannot escape, and gaping mouths with jaws that unhinge to accommodate large fish. Among the most famous of these predators is *Saccopharynx* with its funnel-like throat and tapering body. When the stomach is empty the fish appears slender, but it expands to accept anything the great mouth can swallow. The female angler fish has a dorsal fin modified into a fishing rod with an illuminated lure dangling just above her jaws.

Although fierce and monstrous in appearance, most of these fish are small, between 2 and 30 centimeters (1–12 in) in length. They breathe slowly, and the tissues of their small bodies have a high water and low protein content. These fish go for long periods between feedings and use their food for energy rather than for increased tissue production.

At what depths are most commercial fish found?

How do fish benefit from schooling?

List several characteristics of deep-water fish.

11.13 MAMMALS

Marine mammals are warm-blooded air breathers. They may spend all of their lives at sea, or they may return to land to mate and give birth. In either case, the young are born live and are nursed by their mothers. Included in the group are large and small whales (including porpoises and dolphins), seals, sea lions, walruses, sea otters, and sea cows. See figure 11.22 and table 11.3 for representative whales.

Whales belong to the mammal group called **cetaceans.** Some cetaceans are toothed, pursuing and catching their prey with their teeth and jaws (for example, the killer whale, the sperm whale, and the small whales known as dolphins and porpoises); others have mouths fitted with strainers of **baleen,** or whalebone. Baleen whales gulp water and plankton, then expel the water through the baleen, leaving the krill behind. The mouths of toothed and baleen whales are compared in figure 11.23. The blue, finback, right, sei, gray, and humpback whales are baleen whales. The gray whale feeds mainly on bottom crustaceans and worms.

The "great whales" that have been the focus of the whaling industry are the blue, sperm, humpback, finback, sei, and right whales. The earliest known European whaling was by the Norse between A.D. 800 and 1000 and continued locally, for the next 400 years. In the 1500s, the Basque people of France and Spain crossed the Atlantic to Labrador to hunt whales for their oil, and by 1600 whaling had become a major commercial activity

Table 11.3 Principal Characteristics of the Great Whales

	Distribution	Breeding Grounds	Average Weight (tons)	Greatest Length (m)	Food
Toothed Whales					
Sperm	Worldwide; breeding herds in tropic and temperate regions	Oceanic	35	18	Squid, fish
Baleen Whales					
Blue	Worldwide; large north-south migrations	Oceanic	84	30	Krill
Finback	Worldwide; large north-south migrations	Oceanic	50	25	Krill and other plankton, fish
Humpback	Worldwide; large north-south migrations along coasts	Coastal	33	15	Krill, fish
Right	Worldwide; cool temperate	Coastal	(50)	17	Copepods and other plankton
Sei	Worldwide; large north-south migrations	Oceanic	17	15	Copepods and other plankton, fish
Gray	North Pacific; large north-south migrations along coasts	Coastal	20	12	Benthic invertebrates
Bowhead	Arctic; close to edge of ice	Unknown	(50)	18	Krill
Bryde's	Worldwide; tropic and warm temperate regions	Oceanic	17	15	Krill
Minke	Worldwide; north-south migrations	Oceanic	10	9	Krill

Key: ( ) = estimate.

From K. R. Allen, *Conservation and Management of Whales,* Washington Sea Grant Program, 1980. Reprinted by permission.

Life in the Water

FIGURE 11.22
Relative sizes of baleen and
toothed whales.

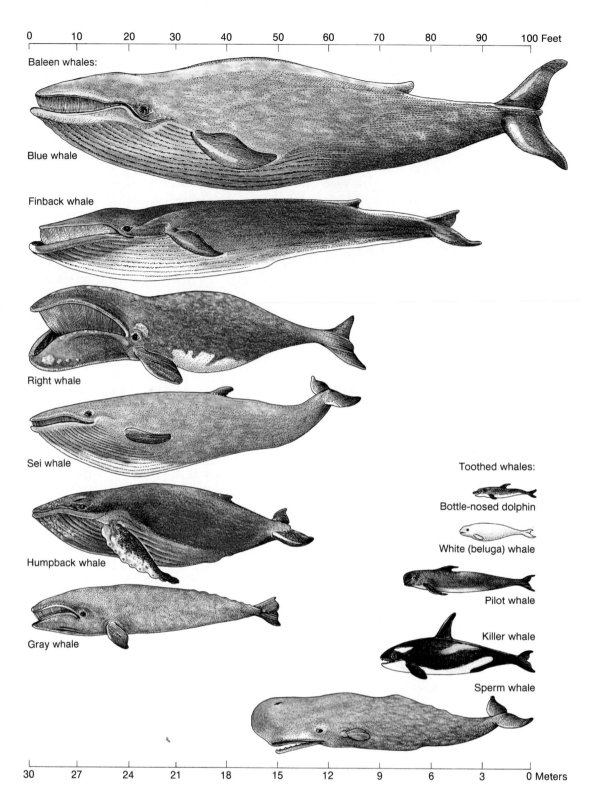

among the Dutch and the British. At about the same time the Japanese began harvesting whales, and in the 1700s and 1800s, whalers from the United States, Great Britain, the Scandinavian countries, and other northern Europeans pursued them far from shore. By the end of the nineteenth century, whales were being hunted with explosive harpoons and high-speed hunting vessels that brought the dead whales to factory ships for conversion to oil. These methods continued into the twentieth century greatly increasing the efficiency of the hunt and rapidly depleting whale stocks.

In the 1930s, the blue whale harvest reduced the population to less than 4 percent of its original numbers, threatening the species with extinction. In 1946, representatives from whaling nations established the International Whaling

(a)

(b)

FIGURE 11.23

(a) The killer whale (*Orcinus orca*) is a toothed whale. (b) Bowhead whale (*Balaena mysticetus*) baleen. This dead whale has been hauled onto the ice and is shown lying on its back.

Table 11.4	World Population Estimates of Great Whales[1]			
Species	Initial	Present	Percent of Initial	Status[2]
Right whale	100,000	>2000	2%	Severely depleted
Blue whale	>196,000	<11,000	<5%	Severely depleted
Fin whale	>464,000	62,000	<13%	Severely depleted
Bowhead whale	>55,000	8000	<15%	Severely depleted
Humpback whale	>120,000	>27,000	<23%	Severely depleted
Sei whale	>105,000	>36,000	34%	Depleted
Sperm whale	2,770,000	1,810,000	65%	At or above optimum sustainable population
Gray whale	<20,000	20,869	100%	Recovered

[1]Population estimates are from *Endangered Whales: Status Update.* 1994. National Marine Mammal Laboratory, National Marine Fisheries Service, National Oceanographic and Atmospheric Administration.

[2]The U.S. Marine Mammal Protection Act defines a stock below its optimal sustainable population as being depleted, and this is thought to occur at or above 60 percent of maximum population size.

Source: Endangered Whales: Status Update, 1994. National Marine Mammal Laboratory, National Marine Fisheries Service, National Oceanographic and Atmospheric Administration.

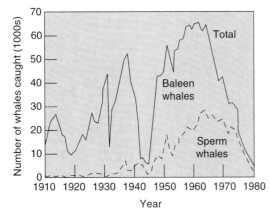

FIGURE 11.24

Total annual catches of baleen and sperm whales in all oceans from 1910 to 1980. A moratorium on commercial whaling began in 1985–86.

Commission (IWC) to set catch quotas and self-regulate the industry. The annual harvest of whales between 1910 and 1980 is given in figure 11.24, and a comparison of original populations and present-day populations is given in table 11.4.

A moratorium on commercial harvesting of whales except dolphin and porpoise, voted by the IWC in 1982, began in 1985–86. Its purpose has been to study the whale populations and assess their ability to recover. Some countries (Japan, Norway, Iceland, Korea, and the former Soviet Union) turned to so-called "scientific" or "research" whaling to assess stocks, but with continued pressure from the IWC all but Japan abandoned their research whaling. Japan continues a research harvest of minke whales, and subsistence whaling by native peoples is carried on in the Arctic.

The 1991 IWC meeting discussed the resumption of commercial whaling, because whale stocks were increasing as a

247

Life in the Water

FIGURE 11.25
Assorted marine mammals;
the elephant seal, harbor
seal, and California sea lion
are pinnipeds.

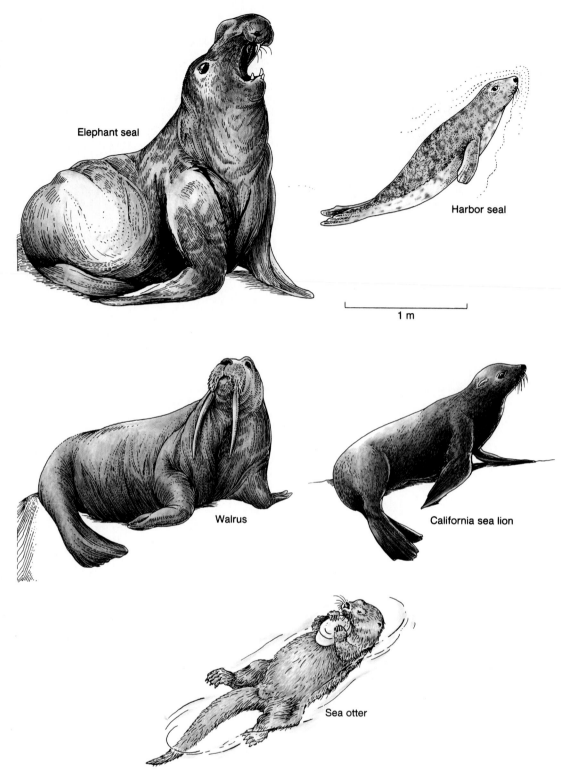

Elephant seal

Harbor seal

1 m

Walrus

California sea lion

Sea otter

result of the moratorium. Whaling nations at this meeting reminded the IWC that the original intent of the organization was to "conserve" whales in order to protect the stocks as harvestable natural resources. The 1992 and 1993 meetings continued the moratorium, but in 1993 the scientific committee concluded that the minke whale population (now about 700,000) was large enough to support small-scale whaling. The IWC declined to accept the scientific committee's recommen-

dation; the committee chairman resigned, and Norway announced its resumption of whaling. In 1994 the IWC voted to establish a whale sanctuary in Antarctic waters below 55°S.

Representative **pinnipeds** ("feather-footed"), named for their four characteristic swimming flippers, are shown in figure 11.25. These animals still retain ties to land, spending considerable time ashore on rocky beaches, ice floes, or in shore caves. They are found from the tropics to the polar seas, ranging

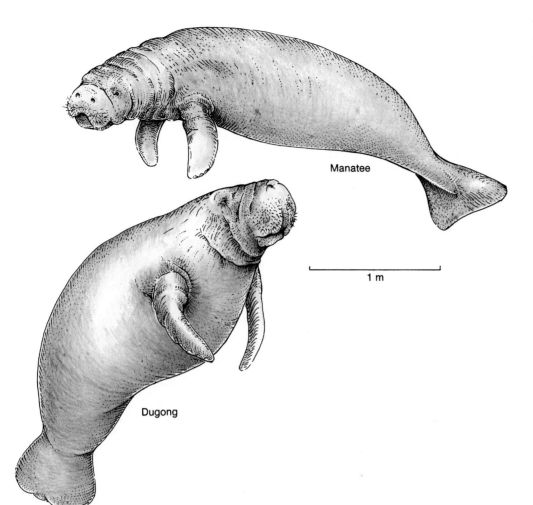

FIGURE 11.26
The tropic manatee of the
Caribbean and the dugong of
Southeast Asia are herbivorous
marine mammals.

Manatee

1 m

Dugong

from the nearly extinct monk seal of the western Hawaiian Islands and Mediterranean area to the fur seals of the Arctic. The common harbor seal, the harp seal of the northwest Atlantic, the leopard seal of the Antarctic, and the elephant seal, the male with its great pendulous snout, are all true seals, or seals without external ears and with torpedo-shaped bodies that require them to use a wriggling motion on land. The northern fur seal and the sea lion are eared seals with longer necks and supple forelimbs tipped with broad flippers, on which the animal can walk and hold its body in a partially erect position when on land.

Seals and sea lions are currently enjoying a period of relative peace, compared to the sealing days of the nineteenth and early twentieth centuries. The northern fur seal's 1870 population of 2–2.5 million was reduced to 300,000 by 1914. The present population is approximately 1.1 million. After a period of decline blamed on the 1982–83 El Niño and entanglement with nets, plastic strapping rings, and other debris, the population appears to have stabilized and is increasing. The 1988 seal pup count was 200,000, up from 169,000 in 1985.

Manatees and **dugongs** or sea cows (fig. 11.26) are slow-moving, docile herbivores. Manatees are found in the brackish coastal bays and waterways of the warm southern Atlantic coasts and in the Caribbean. Dugongs are found in waters around Southeast Asia, Africa, and Australia. At present, the growth of

human populations is putting increasing hunting pressure on the dugong in the southern Pacific. Manatees in the coastal waters of the Caribbean and tropical Atlantic are frequently injured and killed by collisions with the propellers of large and small vessels.

In 1972 the U.S. Congress established the Marine Mammal Protection Act. The act includes a ban on the taking or importing of any marine mammals or marine mammal product. "Taking" is defined by the act as harvesting, hunting, capturing, or killing any marine mammal or attempting to do so. The act covers all U.S. territorial waters and fishery zones. The act effectively removed these animals and their products from commercial trade in the United States. Only under strict permit procedures and with the approval of the Marine Mammal Commission can a few individual marine mammals be caught for scientific research and public display. Since passage of the act, the numbers of some species have increased, and the interactions between people and marine mammals competing for the same resource or habitat are creating problems that are not easily solved. Harbor seal populations have increased 7–10 percent per year along the U.S. west coast. The California sea lion population increases each year, and between 80,000 and 125,000 sea lions consume 100,000–250,000 tons of commercially valuable Pacific hake (whiting) annually along the West Coast of the United States.

Life in the Water

11.14 REPTILES

Few reptiles are found in today's oceans (fig. 11.27). The gavial is a slender-nosed, fish-eating crocodile living in the coastal waters of India. The only modern marine lizard is the big, gregarious marine iguana of the Galápagos Islands. It lives along the shore and dives into the water at low tide to feed on the algae. This iguana has evolved a flattened tail for swimming and has strong legs with large claws for climbing back up on the cliffs.

There are about fifty different kinds of sea snakes found in the warm waters of the Pacific and Indian Oceans; there are no sea snakes in the Atlantic Ocean. Sea snakes are extremely poisonous with small mouths, flattened tails for swimming, and nostrils on the upper surface of the snout that can be closed when the snakes are submerged. The snakes dive as deep as 100 meters (300 ft) and stay submerged as long as two hours. Sea snakes eat fish, and most reproduce at sea by giving birth to live young.

Sea turtles live in the ocean but nest on land. The four large sea turtles are the green, hawksbill, leatherback, and loggerhead turtles. Green sea turtles are herbivores; they weigh 140 kilograms (300 lbs) or more. The hawksbill is a tropic species feeding primarily on reef sponges. The loggerhead, weighing between 70 and 180 kilograms (150–400 lbs), is usually found around wrecks and reefs feeding on crabs, mollusks, and sponges.

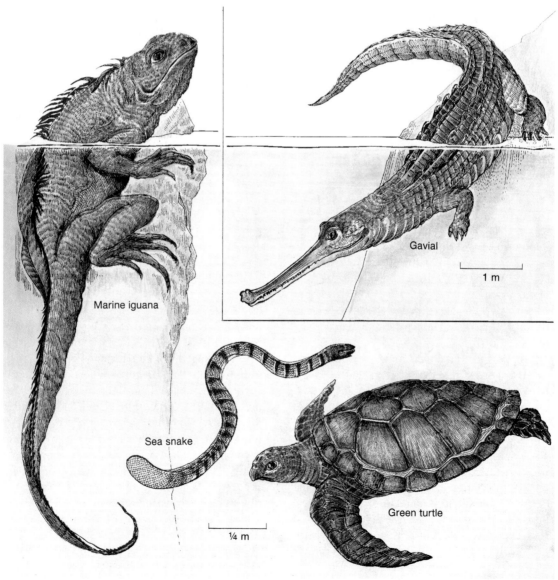

Marine iguana

Gavial

1 m

Sea snake

¼ m

Green turtle

FIGURE 11.27

Marine reptiles.

The giant leatherback (up to 650 kilograms or 1400 lbs) feeds only on jellyfish, following them over large areas of the ocean.

These animals are all migratory and make long sea journeys between their nesting sites and their foraging areas. The female turtle lays a hundred or more eggs in a scooped-out depression in the warm sand, covers them, and returns to the sea. While the eggs incubate, they are an easy prey for humans, dogs, rats, and other carnivores. In more and more cases no eggs survive to replenish the population. If the young turtles hatch, they must make their way across the sand while hungry birds attack and then evade the waiting predatory fish.

The greatest threat to the survival of all sea turtles is humans. Turtle eggs and turtle meat are prized throughout the Pacific, and nests are regularly raided by poachers. Hawksbill turtles are hunted for tortoiseshell, which is used to make combs, boxes, and jewelry. Turtle skins are used for leather and their fat and oils for cosmetics. Although placing sea turtles on the United States' endangered species list prevents the legal importing of turtle products into this country, according to the U.S. Fish and Wildlife Service there is a vast international illegal trade in turtle products that rivals the illegal ivory trade in dollar value. The problem of turtles drowning in shrimp nets is believed to play a significant role in the depletion of sea turtle populations. In 1989 the 15,000 full-time and 40,000 part-time U.S. shrimping trawlers working the Gulf of Mexico were required by federal law to fit their nets with turtle excluding devices (TEDs) that allow the turtles but not the shrimp to escape the nets. Shrimpers resisted the TED requirement, claiming it reduced shrimp yields, but a 1994 statement from a National Marine Fisheries Service (NMFS) spokesperson noted improved compliance, approaching 90 percent. Efforts to protect nesting beaches are growing; some turtle nesting beaches in Florida, Costa Rica, and Mexico have become national parks, where ecotourism has become an important part of the economy. But many nesting beaches are now lined with condominiums and resort communities, and on others the harvesting and poaching continues.

What reptiles inhabit the marine environment?

Why are sea turtles in danger of extinction?

Table 11.5 is a summary of the nekton, their distribution and characteristics.

11.15 COMMERCIAL FISHERIES

World Fish Catches

In 1950, the total world marine fish catch was approximately 21 million metric tons. During the next forty years, as human populations exploded, the fishing effort by all nations intensified, and the technology and gear used to hunt and catch the fish improved dramatically. The catch reached 77 million metric tons in 1970 and rose to 81 million metric tons in 1986. By 1989 the world marine fish catch was 86 million metric tons,

but in 1990, for the first time in forty years, the world's marine harvest dropped to 83 million metric tons. In 1991 it had fallen again, to 81.7 million metric tons.

Fisheries and Overfishing

Around the world too many fishing boats are taking too many fish, too fast. Diminishing fish populations are the victims of relentless overfishing, and harvest regulations fail to keep pace with declining stocks. Fifteen percent of the species currently fished in U.S. waters are near their maximum population potential, and according to the U.S. Office of Fisheries Conservation and Management, 41 percent of the species are overfished. In 1992 Canada banned cod fishing in the waters off Newfoundland and in 1993 and 1994 extended the ban south into the Gulf of St. Lawrence, effectively shutting down its historic cod fishery. Iceland warned its fishing industry that their cod stock could collapse unless the annual catch was cut back by 40 percent, beginning in 1993. North Atlantic swordfish landings in the United States have declined 70 percent from 1980 to 1990, and the average weight of the swordfish has fallen from 115 to 60 pounds. The western Atlantic adult bluefin tuna population dropped from 300,000 to 30,000 in twenty years, and many believe this fish is doomed as a single fish sells for more than $35,000 in Japan. These data mean trouble for people who depend on fish as their source of protein as well as for the hundred million people that make their living by catching, processing, or selling fish.

Today, as overfishing takes its toll of familiar species, the pressure is placed on other species. Until recently, sharks were fished for sport and for local markets, but recent estimates are that more than 200 million sharks are being taken annually, twice the catch of five years ago. Annual U.S. sales of shark steaks and fillets have increased from 3000 metric tons in 1985 to 5000 metric tons in 1990. The Asian demand for shark fin soup supports a shark fishery that amputates the shark fins and throws the carcasses back into the ocean. When commercial shark fishing began in the late 1970s, 21 metric tons of thresher sharks were brought to California docks; by 1982 the thresher catch was 1100 metric tons. In 1989 the thresher fishery collapsed when the catch fell to 300 metric tons; in the 1990s the U.S. east-coast shark fishery also crashed. The worldwide shark harvest dropped from a high of 34,200 metric tons in 1959 to 12,334 metric tons in 1991. Lack of management and lack of information have led to overfishing in all the world's oceans, and the sharks' low reproductive rate and slow growth and maturation rates have intensified the problem. The National Marine Fisheries Service (NMFS) (fig. 11.28) has worked out a management proposal for U.S. shark fishing that reduces commercial fishing by 30 percent and bans cutting off the fins and then dumping the carcass back into the sea. Additional programs are needed, for a worldwide expansion of shark fishing could be devastating to this link in the ocean food web.

Overfishing is sometimes combined with environmental degradation or with environmental change that intensifies its effects. The anchovy fishery concentrated in the upwelling zone off the coast of Peru has produced the world's greatest fish

Life in the Water

Table 11.5 Nekton

Type		Distribution	Characteristics
Mammals			
Whales	Toothed	Worldwide polar to tropic open ocean and coastal waters, some migratory.	Carnivores, prey on fish, squid, seals. Includes sperm and killer whales, dolphins, and porpoises.
	Baleen	Worldwide polar to tropic open ocean and coastal waters, some migratory.	Includes the "great whales"; many populations depleted by whaling industry. Krill feeders.
Pinnipeds	True seals	Mainly polar to midlatitudes; require shore or sea ice areas.	No external ears, torpedo shape, nonrotating flippers, awkward on land. Consume a variety of vertebrates and invertebrates. Harbor seal, elephant seal.
	Eared seals	Mainly polar to midlatitudes; require shore or sea ice areas.	External ears, longer necks, rotating flippers, more agile on land. Consume a variety of invertebrates and vertebrates. Sea lion, northern fur seal.
Walrus		Arctic regions only, needs shore and sea ice areas.	Tusks in males and females. Up to two tons in size, rotating flippers, no external ears. Feeds on shellfish.
Sea cows	Dugongs	Tropical, Australia, Southeast Asia, Africa coastal areas.	Docile, slow-moving herbivores. Heavily impacted by humans.
	Manatees	Tropical, Caribbean, and Gulf of Mexico coastal areas.	Docile, slow-moving herbivores. Heavily impacted by humans.
Sea otters		Subarctic and midlatitude, cold coastal waters.	No blubber layer; fur is the only insulation. Eat and sleep in the water. Feed on shellfish, crabs, and sea urchins.
Fish			
	Cartilaginous (sharks, skates, and rays)	Worldwide, subpolar to tropics, inshore and offshore.	Cartilaginous skeleton. Excellent sensory abilities. Largest sharks and rays are plankton feeders; others are predators and scavengers. Some skates and rays are bottom feeders. Some commercial fishing.
	Bony fish (bottom-living: halibut, flounder, sole, rock fish)	Worldwide, large depth range, associated with the sea floor. Large concentrations in sublittoral.	Bony skeleton, scales and fins. Bottom and near-bottom feeders. Some flattened; many fished commercially.
	Bony fish (pelagic, shallow: tuna, salmon, pollock, anchovies)	Worldwide, upper 200 meters in coastal and upwelling areas.	Bony skeleton, scales and fins. Large variety of fast-swimming types. Some school. Many fished commercially.
	Bony fish (pelagic, deep sea: lantern fish)	Worldwide, 200 to 3000 meters in the sea.	Bony skeleton, scales and fins. Many bioluminescent. Abundant and small. No commercial fisheries.
Reptiles			
	Turtles	Warmer waters worldwide. Require beaches for laying eggs.	Herbivores and carnivores. Migratory. Long-lived but heavily impacted by humans.
	Snakes	Warm waters of Pacific and Indian Oceans.	Poisonous, fish eaters, adapted for swimming and diving.
	Lizards	Galápagos Islands, Pacific.	Marine-swimming iguana, herbivore.

catches for any single species. Anchovies are small, fast-growing fish that feed directly on the phytoplankton in the upwelling and travel in dense schools that are easy to net in large quantities. These fish are converted to fish meal that is exported as feed for domestic animals.

The fishery began in 1950 with a harvest of 7000 metric tons; by 1970, the peak year, about 12.3 million metric tons were taken. As the fishing increased, the average size of the fish taken decreased, and it took more and more fish (smaller and younger) to make up the same catch. During this time the

FIGURE 11.28
National Marine Fisheries scientists sample shark populations to evaluate commercial fisheries impacts.

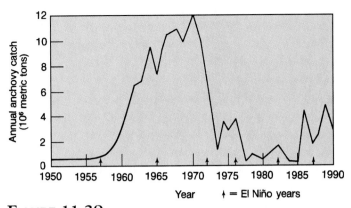

FIGURE 11.29
The anchovy catch of Peru and Ecuador by calendar year given in millions of metric tons. Years of El Niño are 1957, 1965, 1972, 1976, 1982, and 1986.

Peruvian coast was visited three times by El Niño (refer to Chapter 6). The heavy fishing of 1970–71 was followed by a severe El Niño in 1972, and only 2 million metric tons of anchovies were harvested in 1973. After 1973, government quotas were placed on the catch, and El Niño struck again in 1976 and 1982–83. The 1983–85 catches were kept below 150,000 metric tons. Since then the catch has stabilized with harvests between 1988 and 1990 being 3.6, 5.4, and 3.8 million metric tons, respectively. The variations in catch since 1950 and the El Niño years are shown in figure 11.29.

The wild salmon runs of the Pacific Northwest and Alaska are becoming smaller, and the fishing is now tightly controlled. The fishing season is shortened and the catches decrease, causing profits for those who fish to fall; management of fish stocks requires more and more regulation in attempts to ensure a returning population for the next year. The numbers of coho and chinook salmon caught off the Oregon coast in 1992 were 62 percent below the 1991 catch; a fishing season that once extended from May to October was opened for only two weeks in 1993. In 1994 Washington and Oregon closed their coastal and estuarine salmon fisheries for the first time.

The degradation of the freshwater environment to which the wild fish must return for spawning adds to their difficulties. Salmon need high-quality fresh water and clean, gravel-bottomed, shady, cool streams to spawn. But rivers are routinely dammed for power and flood control, cutting off salmon from their home streams. Trees are harvested to the stream edges, removing shade and allowing mud and silt eroded from the exposed land to cover the streambed. Wastes in the stream degrade the water quality and damage the juvenile fish and the returning runs. Although these activities are being corrected in many areas, the populations of naturally spawning fish continue to decline. However, variations in the return of combined wild and hatchery salmon that mature in one, two, or three years can still produce high fishing yields. For example, the 1991

Alaska pink salmon harvest set a record of 175 million fish, surpassing the 1990 record of 155 million fish.

Even with increased management and regulation of fish stocks, the general trend of ocean fishing is down and the costs are up. One way to increase the fish harvest is by developing new fisheries such as the North Pacific bottom fishery and the South African pilchard fishery, and another is to foster an increased consumer demand for new fish and fish products. For example, Alaskan pollock, a bottom fish, is processed to remove the fats and oils that give the fish its flavor, and a highly refined fish protein called **surimi** is processed and flavored to form artificial crab, shrimp, and scallops. Surimi is a major fish product in Japan and has a growing market in the United States.

In all cases the fishing boat and gear, the crew's wages, and the fuel required are increasing the cost of commercial fishing (fig. 11.30). In the United States, those who fish are usually independent operators who sell their catch directly to the processor. The U.S. consumer prefers fish fillets and fish steaks, not low-cost minced fish products; therefore, the fishing is for high-cost fish such as salmon, swordfish, and halibut.

In every fishery large numbers of unwanted fish are caught incidentally while fishing for other species. These "trash fish" are referred to as **incidental catch** or **by-catch** and represent a tremendous waste of marine resources. In the United States, if incidental catch does not bring a high enough price and if processors are not available, these fish will be returned to the sea, usually dead. In other areas of the world, most of the by-catch is marketed locally. For example, the world's shrimp fishery is estimated to have an annual catch of 1.1–1.5 million tons; the associated by-catch of finfish is 5–21 million tons, of which 3–5 million tons are discarded. Alaskan trawlers for pollock and cod throw back to the sea some 25 million pounds of halibut, worth about $30 million, as well as salmon and king crab because they are prohibited from keeping or selling this incidental catch. Another 550 million pounds of ground fish are discarded in Alaskan waters to save space for larger or more valuable fish.

Life in the Water

FIGURE 11.30
A salmon purse seiner hauling in its net.

Fish Farming

An alternative way to increase the harvest is by fish farming. Farming the water, known as **mariculture** or **aquaculture,** began in China some 4000 years ago. In China, Southeast Asia, and Japan, fish farming in fresh and salt water has continued to the present as a practical and productive method of raising large quantities of fish. The methods are labor intensive, and most fish farms are small, family-run operations. In **monoculture,** fish are raised as a single species; in **polyculture,** surface feeders and bottom feeders are raised together, making use of the total volume of the pond.

Fish farming in the United States produces only 2 percent of our fishery products, but that amount includes 50 percent of the catfish and nearly all of the trout. To succeed in the United States, fish farming requires a proven market and a large-scale operation to keep costs down. To be economically productive, a farmed species must have juvenile forms that survive well under controlled conditions, gain weight rapidly, eat a cheap and available food, and adults that fetch a high market price and reproduce in captivity.

Salmon, shrimp, and oyster culture have grown steadily in the last twenty years. In 1990, salmon pens around the world produced 242,000 metric tons of salmon, a 58 percent increase over 1988. The two largest producers were Norway, harvesting 150,000 metric tons, and Scotland, producing 35,000 metric tons followed by Japan at 26,000 metric tons and Canada with 23,300 metric tons. The world's largest salmon exporters in 1992 were Norway, Chile, and Canada. Although pen-rearing of salmon has been tried in the protected waters of Puget Sound in Washington state (fig. 11.31), it is an expensive operation that raises problems in the Sound's multiple-use waterways.

FIGURE 11.31
An aerial view of salmon pens in which fish are raised to market size by Sea Farm Washington, Inc.

Life in the Water

FIGURE 11.32
The ocean-ranching release and recapture facility at Coos Bay, Oregon, is operated by Anadromous, Inc., of Corvallis, Oregon.

Salmon pens are not attractive to those with shoreside homes, and the pens interfere with recreational boating and water sports. Also the fish excrete wastes into the water.

Aquaculture research is focusing on several other fish species for marketing by the year 2010. In Hawaii pen-reared mahi mahi now grow to 1.5 kilograms (3 lbs) size in 150 days, and are expected to be the first to enter the market. Norway has the technology to grow Atlantic cod, but the cost is still too high for the commercial market. The United States and Canada are experimenting with the Atlantic halibut.

Hatchery rearing and release of seagoing salmon and steelhead trout are called **ocean ranching** or **sea ranching.**

Federal and state hatcheries provide millions of fish, which are caught in both the commercial and recreational fisheries along the coasts of Oregon, Washington, and Alaska. The release from private hatcheries of salmon that will return to the hatcheries to be harvested is being tried in Oregon (fig. 11.32). The idea of private ocean ranching has not proved popular with those who fish commercially, because they fear the intrusion and control of the large companies and a possible drop in salmon prices. Most economists and fishery biologists believe that an ocean-ranching program in the private sector would add fish to the harvest without any great effect on market price.

SUMMARY

Marine organisms are grouped as plankton, nekton, and benthos. Phytoplankton are plants and zooplankton are animals. The marine environment is subdivided into the benthic and pelagic zones; there are subdivisions of each of these zones.

Organisms of the sea are buoyed and supported by the seawater. Adaptations for staying afloat include floats, swim bladders, oil and fat storage, and extended surface-area appendages.

Most marine fish lose water to their environments. They drink continually and excrete salt to prevent dehydration. Salinity is a barrier to some organisms; others can adapt to large salinity changes.

Animals use color for concealment and camouflage, and to warn predators of poisonous flesh and bitter taste. Barriers for marine organisms include water properties, light intensity, and seafloor topography.

The plankton are the drifting organisms of the oceans. Phytoplankton are single cells or chains of cells. *Sargassum* is the only large planktonic seaweed. Diatoms are round or elongate golden-brown algae with silica coverings. They reproduce rapidly by cell division and make up the first trophic level of the open ocean.

Heavy blooms of some dinoflagellates produce red tides; some are toxic and others are not. Dinoflagellate toxin is concentrated in shellfish and produces several types of poisoning in humans. Red tides are apparently triggered by certain combinations of environmental factors and the disturbance of dormant dinoflagellates. Domoic acid, produced by diatoms, is a newly discovered toxin. Ciguatoxin is another dinoflagellate toxin that affects humans and hampers fishery development.

Concentrations of zooplankton are found at convergence zones and along density boundaries. Some zooplankton migrate toward the sea surface at night and away from it during the day.

The calcareous-shelled foraminiferans and the silica-shelled radiolarians are single cells. The copepods and euphausiids are the most abundant zooplankton. Euphausiids, also known as krill, are the basic food of the baleen whales. Other small members of the plankton are the carnivorous arrowworms and the swimming snails or pteropods. Large zooplankton include the nearly transparent comb jellies and jellyfish.

Zooplankton that spend their entire lives in the plankton are holoplankton. Meroplankton are juvenile (or larval) stages of nonplanktonic adults including fish eggs, very young fish, and the larvae of barnacles, snails, crabs, and starfish. Marine bacteria are also planktonic and exist on any available surface.

Plankton are sampled with nets and pumps. Various techniques are used to determine the kinds and numbers of organisms.

Nekton swim freely. Squid are fast-swimming invertebrate nekton, but fish dominate the nekton. Sharks, rays, and skates are fish with cartilaginous skeletons. All other fish have bony skeletons, including the commercially fished species and the highly specialized types of the deeper sea.

Marine mammals include whales, seals, sea lions, and sea cows. The whalers hunted the great baleen whales and the toothed sperm whale. As hunting techniques changed, harvesting became more intensive, and some species of whales have been threatened with extinction. The International Whaling Commission regulates whaling on a voluntary basis; its moratorium on commercial whaling continues. Norway has broken the moratorium, and other nations have harvested whales for population assessment. The IWC has established a whale sanctuary in Antarctica.

Pinnipeds are widespread; fur seals were hunted heavily during the nineteenth and early twentieth centuries but are now protected, and their populations are increasing. Manatees and dugongs are found in the warm waters of the Indian and Atlantic Oceans. The Marine Mammal Protection Act protects all marine mammals in U.S. waters and prohibits U.S. commercial trade in marine mammal products.

Sea snakes, the marine iguana, the gavial, and sea turtles are reptile members of the nekton. Sea turtle populations are under great pressure from hunters and poachers.

World fish catches increased dramatically between 1950 and the 1980s due to increased fishing efforts and improved technology, but declined in 1990. Overfishing has been relentless. Atlantic cod fisheries and North Pacific coastal salmon fisheries have been cut back and some closed in the 1990s. The Peruvian anchovy fishery nearly collapsed with overfishing and El Niño events. The West Coast shark fishery is threatened with collapse due to its recent explosive growth. Declining fish catches, increased regulation, incidental catch, and increased costs are worldwide problems.

Aquaculture has proved to be practical and productive in Asia, but in the United States, it is a small industry that requires large-scale operations, proven markets, and cost-cutting technology. Salmon farming has increased, especially in Norway, Chile, and Scotland. Ocean ranching is the release of hatchery-raised seagoing fish to increase fishing stocks.

KEY TERMS

plankton	buoyancy	copepod	cetaceans
phytoplankton	osmosis	euphausiid	vertebrate
zooplankton	bioluminescence	krill	baleen
benthos	algae	pteropod	pinniped
nekton	diatom	mollusk	manatee
pelagic	dinoflagellate	ctenophore	dugong
benthic	bloom	cilia	surimi
neritic	flagella	coelenterate	incidental catch/by-catch
oceanic	red tide	colonial	mariculture
photic	deep scattering layer (DSL)	holoplankton	aquaculture
aphotic	foraminiferan	meroplankton	monoculture
supralittoral	radiolarian	larvae	polyculture
littoral/intertidal	crustacean	invertebrate	ocean ranching/sea ranching
sublittoral/subtidal			

Life in the Water

SUGGESTED READINGS

Anderson, D. M. 1994. Red Tides. *Scientific American* 271 (2):62–68.

Conniff, R. 1993. Disappearing Shadows in the Surf. *Smithsonian* 24 (2):32–43. Worldwide drops in shark populations and conservation methods.

Costa, D. P. 1993. The Secret Life of Marine Mammals. *Oceanography* 6 (3):120–27.

Gentry, R. 1987. Seals and Their Kin. *National Geographic* 171 (4):475–501.

Horn, M. H., and R. N. Gibson. 1988. Intertidal Fisheries. *Scientific American* 258 (1):64–70.

Kane, H. 1993. Growing Fish in Fields. *World Watch* 6 (5):20–27.

Klimley, A. P. 1994. The Predatory Behavior of the White Shark. *American Scientist* 82 (2):122–33.

Life in the Sea, Readings from Scientific American. 1982. W. H. Freeman, San Francisco, 248 pp.

Lohmann, K. J. 1992. How Sea Turtles Navigate. *Scientific American* 266 (1):100–6.

Minton, S. A., and H. Heatwole. 1978. Snakes and the Sea. *Oceanus* 11 (2):53–56.

Nelson, K., and C. Arneson. 1985. Marine Bioluminescence: About to See the Light. *Oceanus* 28 (3):13–18.

Oceanus. Spring 1989. 32 (1). Issue devoted to whales.

O'Shea, T. J. 1994. Manatees. *Scientific American* 271 (1):66–72.

Ritchie, T. 1989. Marine Crocodiles. *Sea Frontiers* 35 (4):212–19.

Roper, C. F. E., and K. J. Boss. 1982. The Giant Squid. *Scientific American* 246 (4):96–105.

Rudloe, A., and J. Rudloe. 1994. Sea Turtles: In a Race for Survival. *National Geographic* 185 (2):94–121.

Sanderson, S. L., and R. Wassersug. 1990. Suspension-Feeding Vertebrates. *Scientific American* 262 (3):96–101. Plankton-feeding sharks and whales.

VanDyk, J. 1990. The Long Journey of the Pacific Salmon. *National Geographic* 178 (1):3–37.

Wacker, R. 1994. Strip Mining the Seas. *Sea Frontiers* 40 (3):14–17 and 60.

Ward, P., L. Greenwald, and O. E. Greenwald. 1980. The Buoyancy of the Chambered Nautilus. *Scientific American* 243 (4):190–203.

Wrobel, D. J. 1990. Transient Jewels. *Sea Frontiers* 36 (2):8–17. Jellyfish and comb jellies.

Wu, N. 1990. Fangtooth, Viperfish and Black Swallower. *Sea Frontiers* 36 (5):32–39. Deep-sea fish.

Wursig, B. 1988. The Behavior of Baleen Whales. *Scientific American* 258 (4):102–7.

Marine Birds

There are an estimated 8600 species of birds, but only about 3 percent of these are considered marine species. Marine birds include the oceanic albatross, the flightless penguin, the large pelicans and cormorants, the diving puffins, the familiar gulls, and the shorebirds such as herons, egrets, sandpipers, sea ducks, and oystercatchers. Their association with the sea may be periodic or continuous, but all spend a large portion of their lives there. Some are so well adapted to oceanic life that they rarely come ashore; others move daily into coastal waters to feed, but all return to shore to nest. They often congregate in large groups, thousands and hundreds of thousands of birds crowded together along shore cliffs and beaches or on islands. Since a very large portion of the population may be in one small area at the same time, they are vulnerable to human predation and pollution. Most have definite breeding seasons, and they may migrate thousands of kilometers as they travel from feeding ground to breeding area.

Seabirds

The wandering albatross of the southern oceans is the most truly oceanic of marine birds; it has the largest wing spread of all birds, 3.5 meters (11 ft). These great white birds with black-tipped wings spend four to five years at sea before returning to their nesting sites. The smallest of the oceanic birds, Wilson's petrel, is a swallowlike bird that breeds in Antarctica and flies 16,000 kilometers (10,000 mi) along the Gulf Stream to Labrador during the Southern Hemisphere winter, returning to the Southern Hemisphere for the southern summer, another 16,000 kilometers.

Penguins (fig. 1) are gregarious birds living in crowded breeding areas or rookeries; they swim in flocks using their wings for propulsion and steering with their feet. Their underwater swimming speed is almost 16 kilometers per hour (10 mph). They feed on fish, krill, squid and shellfish. All but the Galápagos penguin are found in the Southern Hemisphere, and two species, the emperor and the adelie, are found in Antarctica. The emperor is the largest of the penguins; it is 1.3 meters (4 ft) tall and weighs 45 kilograms (100 lbs). This penguin can remain under water for more than 15 minutes and dive to depths of 265 meters (900 ft). The emperor penguin breeds on the Antarctic ice sheet, keeping its egg warm by holding it on its feet and covering it with a fold of skin.

FIGURE 1

These Magellanic penguins are part of a 500,000-bird colony at Punta Tombo, Argentina.

Pelicans and cormorants are large fishing birds with big beaks. They are strong fliers found mostly in coastal areas, but some venture far out to sea. A pelican has a particularly large beak from which hangs a pouch used in catching fish. White pelicans of North America nest inland but winter in California, in Florida, and along the Gulf Coast. They fish in groups, herding small schools of fish into shallow water and then scooping them up in their large pouches. The brown pelican (fig. 2) lives along the Pacific and southeastern coasts. It does a spectacular dive from up to 10 meters (30 ft) above the water to capture its prey. Cormorants are black, long-bodied birds with snakelike necks and moderately long bills that are hooked at the tip. They are found along the coasts of all the world's oceans. Cormorants settle on the water and make repeated dives from the surface. Swimming primarily with their feet but also using their wings, they chase and catch fish under water. Cormorants float low in the water with only their necks and heads above the surface, for they do not have the water repellent feathers of other seabirds and must return to land periodically to dry out. There is a flightless cormorant in the Galápagos Islands.

continued . . .

FIGURE 2
A brown pelican takes off from the water.

FIGURE 4
A mixed flock of dunlins and sanderlings takes flight along a sandy beach.

FIGURE 3
A western gull, one of the large gulls of North America's Pacific coast.

Gulls and terns are found all over the world except in the South Pacific between South America and Australia. Gulls (fig. 3) are strong flyers and feed on anything and everything, foraging over beach and open water. Most are white or white with some black or gray. The terns are smaller and more graceful with a slender bill and forked tail; they plunge into the water to catch fish. The Arctic tern breeds in the Arctic and in winter migrates south of the Antarctic Circle, a round-trip of 35,000 kilometers (20,000 mi).

The puffins, murres, and auks are heavy-bodied, short-winged, and short-legged diving birds. They feed on fish, crustaceans, squid, and some krill. All are limited to the North Atlantic, North Pacific, and Arctic areas where most nest on isolated cliffs and islands. In prehistoric times great auks were abundant on both sides of the North Atlantic. The great auk was a large, slow, flightless bird, 0.6 meters (2 ft) high that provided food for generations of sea travelers. It was easily killed for both its meat and its feathers (for stuffing mattresses), and as its numbers dwindled museums and private collectors paid more and more for each bird. The last two were killed on a small island off Iceland on June 2, 1844.

Shorebirds

Most shorebirds are migratory and are present in very large numbers (thousands and hundreds of thousands) only at certain times of the year. Large flocks arrive on intertidal sands and mudflats, but there is no severe competition for food, because the different kinds of birds arrive at different times of the year. There is also no extreme depletion of food, for areas have time to recover when a migratory population moves on. Different birds select different types of shoreland; for example, herons, egrets, and dowitchers wade in the calm waters of salt marshes, but sandpipers prefer more exposed beaches (fig. 4).

Shorebirds venture into coastal waters to feed, and each bird exploits a different food resource related to the length of its neck, bill, and legs. The varying lengths and shapes of their bills allow different kinds of birds to specialize in food at different depths of mud or sand. Herons and egrets are waders; their long necks enable them to strike at small fish and insects in the water. Sandpipers probe the sand and mud with their bills to locate worms and other organisms; plovers use their eyesight to detect slight movements on the mud surface, allowing them to find their prey. Oystercatchers feed on shellfish and crabs in the rocky intertidal zone; ducks, terns, and gulls swim and dive in the estuaries and rest on the beaches and mudflats.

Today and Tomorrow

Marine birds compete with humans for space and food resources along the world's coasts. All over the world increasing numbers of people choose to live or vacation along coasts and beaches, and isolated shore areas available to birds are rapidly decreasing. Wetlands are diked and filled removing food sources and nesting sites; vacation homes, harbors, and resorts are built along once lonely beaches; runoffs from town, industry, and agriculture contaminate estuaries; accidental spills and industrial dumping pollute inshore waters. The intensity of fishing has increased in all the oceans, and as fish populations have diminished, there is less fish for both humans and birds. The flocks decrease until at last they no longer visit beaches that once supported them by the hundreds of thousands. Even in still secluded areas, increasing numbers of people and pets strolling the beaches and boaters visiting previously undisturbed bays and rocky islets unintentionally interfere with the habitat. Most recreational uses of shore areas are in the late spring and early summer, and this is also the time that most marine birds look for isolation to nest and rear their young (fig. 5).

FIGURE 5
Kittiwakes (gull-like birds) and common murres nesting on an island in Kachemak Bay, Alaska.

Some birds do find ways to use the changes humans make, and these populations may thrive. Birds use jetties and docks as nesting sites and lookouts for prey. Cormorants sit on old wharves and net sheds to dry their wings, and gulls nest on the flat roofs of commercial buildings. Artificially constructed and maintained beaches become feeding and nesting areas. The submerged portions of piers and pilings provide a home for many organisms that are an important part of some marine birds' diets.

Natural disasters such as volcanic eruptions and landslides disturb nesting areas, and events such as El Niño, with its disruption of the fish populations, depress the great bird populations along the west coast of South America (refer to sections 6.14 and 11.15). However these disruptions are usually cyclical; a population decrease in one year builds up again over the next series of unaffected years. Human changes are generally more permanent, and the flocks of migratory seabirds that were described by naturalists of the seventeenth, eighteenth, and nineteenth centuries may never be seen again.

Life in the Water

C H A P T E R

12

Life on the Sea Floor

Outline

Learning Objectives

After reading this chapter, you should be able to

■ Describe a typical alga and explain its structure.

■ List the major groups of seaweeds and explain where they are found.

■ Understand how benthic plants modify the marine environment for other marine organisms.

■ Recognize and appreciate the diversity of benthic animals.

■ Discuss factors that modify the distribution of benthic animals.

■ Compare intertidal zonation on rocky beaches with intertidal zonation on sandy or mud beaches.

■ Give examples of adaptations that allow various benthic animals to colonize different seafloor environments.

■ Explain symbiotic relationships and give examples of the different kinds.

■ Discuss the conditions required for a stony coral reef and the interrelationships between its organisms.

■ Describe the present state of commercial benthos harvests: wild and farmed.

■ Explain how the new techniques of genetics are being used to improve farmed fish and shellfish.

◀ Red sea urchins and kelp in shallow water. British Columbia, Canada.

Most marine animal species and nearly all of the larger marine plants are benthic. The plants are found along the sunlit, shallow coastal areas and in the intertidal zone, while the benthic animals are found at all depths on the sea floor or in the sediments. The benthos includes a remarkably rich and diverse group of organisms, among them the luxuriant, colorful tropical coral reefs, the great cold-water kelp forests, and the hidden life below the surface of the mudflat and sandy beach. Benthic organisms are important food resources and provide valuable commercial harvests (for example, clams, oysters, crabs, and lobsters).

12.1   THE SEAWEEDS

Seaweeds are members of the large group called algae. Because algae photosynthesize, some biologists consider them members of the plant kingdom; other biologists consider them plantlike but prefer to place them in other categories because of their body form, reproduction, and biochemistry. Here the algae will be considered primitive members of the plant kingdom. Algae have simple tissues; they do not produce flowers or seeds; and their pigments and storage compounds vary from group to group.

Seaweeds grow attached to rocks, shells, or any solid object. Those found floating at the water's edge or thrown up on the shore are benthic organisms that have been dislodged from the bottom. Seaweeds are attached by a **holdfast** that anchors the plant firmly to a solid base or **substrate.** The holdfast is not a root; it does not absorb water or nutrients. Above the holdfast is a stemlike portion known as the **stipe.** The stipe may be so short that it is barely identifiable, or it may be up to 35 meters (115 ft) in length. It acts as a flexible connection between the holdfast and the **blades,** the plant's photosynthetic organ. Seaweed blades serve the same purpose as a leaf but do not have the specialized tissues and veins of leaves. The blades may be flat, ruffled, feathery, or even encrusted with calcium carbonate.

Seaweeds grow attached to rocky substrates; they are not found in areas of mud or sand where their holdfasts have nothing to which they can attach. Because the benthic algae are dependent on sunlight, they are confined to the shallow, sunlit depths of the ocean where they are surrounded by water with its dissolved carbon dioxide and nutrients. They are efficient primary producers, exposing a large blade area to both the water and the sun. The general characteristics of a benthic alga are shown in figure 12.1.

Algae can be divided into groups based on their pigments. Green algae are moderate in size and may form fine branches or thin, flat sheets. Brown algae range from microscopic chains of cells to the **kelps,** which are the largest of the algae. Kelps have more structure than most algae, with strong stipes and holdfasts that allow them to grow in fast currents and heavy surf, or their holdfasts may attach below the depth of wave action and their blades float at the surface supported by gas-filled floats. Kelps are especially abundant along the coasts of Alaska, British Columbia, Washington, California, Chile, New Zealand, northern Europe, and Japan. Their brown color comes from a pigment that masks their green chlorophyll. Red algae are the most abundant and widespread of the large marine algae. Their body forms are varied, flat, ruffled, lacy, or intricately branched. They also contain pigments that mask the green chlorophyll.

Although the algae are commonly classified by color, the visible color can be misleading. Some red algae appear brown, green, or violet, and some brown algae appear black or greenish. Some representative algae are shown in figure 12.2.

The characteristic pattern for seaweed growing on a rocky shore is first the green algae, then the brown algae, and last the red algae as depth increases. Remember that the quality as well as the quantity of light changes with depth in the sea (see Chapter 5). Green algae grow in shallow water because their chlorophyll absorbs light from the visible light spectrum

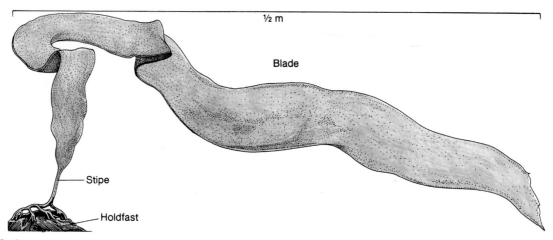

½ m

Blade

Stipe

Holdfast

FIGURE 12.1

Benthic algae are attached to the sea floor by a holdfast. A stipe connects the holdfast to the blade. *Laminaria* is a kelp and a member of the brown algae.

FIGURE 12.2

Representative benthic algae. *Ulva* and *Codium* are green algae. *Postelsia*, *Nereocystis*, and *Macrocystis* are kelps. The kelps and *Fucus* are brown algae. The red algae are *Corallina*, *Porphyra*, and *Polyneura*. *Corallina* has a hard, calcareous covering.

available at the sea surface. Brown algae are found at moderate depths; their brown pigment is more efficient at trapping the shorter wavelengths of light available at these depths. At maximum growing depths, the algae are red, for the red pigment can best absorb the remaining blue-green light.

Seaweeds provide food and shelter for many animals. They act in the sea much as the trees and shrubs do on land. Some fish and other animals, such as sea urchins, limpets, and some snails, feed directly on the algae; other animals feed on shreds and pieces that settle to the bottom. Some organisms use large seaweeds as a place of attachment; some of the smaller algae grow on the large kelps. Algae that produce calcareous outer coverings are important in the building of tropical coral reefs, discussed in section 12.10.

Life on the Sea Floor

Although the diatoms have been discussed as part of the plankton in the previous chapter, there are benthic diatoms as well. Benthic diatoms are usually elongate and grow on rocks, muds, and docks, where they produce a slippery brown coating.

Sketch a large benthic alga (seaweed). Label its parts and give the function of each part.

How do seaweeds differ from land plants?

What are the three general classes of seaweeds?

How are seaweeds distributed with depth and why are they distributed this way?

What is a kelp? How are kelps modified for their habitat? (see figure 12.2, *Nereocystis, Macrocystis,* and *Postelsia*)

What roles are played by large algae in the marine environment?

12.2 OTHER MARINE PLANTS

There are a few flowering plants with true roots, stems, and leaves that have made a home in the sea. Eelgrass, with its strap-shaped leaves, is found growing on mud and sand in the shallow, quiet waters of bays and estuaries along the Pacific and Atlantic coasts, and turtle grass is common along the Gulf Coast. Surf grass flourishes in more turbulent areas exposed to waves and surge action. These sea grasses are rich sources of food and shelter for animals and act as attachment surfaces for some algae.

In the tropics, mangrove trees grow in swampy intertidal areas (refer to fig. 9.2f), while midlatitude salt marshes are dominated by marsh grasses able to tolerate the salty water (refer to fig. 9.2g). The intertwining roots of the mangrove trees provide shelter and also trap sediments and organic material, helping eventually to fill in the swamps and move the shore seaward. Marsh grasses are partly consumed by marsh herbivores, but much of the grass breaks down in the marsh and is washed into the estuaries by the tidal creeks. There the marsh remains are broken down further by bacteria that release nutrients to the water to be recycled by the marsh grasses and other plant forms.

How are marine habitats modified by each of these plants: (a) sea grasses, (b) mangroves, (c) salt marsh grasses.

12.3 THE ANIMALS

Benthic animals are found at all depths and are associated with all substrates. There are more than 150,000 benthic species as opposed to about 3000 pelagic species. About 80 percent of the benthic animals belong to the **epifauna;** these are the animals that live on or are attached to rocky areas or firm sediments. Animals that live buried in the substrate belong to the **infauna** and are associated with soft sediments such as mud and sand.

Some animals of the sea floor are **sessile,** attached to the sea floor as adults (for example, barnacles, sea anemones, and oysters), while others are **motile,** or free-moving, all their lives (for example, crabs, starfish, and snails). Most benthic forms produce motile larvae that spend a few weeks of their lives as meroplankton (Chapter 11), allowing the species to avoid overcrowding and to colonize new areas. Sessile adults must wait for their food to come to them, either under its own power or carried by waves, tides, and currents. Motile organisms are able to pursue their prey, scavenge over the bottom, or graze on the seaweed-covered rocks.

The distribution of benthic animals is controlled by a complex interaction of factors, creating living conditions that are extremely variable. The substrate may be solid rock, shifting sand, or soft mud. Temperature, salinity, pH, exposure to air, oxygen content of water, and water turbulence change abruptly in the shallow intertidal zone; the same factors are nearly constant in deep water. Benthic animals exist at all depths and are as diverse as the conditions under which they live.

Compare infauna and epifauna. Give examples of each.

How do the food-gathering strategies of sessile and motile animals differ?

12.4 ANIMALS ON ROCKY SHORES

The rocky coast is a region of rich and complex plant and animal communities living in an area of environmental extremes. At the top of the littoral zone, organisms must cope with long periods of exposure to air as well as heat, cold, rain, snow, and predation by land animals and seabirds, also the waves and turbulence of the returning water. At the littoral zone's lowest reaches, the plants and animals are rarely exposed but have their own problems of competition for space and predation by other organisms. The exposure to air endured by marine life at different levels in the littoral zone is shown in figure 12.3.

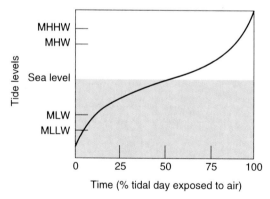

FIGURE 12.3

The time exposure to air for intertidal benthic organisms is determined by their location above and below mean sea level and by the tidal range. (MLLW = mean lower low water range; MLW = mean low water; MHW = mean high water; MHHW = mean higher high water.)

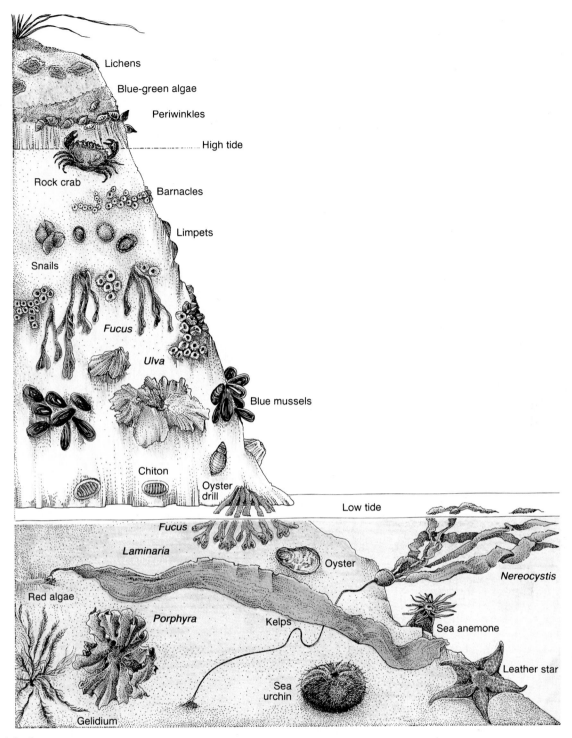

FIGURE 12.4

A typical distribution of benthic plants and animals on a rocky shore at midlatitudes. Vertical intertidal zonation is the result of the relationships of the organisms to their intertidal environment.

The distribution of the plants and animals is governed by their ability to cope with the stresses that accompany exposure, turbulence, and loss of water. Along the rocky coasts of such areas as North America, Australia, and South Africa, biologists have noted that patterns form as the plants and animals sort themselves out over the intertidal zone. This grouping is called **intertidal zonation** and is shown in figure 12.4. Zones vary with local conditions; they are generally narrow where the shore is steep or the tidal range is small, and wide where the beach is flat and the range of tides is large. The distribution of seaweeds, with the green algae in shallow water, the brown algae in the intertidal zone, and the red algae in the subtidal area, is an example of vertical zonation.

Life on the Sea Floor

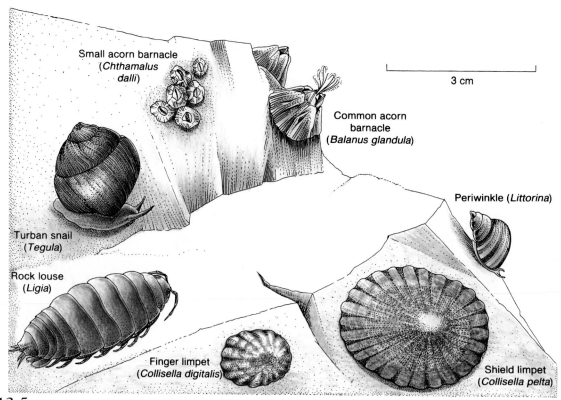

FIGURE 12.5

Organisms of the supralittoral zone. The limpets and the snail, *Littorina*, are herbivores. The barnacles feed on particulate matter in the water. *Ligia* is a scavenger.

In the supralittoral (or splash) zone, which is above the high water level and covered with water only during storms and the highest tides, the animals and plants occupy an area that is as nearly land as it is ocean bed. At the top of this area, patches of dark lichens and algae appear as crusts on rocks. Scattered tufts of algae provide grazing for small herbivorous snails and limpets (fig. 12.5). The small acorn barnacles filter food from the seawater and are able to survive even though they are covered with water only briefly during spring tides. The width of the supralittoral zone varies with the slope of the rocks, variations in light and shade, exposure to waves and spray, tidal range, and the frequency of cool days and damp fogs.

Conspicuous members of the midlittoral (fig. 12.6) include several other species of barnacles, limpets, snails, mussels, and chitons. Chitons and limpets are grazers that scrape algae from hard surfaces, while mussels filter organic material from the water. A muscular foot anchors chitons and limpets to the rocks; strong cement secures barnacles; and special threads attach mussels. Tightly closed shells protect many of these organisms from drying out during low tide, and their rounded profiles present little resistance to the breaking waves. Species of brown algae found here have strong holdfasts and flexible stipes. Mussel beds provide shelter for less conspicuous

animals such as sea worms and small crustaceans. Shore crabs of varied colors and patterns are found in the moist shelter of the rocks, and small sea anemones huddle together in large groups to conserve moisture. The area is crowded, and the competition for space is extreme. New space in an inhabited area becomes available by predation. Seasonal die-offs of algae and the battering action of strong seas and floating logs also clear space for newcomers.

A selection of organisms from the lower littoral is found in figure 12.7. The larger flowerlike anemones attach firmly to the rocks and spread their tentacles to grasp and paralyze any prey that touches them. Starfish of many colors and sizes make their home in this zone preying on shellfish, sea urchins, and limpets. The filter-feeding sponges encrust the rocks, and on a minus tide delicate, free-living flatworms and ribbon worms are found keeping moist under the mats of algae. Colorful sea slugs are active predators, feeding on sponges, anemones, and the spawn of other organisms. Although soft-bodied, sea slugs have few if any enemies because they produce poisonous acid secretions. Species of chitons, limpets, and sea urchins graze on the algae covering the rocks. Sea cucumbers are found wedged in cracks and crevices. Tube worms secrete the leathery or calcareous tubes in which they live and extend only their feathery tentacles to strain their food from the water.

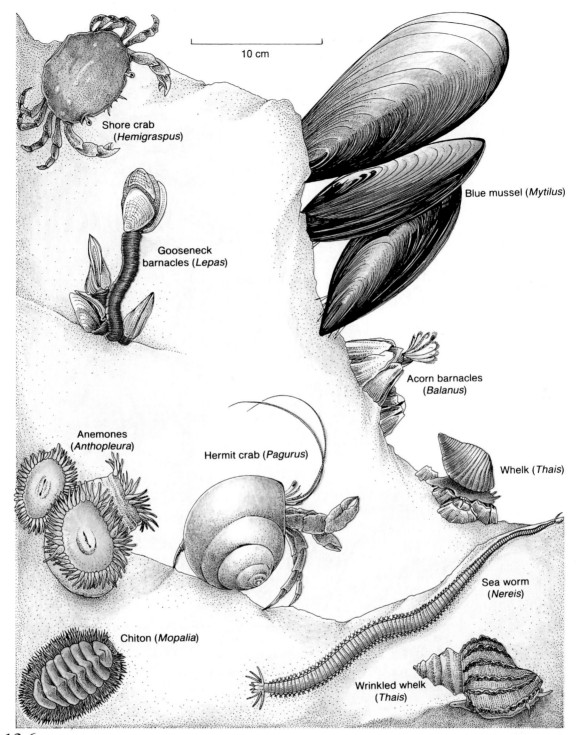

FIGURE 12.6

Representative organisms from the midlittoral zone. The mussels and barnacles filter their food from the water. The chitons graze on the algae covering the rocks. *Thais*, a snail, is a carnivore. Small shore crabs and hermit crabs are scavengers. *Balanus cariosus* is a larger and heavier barnacle than the barnacles of the supralittoral zone. *Nereis* is often found in the mussel beds. The anemones huddle together to keep moist when exposed.

Octopuses are seen occasionally from shore on very low tides. They feed on crabs and shellfish, live in caves or dens, and are known for their ability to flash color changes and move gracefully and swiftly over the bottom and through the water. They are not aggressive, although they are curious and have been shown to have learning ability and memory. The world's

Life on the Sea Floor

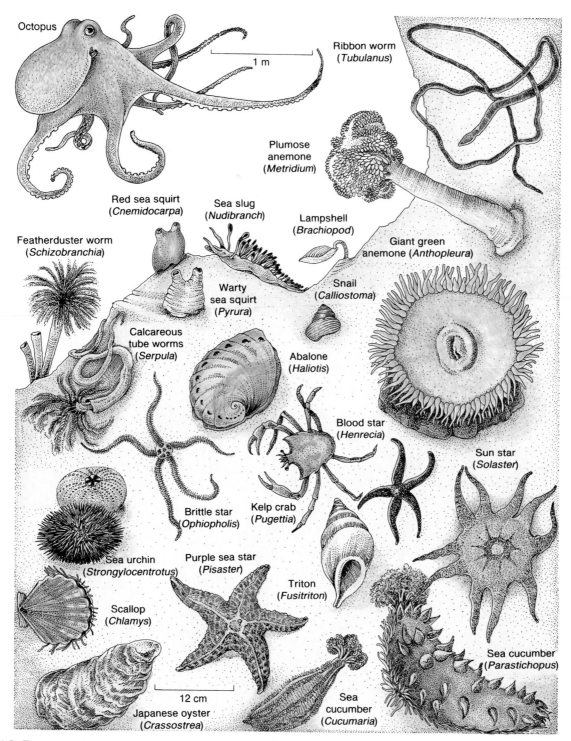

Octopus

Ribbon worm
(*Tubulanus*)

1 m

Plumose anemone (*Metridium*)

Red sea squirt (*Cnemidocarpa*)

Sea slug (*Nudibranch*)

Lampshell (*Brachiopod*)

Featherduster worm (*Schizobranchia*)

Giant green anemone (*Anthopleura*)

Warty sea squirt (*Pyrura*)

Snail (*Calliostoma*)

Calcareous tube worms (*Serpula*)

Abalone (*Haliotis*)

Blood star (*Henrecia*)

Sun star (*Solaster*)

Brittle star (*Ophiopholis*)

Kelp crab (*Pugettia*)

Sea urchin (*Strongylocentrotus*)

Purple sea star (*Pisaster*)

Triton (*Fusitriton*)

Scallop (*Chlamys*)

Sea cucumber (*Parastichopus*)

12 cm

Japanese oyster (*Crassostrea*)

Sea cucumber (*Cucumaria*)

FIGURE 12.7

Lower littoral zone organisms. The starfish feed on the oysters; the sea urchins are herbivores; and the sea cucumbers feed on detritus suspended in the water. Among the mollusks are the oysters, scallops, snails, abalone, sea slugs, and octopuses. The oysters and scallops are filter feeders. *Calliostoma* and the abalone are grazers; the sea slugs, the octopus, and the triton snail are predators.

largest octopus, found in the coastal waters of the eastern North Pacific, commonly measures up to 3 meters (16.5 ft) from tip to tip and weighs 20 kilograms (45 lbs); specimens in excess of 7 meters (23 ft) and 45 kilograms (100 lbs) have been observed.

The bottom of the littoral zone merges into the beginning of the sublittoral zone extending across the continental shelf. If the shallow areas of the subtidal zone are rocky, many of the same lower littoral zone organisms will be found. When soft sediments begin to collect in protected areas or deeper water, the population types change, and animals appear that are commonly found on mud and sand substrates.

Define intertidal zonation and explain what causes it.

Give examples of animals living in the supralittoral, the littoral, and the sublittoral zones along a rocky shore. What are the survival strategies and adaptations of the animals at each level?

12.5   TIDE POOLS

A tide pool, left by the receding tide in a rock depression or basin, provides a habitat for animals common to the lower littoral zone. Some small, isolated tide pools provide a very specialized habitat of increased salinity and temperature due to solar heating and evaporation. Other tide pools act as catch basins for rainwater, lowering the salinity and temperature in fall and winter. Isolated tide pools often support blooms of microscopic algae that give the water the appearance of pea soup.

The deeper the tide pool and the greater the volume of water, the more stable its environment when isolated by a falling tide. The larger the tide pool, the more slowly it changes temperature, salinity, pH, and carbon dioxide-oxygen balance. Animals such as starfish, sea urchins, and sea cucumbers require large, deep pools. Fish species found in tide pools are patterned and colored to match the rocks and the algae within the pool and spend much of their time resting on the bottom, swimming in short spurts. Each tide pool is a specialized environment populated with organisms that are able to survive under the conditions established in that particular pool.

What happens to water isolated in a tide pool?

How does this affect the organisms that live in a tide pool?

12.6   ANIMALS OF THE SAND AND MUD

Along exposed gravel and sandy shores, waves produce an unstable benthic environment. Few plants can attach to the shifting substrate, and therefore few grazing animals are found. When currents deposit sand and mud in quiet coves and bays, the habitat is more stable. Here the size and shape of the sediment particles and the organic content of the sediment determine the quality of the environment. The size of the spaces between particles regulates the flow of water and the availabil-

ity of dissolved oxygen. Beach sand is fairly coarse and porous, gaining and losing water quickly, while fine particles of mud hold more water and replace the water more slowly. The finer the mud particles, the tighter they pack together and the slower the exchange of water; oxygen is not resupplied quickly, and wastes are removed slowly. Deeper into the sediments, the decomposition of organic material produces hydrogen sulfide with its rotten-egg odor, and lack of oxygen restricts the depth to which infauna can be found. Animals such as clams live below the oxygen level but use their siphons to obtain food and oxygen from the water at the surface. In locations protected from waves and currents, eelgrass and surf grass help stabilize the small-particle sediments and provide shelter, substrate, and food, creating a special community of plants and animals.

Most sand and mud animals are **detritus** feeders, and most detritus is former plant material that is degraded by bacteria and fungi. The sand dollar feeds on detritus particles found between the sand grains. Clams, cockles, and some worms are filter feeders, feeding on the detritus and microscopic organisms suspended in the water. Other animals are deposit feeders that engulf the sediments and process them in their guts to extract organic matter (for example, burrowing sea cucumbers). Areas of mud that are high in organic detritus support large quantities of bacteria that are a food source for many small organisms.

The intertidal area of a soft-sediment beach shows some zonation of benthic organisms, but because of the low slope and overlapping boundaries between zones, it is not nearly as clear-cut as the zonation along a rocky shore. The distribution of life in soft sediments is shown in figure 12.8, and a selection of animals from this region is found in figure 12.9.

Compare the habitats of a protected mudflat and an exposed sandy beach. What plants and animals would you expect to find in each of these environments? What plants and animals would you *not* expect to find in each environment?

What is detritus, and what is its importance in marine food webs?

12.7   ANIMALS OF THE DEEP SEA FLOOR

The deep sea floor includes the flat abyssal plains, the trenches, and the rocky slopes of seamounts and mid-ocean ridges. The seafloor sediments are more uniform and their particle size is smaller than those of the shallow regions close to land sources, and the environment is uniformly cold and dark.

The stable conditions of the sea floor appear to have favored deposit-feeding infaunal animals of many species. Many members of the deep-sea infauna are very small and measure 2 millimeters or less; dominant members include detritus-eating worms, burrowing crustaceans, and burrowing sea cucumbers.

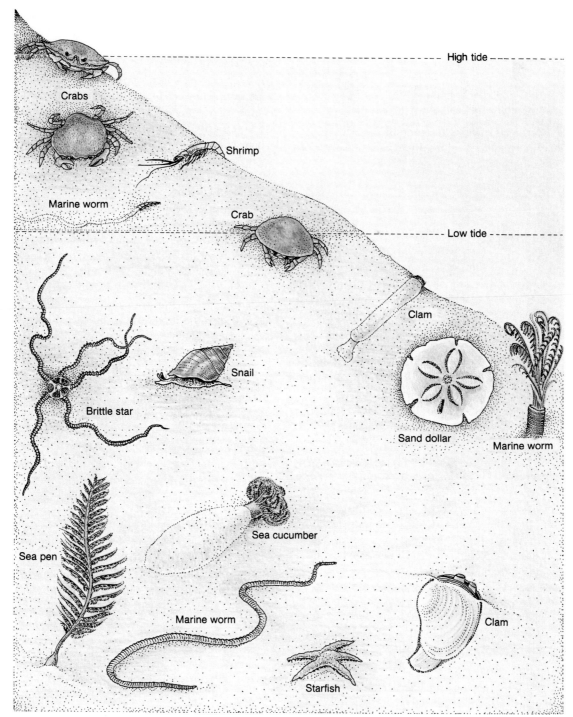

FIGURE 12.8
Zonation on a soft-sediment beach is less conspicuous than that found on a rocky beach. Animals living at the higher tide levels burrow to stay moist.

The deep sediments are continually disturbed and reworked by sea cucumbers and worms as they extract organic matter. This process, called **bioturbation,** results in a well-mixed, uniform sediment layer that covers vast areas of the ocean floor.

Single-celled protozoans are abundant and widely distributed. Glass sponges attach to the scattered rocks on oceanic ridges and seamounts; so do sea squirts and sea anemones (fig. 12.10). Their stalks lift them above the soft sediments into the water, where they feed by straining out organic matter. Stalked barnacles attach to glass sponges as well as to shells and boulders. Tube worms are common, ranging in size from a few millimeters to 20 centimeters (8 in), and sea

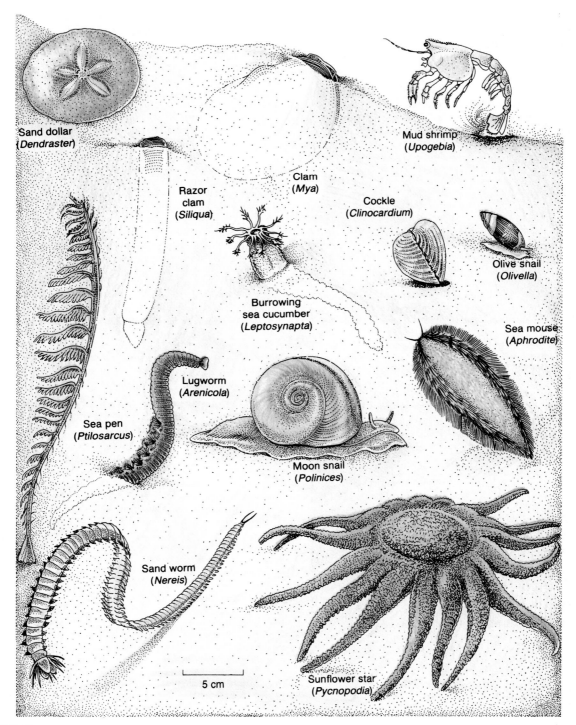

FIGURE 12.9

Organisms of the soft sediments. Infauna types include the mud shrimp, the lugworm, the clam, the cockle, and the burrowing sea cucumber. The sand dollar feeds on detritus; the moon snail drills its way into shellfish; the sea pens feed from the water above the soft bottom. The sea mouse, like *Nereis*, is a polychaete worm.

spiders, brittle stars, and sea cucumbers are found at depths down to 7000 meters (23,000 ft). Snails are found to the greatest depths; some in the deepest trenches have no eyes or eye stalks. The specialized communities of the hydrothermal vents are discussed in Chapter 10.

Describe the environment of deep-sea benthos. What kinds of organisms are found in this habitat?

What produces bioturbation, and what is its effect?

Life on the Sea Floor

(a)

(b)

(c)

FIGURE 12.10

(a) A vase-shaped glass sponge 640 meters (2100 ft) under the surface of the Brown Bear Sea Mount in the northeast Pacific Ocean. (b) A deep-sea crab photographed at a depth of 2000 meters (6550 ft) on the Juan de Fuca Ridge, in an area that has very little life. (c) A group of deep-sea sponges and an anemone are seen at 684 meters (2244 ft) on the Brown Bear Sea Mount.

12.8 FOULING AND BORING ORGANISMS

Organisms that settle and grow on pilings, docks, and boat hulls are said to foul these surfaces. Fouling organisms include barnacles, anemones, tube worms, sea squirts, and algae. They make it difficult to operate vessels and equipment in the ocean environment, and much research goes into experiments with paints and metal alloys to discourage and control these fouling organisms.

Other organisms naturally drill or bore their way into the substrate. Sponges bore into scallop and clam shells; snails bore into oysters; some clams bore into rock. Organisms that bore into wood are costly, for they destroy harbor and port structures as well as wooden boat hulls. The shipworm is a wormlike mollusk with one end covered by a two-valved shell (fig. 12.11). Shipworms secrete enzymes that break down and partially digest the wood fibers, which are scraped away by the shell's rocking and turning movement. This motion allows the animal to bore deep and destructive holes all through a piece of wood. The gribble, a small crustacean, gnaws more superficial and smaller burrows but is also very destructive (fig. 12.11).

Give examples of fouling and boring organisms.

Why are these organisms a problem for humans? How do humans combat them? What are the consequences of this human activity to the marine environment?

(a)

(b)

12.9 INTIMATE RELATIONSHIPS

Competition for food and space and predator-prey relationships are common in the marine environment, but relationships can also be cooperative. The beautiful green anemone of the Pacific coast is green because a small, single-celled alga grows in the animal's cells. This mutually beneficial relationship or **mutualism,** in which the alga receives protection as well as carbon

dioxide and nitrogenous compounds from the anemone, and the anemone acquires organic compounds and some oxygen from the alga, is a type of symbiotic relationship. **Symbiosis** refers to a state in which two dissimilar organisms live in a close, intimate relationship.

Another interesting symbiotic relationship is found in tropical waters between the sea anemone and the clownfish (fig. 12.12). The clownfish acquires protection by nestling

275

Life on the Sea Floor

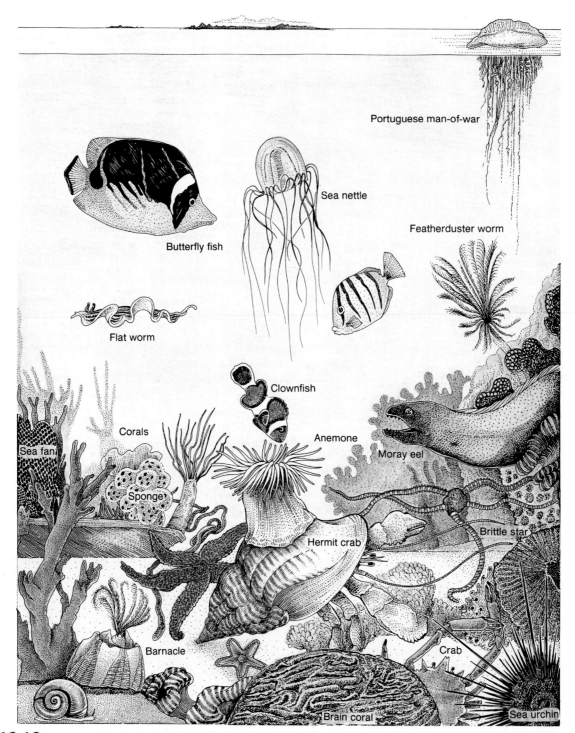

FIGURE 12.12

A coral reef is a complex, interdependent, but self-contained community. Members include corals, clams, sponges, sea urchins, anemones, tube worms, algae, and fish.

among the anemone's stinging tentacles while acting to lure other fish within the anemone's grasp. Shellfish of all types play host to various worms and small crustaceans. Rather than being mutually beneficial, these relationships are often beneficial to one partner and of no harm to the other, a relationship known as **commensalism.**

A relationship in which one partner is harmed by the other is **parasitism.** Parasitic flatworms, roundworms, and bacteria infect marine animals in the same way that land animals are infected. Parasitic relationships are not confined to benthic organisms; both fish and sea mammals are often heavily parasitized.

Compare the three types of symbiosis. Give an example of each.

12.10 TROPICAL CORAL REEFS

Coral reefs are the most luxuriant and complex of all benthic communities (fig. 12.12). The largest coral reef in the world, the Great Barrier Reef, stretches more than 2000 kilometers (1250 mi) from New Guinea southward along the east coast of Australia. Corals are colonial animals, and individual coral animals are called **polyps.** A coral polyp is very similar to a tiny sea anemone with its stinging tentacles, but unlike the anemone, a coral polyp extracts calcium carbonate from the water and builds within its tissues a calcareous skeletal cup. Large numbers of polyps grow together in colonies of delicately branched forms or rounded masses.

Tropical reef-building corals have specialized requirements; they require warm, clear, shallow, clean water and a firm substrate to which they can attach. Because the water temperature must not go below 18°C and the optimum temperature is 23° to 25°C, their growth is restricted to tropical waters between 30°N and 30°S and away from cold-water currents. Waters at depths greater than 50 to 100 meters (150 to 300 ft) are also too cold for significant secretion of calcium carbonate. Most Caribbean corals are found in the upper 50 meters of lighted water, whereas Indian and Pacific corals are found to depths of 150 meters (400 ft) in the more transparent water of these oceans.

Within the tissues of the coral polyps are masses of single-celled dinoflagellate algae called **zooxanthellae.** Polyps and zooxanthellae have a symbiotic relationship in which the coral provides the algal cells with a protected environment, carbon dioxide, and nitrate and phosphate nutrients, and the algal cells photosynthesize, return oxygen, remove waste, and produce carbon compounds which help to nourish the coral. Some coral species receive as much as 60 percent of their food from their algae. Zooxanthellae also cause the coral to produce more calcium carbonate and increase the growth of their calcareous skeleton. The polyps feed actively at night, extending their tentacles to catch zooplankton, but during the day, their tentacles are contracted, exposing the outer layer of cells containing zooxanthellae to the sunlight.

Corals are slow-growing organisms; some species grow less than 1 centimeter (0.4 in) in a year, and others add up to 5 centimeters (2 in) each year. The same coral may be found in different shapes and sizes, depending on the depth and the wave action of an area. Environmental conditions vary over a reef, forming both horizontal and vertical zonation patterns as shown in figure 12.13. On the sheltered (or lagoon) side of the reef, the shallow reef flat is covered with a large variety of branched corals and other organisms. Fine coral particles broken off from the reef top produce sand, which fills the sheltered lagoon floor. On the reef's windward side, the reef's highest point, or reef crest, may be exposed at low tide and is pounded by the breaking waves of the surf zone. Here the more massive rounded corals grow. Below the low tide line, to a depth of 10 to 20 meters (30 to 60 ft) on the seaward side, is a zone of steep, rugged buttresses, which alternate with grooves in the reef face. Masses of large corals grow here, and many large fish frequent the area. The buttresses dissipate the wave energy, and the grooves drain off fine sand and debris, which would smother the coral colonies. At depths of 20 to 30 meters (60 to 100 ft), there is little wave energy, and the light intensity is only about 25 percent of its surface value. The corals are less massive at this depth, and more delicately branched forms are found. Between 30 and 40 meters (100 and 130 ft), the slope is gentle and the level of light is very reduced; sediments accumulate at this depth, and the coral growth becomes patchy. Below 50 meters (150 ft), the slope drops off sharply into the deep water.

Coral reefs are complex assemblages of many different types of plants and animals, and competition for space and food is intense. It has been estimated that as many as 3000 animal species may live together on a single reef. The giant clam, measuring up to a meter in length and weighing over 150 kilograms (330 lbs), is among the most conspicuous. These clams also possess zooxanthellae in large numbers in the colorful tissues that line the edges of the shell. Crabs, moray eels, colorful reef fish, poisonous stonefish, long-spined sea urchins, sea horses, shrimp, lobsters, sponges, and many more organisms are all found living here together. The outer calcareous coverings of encrusting algae, shells, the tubes of worms, and the spines and plates of sea urchins are compressed and cemented together to form new places for more organisms to live. At the same time, some sponges, worms, and clams bore into the reef; some fish graze on the coral and the algae, and the sea cucumbers feed on the broken fragments, reducing them to sandy sediments.

Coral reefs around the world are showing signs of stress. Episodes of coral bleaching, in which the corals expel their zooxanthellae and turn white, have been occurring more frequently and with greater severity. In the tropical eastern Pacific in 1983 large amounts of bleaching, thought to have been associated with the 1982–83 El Niño event (see section 6.14), occurred in reefs off Costa Rica, Panama, and the Galápagos Islands. Between 70 and 90 percent of the corals in Panama and Costa Rica and more than 95 percent of the corals in the Galápagos were destroyed. Bleaching also occurred in the Society Islands and along the Great Barrier Reef. Shallow reefs of the Java Sea lost 80 to 90 percent of their coral cover, and five years later coral cover was only 50 percent of its former level. Mass bleaching in the Atlantic associated with elevated water temperatures began in 1987 in the Florida Keys and eventually stretched from Bermuda south to the Bahamas, affecting coral reefs throughout the Caribbean. In 1990–91 the water warmed again, and the corals bleached; French Polynesia also reported large-scale bleaching in 1991.

No one is sure why massive bleaching episodes occur. One possibility is that when the water warms it causes the algae to produce more oxygen (a by-product of photosynthesis) than the corals can stand; another theory is that stressed coral polyps provide fewer nutrients to the algae. Some researchers have attempted to link the problem to global warming, but there is little evidence so far of a permanent warming trend in the oceans. Whatever the cause, when corals bleach the delicate balance between the zooxanthellae and the corals is destroyed, leaving them susceptible to disease and other stresses.

Periodically a dramatic population increase of the crown-of-thorns sea star, which feeds on coral polyps, occurs. No one knows why the crown-of-thorns population increases so rapidly;

Life on the Sea Floor

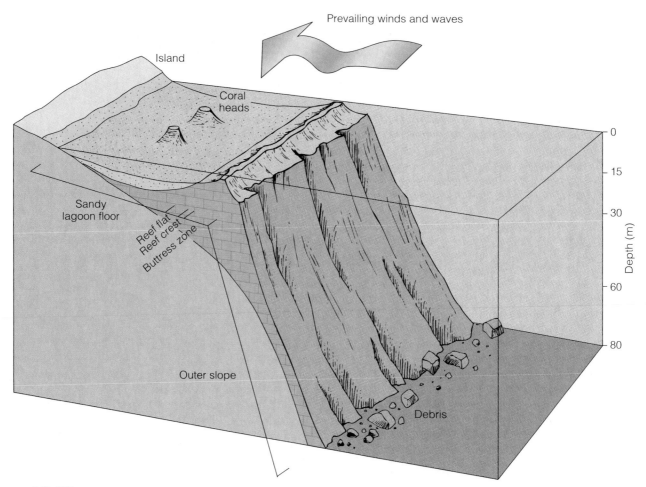

FIGURE 12.13

Coral reef zonation. Water depths and wave action vary over the reef. The reef crest may be exposed at low tide. Sand from coral debris fills the lagoon and drains off down the grooves of the outer slope, where buttresses dissipate wave energy.

evidence points to a correlation between rainy weather (low salinity) and runoff (increased nutrients), allowing large numbers of larvae to survive. The reefs have been able to regenerate in most cases when the crown-of-thorns starfish population dies back; the opening up of grazed areas on the reef may even allow slower growing species to expand, but the outbreaks of starfish do appear to be occurring more frequently and sometimes become chronic.

In addition, coral reefs are mined for building materials, despoiled by shell and aquarium collectors, and dynamited to harvest reef fish. They are damaged by careless sport divers trampling delicate corals and by boats grounding or dragging their anchors. Increased fertilizers and sewage flowing into coastal waters from increasing farming and advancing industrial and recreational complexes also damage the reefs. Too little oxygen and the animals die; too many nitrates and the algae may overgrow and smother the corals. If the grazers decline, the predators decline, upsetting the balance again. To ensure the continued existence of these beautiful and productive areas requires

both an increased understanding of the complex nature of reef communities and the development of policies designed to protect them from human interference.

What conditions are required for the growth of a tropical stony coral reef?

Explain the symbiotic relationship between coral polyps and zooxanthellae.

How do reef growth and wave action combine to create basic reef features?

Discuss the effect of the crown-of-thorns starfish and episodes of bleaching on coastal reefs.

How do human activities impact these reefs?

Table 12.1 is a summary of the benthos, their distribution and characteristics.

Table 12.1 Benthos

Type		Distribution	Characteristics
Plants			
	Diatoms	Cool waters, attached to rocks, mud, other surfaces. Photic zone.	Elongate diatoms with bilateral symmetry.
	Seaweeds	Polar to tropic waters; most abundant at midlatitudes. Limited to photic zone. Vertical zonation associated with pigments, light, and substrate.	Algae with holdfast, stipe, and blades; no flowers or seeds. Nutrients and gases absorbed from the water. Require hard substrate for attachment. Kelps reach 35 meters in length.
	Seed plants	Midlatitudes and tropics. Quiet shallow water; mud and sand substrates.	Plants with flowers and seeds, roots, stems, and leaves. Salt tolerant. Salt marshes and aquatic grass areas in bays. Special mangrove tree habitat in tropics.
Animals			
Littoral zone			
	Epifauna	Intertidal of all latitudes. Animals live on or attached to rocks or firm sediments. Vertical zonation governed by exposure, predation, competition, and substrate.	Carnivores and herbivores. Some motile, predators and scavengers. Some sessile, filter feeders or contact feeders. Adapted to withstand desiccation.
	Infauna	Intertidal of all latitudes. Animals live buried in substrate. Little zonation.	Animals buried in substrate. Some filter feeders; some process substrate to obtain organics.
Sublittoral			
	Epifauna	Animals live on or attached to rocks or firm sediments. No zonation; substrate controls distribution. Populations decrease with depth and availability of food.	Animals similar to lower littoral zone. Less environmental stress.
	Infauna	Animals live buried in substrate. No zonation; substrate controls distribution. Populations decrease with depth and availability of food.	Animals similar to lower littoral zone.
Chemosynthetic communities (See section 10.7)		Isolated communities associated with deep-sea hydrothermal vents and continental shelf gas and oil seeps.	Life-forms feed on chemosynthetic bacteria or harbor bacteria within their tissues. Bacteria at vents use hydrogen sulfide gas as energy source. Hydrocarbons and methane used at seeps. Communities include large clams, large tube worms, crabs, mussels, and barnacles.
Coral reefs		Tropical. Warm, shallow, clean water between 30°N and 30°S. Vertical zonation and reef profile are products of wave action and water depth.	Luxuriant and complex benthic community. Includes reef-building corals, giant clams with internal photosynthetic zooxanthellae, and many other algae and animals.

12.11 HARVESTING THE BENTHOS

Benthic animals are a valuable part of the seafood harvest that includes crabs, shrimp, prawns, and lobsters (crustaceans) and clams, mussels, and oysters (mollusks or shellfish). World estimates for 1991 include catches of 7.8 million metric tons of shellfish and 4.5 million tons of crustaceans. Because of the demand for shellfish and crustaceans, the catches are much more important in dollar value than the weight of the catches would suggest.

Many of the problems of the finfish fisheries are repeated in the benthic fisheries. For example, between 1975 and 1980, more and more boats entered the U.S. king crab fishery of the Bering Sea. In 1979 and 1980, more than 45,000 metric tons of crab meat were harvested (fig. 12.14), but in 1981 the catch slipped to 21,000 tons and then rapidly decreased year by year to 4000 tons in 1985. The rapid drop in catch is apparently due to the classic ills of overfishing and to an insufficient knowledge of the crab's natural history. Under tight control, the

FIGURE 12.14
Alaska king crab being unloaded from a crab pot.

fishery has slowly increased since 1985, reaching 15,000 metric tons in 1990. It is unlikely that the catches of 1979 and 1980 will ever be repeated, but a regulated fishery may ensure years of future fishing if controls are maintained.

Attempts to increase the harvests of crustaceans and mollusks have focused on aquaculture, or mariculture. Oysters, mussels, and clams are raised on aquaculture farms around the world (fig. 12.15). The world harvest of shrimp produced in aquaculture ponds was 500,000 tons in 1990.

Aquaculture projects for benthic species in the United States are faced with the same difficulties as those facing fish farms. High costs, strict licensing policies, the need for technology to replace hand labor, and the need for research to improve diet and disease control require considerable attention if we are to increase our seafood harvest in this way.

Seaweeds are gathered from the wild in northern Europe, Japan, China, and Southeast Asia; they are also an important part of the Japanese aquaculture industry. The 1990 world estimate for total seaweed harvest was 4 million tons. Certain species of green algae, known as sea lettuces, are used in seaweed soups, in salad, and as a flavoring in other dishes. Kelp blades, fresh, dried, pickled, and salted, are used in soups and stews. Japanese **nori** is used in soups and stews and is rolled around portions of rice and fish for flavor. Historically in coastal areas, kelp was used as winter fodder for sheep and cattle and to mulch and fertilize the fields.

There are also important industrial uses for some algal products. **Algin** is extracted from brown algae, and **agar** and **carrageenan** are obtained from red algae. Algin derivatives are used as stabilizers in dairy products and candies as well as in paints, inks, and cosmetics. Agar is used as a medium for bacterial culture in laboratories and hospitals; in addition, it serves as an ingredient in desserts and in pharmaceutical products. Algae rich in carrageenan are gathered in the wild in New England and northeastern Canada; carrageenan is a stabilizer and emulsifier used to prevent separation in ice cream, salad dressing, soup, pudding, cosmetics, and medicines. About 500 metric tons of agar and 5000 metric tons of carrageenan are used in the United States each year.

FIGURE 12.15
Raft culture of bay mussels in Puget Sound. Mussels adhere to ropes suspended from floating rafts.

Assess the importance of benthic organisms in world food harvests.

Why have some harvests declined?

How are harvests being increased?

What are the uses of the seaweed harvest?

12.12 GENETIC MANIPULATION

In 1990, worldwide sales of fish raised in ponds and sea pens reached $22 billion and accounted for 15 percent of the fish consumed worldwide. By the turn of the century, the United Nations Food and Agricultural Organization estimates that farm-raised fish will make up 20 percent of the world supply. Fisheries scientists in North America, Japan, and northern Europe are using gene transfer and chromosome manipulation techniques to keep fish farms stocked with species that will reach market weight faster and be more resistant to disease and freezing.

Aquaculturalists are transferring genes that boost the production of natural growth hormone into fish eggs. Although the foreign genes do not always function, when they do, the gene is passed on to the next generation. Transgenic fish and their offspring have shown from 20 to 46 percent faster growth than the wild stock. Another method is to transfer genes that will increase the immunity of the fish and so boost their survival. Many farm-raised halibut and Atlantic salmon die in winter when their body fluids freeze. A gene from the Arctic winter flounder has been transferred into Atlantic salmon, and preliminary results indicate that these fish survive very cold temperatures better than normal fish.

Another technique produces fish carrying an extra set of chromosomes, or **triploids.** Triploid salmon and trout are produced by exposing the eggs to temperature, pressure, or chemical shock shortly after fertilization. The shock interferes with the division of the egg nucleus, and the egg retains a third set of chromosomes. Triploid salmon and trout do not mature sexually but continue to eat and reach a larger size since they are not spending energy on egg or sperm production. This improved growth and survival has led to widespread production of farmed triploid trout in the United Kingdom.

The gender of fish can also be controlled by chromosome manipulation. Researchers expose fish sperm to ultraviolet light, denaturing its chromosomes. Eggs fertilized with the inactivated chromosome sperm are briefly chilled, causing the eggs to develop with only the female's chromosomes. The result is all female offspring. Many female food fish grow bigger and live longer than males. Female flounder grow to twice the size of males; female Coho salmon have firmer and more flavorful flesh, and all-female sturgeon populations would bring higher profits to caviar producers. It is also possible to inactivate the egg chromosomes and double the sperm chromosomes to produce an all-male population of fish.

Shellfish are also being genetically engineered to produce better commercial forms. Normal oysters enter a summer reproductive phase during which their meat is poor in quality and the oysters are unmarketable. Triploid oysters do not produce sperm and egg but continue to grow steadily, becoming significantly larger than normal oysters and ready for the summer market. Currently triploid oysters are being farmed on about 450 acres of tidelands in Washington and northern California; the 1994 harvest is expected to total about 540,000 gallons with a wholesale value of about $16.2 million. The blue mussel is also ready to market in triploid form.

What research techniques are being used to improve farmed fish and shellfish?

Life on the Sea Floor

Summary

Benthic algae grow attached to a solid surface. These algae have a holdfast, a stipe, and photosynthetic blades but no roots, stems, or leaves. Green algae grow nearest the surface; brown algae including kelps grow at moderate depths; and red algae are found primarily below the low tide level. Each group's pigments trap the available sunlight at these depths. Seaweeds provide food, shelter, and substrate for other organisms. There are benthic diatoms and a few flowering plants, including marine grasses and mangroves.

Benthic animals are subdivided into the epifauna, which live on or attached to the bottom, and the infauna, which live buried in the substrate. Animals that inhabit the rocky littoral region are sorted by the stresses of the area into a series of zones. Organisms that live in the supralittoral zone spend long periods of time out of water. The animals of the midlittoral zone experience nearly equal periods of exposure and submergence; they have tight shells or live close together to prevent drying out. The lower littoral zone, a less stressful environment, is home to a wide variety of animals. Tide pools provide homes for some organisms in the littoral zone.

The littoral is crowded, and competition for space is great. Organisms of the littoral zone are herbivores and carnivores, each with its specialized lifestyle and adaptations for survival.

Sand, and gravel areas are less stable than rocky areas; these substrates show little zonation. The size of the spaces between the substrate particles determines the water and oxygen content of the substrate. Some sediments have a higher organic content than others. Few algae can attach to soft sediments, so few grazers are found here. Eelgrass and surf grass provide food and shelter for specialized communities. Most organisms that live associated with soft sediments are detritus feeders or deposit feeders.

The environment of the deep sea floor is very uniform. Most animals are infaunal deposit feeders. Large burrowers like sea cucumbers continually rework the sediments.

Fouling organisms settle on pilings, docks, and boat hulls; borers drill holes into other organisms. Gribbles and shipworms bore into wood and cause great damage. Symbiotic relationships are intimate relationships between organisms. Mutualism, commensalism, and parasitism are types of symbiosis found in the marine environment.

Tropical coral reefs are specialized, self-contained systems. The coral animals require warm, clear, clean, shallow water and a firm substrate. Photosynthetic dinoflagellates, called zooxanthellae, live in the cells of the corals and the giant clams. The reef exists in a complex but delicate biological balance, which can be easily upset. Reefs have a typical zonation and structure associated with depth and wave exposure.

Shellfish are valuable world food resources. Aquaculture can be used to increase the shellfish and shrimp harvests. Algae are gathered in many countries and are cultivated in Japan. Some algae are used directly as food; others yield substances that are used as stabilizers and emulsifiers in foods and other products.

Gene transfers are used to improve fish growth and increase immunity of stocks for fish farming. Triploid fish and shellfish, because they do not mature sexually, continue to grow to larger than normal size.

KEY TERMS

holdfast

substrate

stipe

blade

kelp

epifauna

infauna

sessile

motile

intertidal zonation

detritus

bioturbation

mutualism

symbiosis

commensalism

parasitism

polyp

zooxanthellae

nori

algin

agar

carrageenan

triploid

SUGGESTED READINGS

Birkeland, C. 1989. The Faustian Traits of the Crown of Thorns Starfish. *American Scientist* 77 (2):154–63.

Brown, B. E., and J. C. Ogden. 1993. Coral Bleaching. *Scientific American* 268 (1):64–70.

Childress, J., H. Felbeck, and G. Somero. 1987. Symbiosis in the Deep Sea. *Scientific American* 256 (5):114–20.

Fischetti, M. 1991. A Feast of Gene-Splicing Down on the Fish Farm. *Science* 253 (5019):512–13.

Golden, F. 1991. Reef Raiders. *Sea Frontiers* 37 (1):18–25.

Grall, G. 1992. Life on a Wharf Piling. *National Geographic* 182 (1):95–115.

Leal, J. H. 1991. Australia's Vast Offshore Maze. *Sea Frontiers* 37 (4):48–51. Great Barrier Reef.

Life in the Sea, Readings from Scientific American. 1982. W. H. Freeman, San Francisco, 248 pp.

Oceanus. Summer 1986. 29 (2). Issue devoted to coral reefs.

Peterson, C. H. 1991. Intertidal Zonation of Marine Invertebrates in Sand and Mud. *American Scientist* 79 (3):236–49.

Sissenwine, M. P., and A. A. Rosenberg. 1992. U.S. Fisheries-Status, Long-Term Potential Yields and Stock Management Ideas. *Oceanus* 36 (2):48–54.

Standish, K. A., Jr. 1988. Triploid Oysters Ensure Year-Round Supply. *Oceanus* 31 (3):58–63.

Tennesen, M. 1992. Kelp: Keeping a Forest Afloat. *National Wildlife* 30 (4):4–11.

Torrance, D. C. 1991. Deep Ecology: Rescuing Florida's Reefs. *Nature Conservancy* 41 (4):8–17.

Ward, F. 1990. Florida's Coral Reefs Are Imperiled. *National Geographic* 178 (1):115–32.

APPENDIX A
Latitude and Longitude

To determine a location on the surface of the earth, a grid of surface lines that cross each other at right angles is used. These lines are latitude and longitude. Lines of latitude begin at the equator, which is created by passing a plane through the earth halfway between the poles and at right angles to the earth's axis. The equator is marked as 0° latitude, and other latitude lines are drawn around the earth parallel to the equator, north to 90°N, the North Pole, and south to 90°S, the South Pole (fig. A.1a). Lines of latitude are also termed parallels because they are parallel to the equator and to each other. Lines of latitude describe smaller and smaller circles as the poles are approached. All parallels of latitude must be designated as an angle either north or south of the equator. The latitude value is determined by the internal angle (φ or phi) between the latitude line, the earth's center, and the equatorial plane (fig. A.1b).

Lines of longitude, also called meridians, are formed at right angles to the latitude grid (fig. A.2a). Longitude begins at 0°, a line on the earth's surface extending from the North Pole to the South Pole that passes directly through the Royal Naval Observatory in Greenwich, England. The 0° longitude line is known as the prime meridian. Longitude lines are identified by the angular displacement (θ or theta) to the east and west of 0° longitude (fig. A.2b). Directly opposite the prime meridian, on the other side of the earth, 180° longitude approximates the international date line. All meridians are the same size; they mark the intersection of the earth's surface with planes passing through the earth's center at right angles to the parallels of latitude. Any circle that passes through the earth's center is called a "great circle," and all longitude lines form great circles; only the equator is a great circle of latitude. A great circle connecting any two points on the earth's surface defines the shortest distance between them.

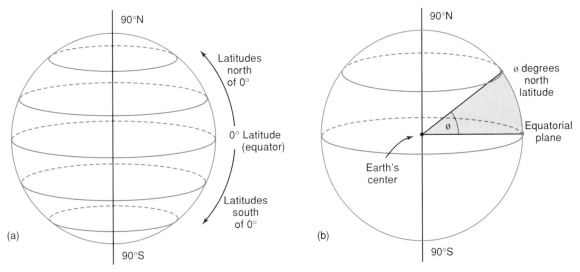

FIGURE A.1
(a) Latitude lines are drawn parallel to the equatorial plane. (b) The value of a latitude line is expressed in angular degrees determined by the angle formed between the equatorial plane and the latitude line to the earth's center. This is the angle φ (phi). The degree value of φ must be noted as north or south of the equator.

To identify any location on the earth's surface, use the crossing of the latitude and longitude lines; for example, 158°W, 21°N is the location of the Hawaiian Islands, and 20°E, 33°S identifies the Cape of Good Hope at the southern tip of Africa. For greater accuracy, one degree of latitude or longitude is divided into 60 minutes of arc; each minute is divided into 60 seconds of arc. One minute of arc length of latitude or longitude at the equator is equal to one nautical mile (1.852 km or 1.15 land miles).

FIGURE A.2

(a) Longitude lines are drawn with reference to the prime meridian. (b) The value of a longitude line is expressed in angular degrees determined by the angle formed between the prime meridian and the longitude line to the earth's center. This is the angle θ (theta). The value of θ is given in degrees east or west of the prime meridian.

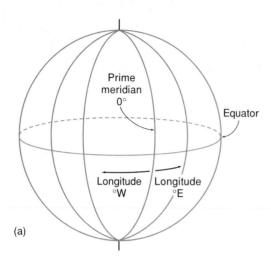

(a)

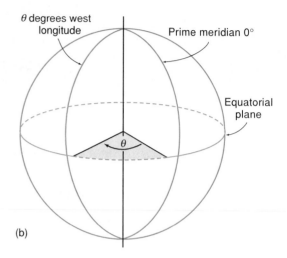

(b)

APPENDIX B
Classification Summaries

Classification Summary of the Plankton

The types of plankton can be categorized in kingdoms, divisions, phyla, and classes, as in the following list.

I. **Kingdom Monera:** cells, simple and unspecialized; single cells, some in groups or chains.
   A. *Bacteria:* single cells, in chains or groups; autotrophic and heterotrophic, aerobic and anaerobic; important as food source and in decomposition. *Cyanobacteria:* blue-green algae; autotrophic single cells, in chains or groups; produce some red blooms in sea; phytoplankton.

II. **Kingdom Protista:** convenience grouping of microscopic and mostly single-celled organisms; autotrophs (algae) and heterotrophs (protozoa).
   A. *Division Chrysophyta:* golden-brown algae; yellow to golden autotrophic single cells, in groups or chains; contribute to deep-sea sediments; phytoplankton.
      1. *Class Bacillariophyceae:* diatoms.
      2. *Class Chrysophyceae:* coccolithophores and silicoflagellates.
   B. *Division Dinoflagellata:* fire algae; single cells with flagella; produce most red tides; bioluminescence common; usually considered phytoplankton.
   C. *Phylum Sarcodina:* microscopic heterotrophs, moving with pseudopodia; radiolarians and foraminiferans.
   D. *Phylum Ciliophora:* microscopic heterotrophs, moving with cilia; ciliates.

III. **Kingdom Plantae:** plants; primarily nonmotile, multicellular, photosynthetic autotrophs.
   A. *Division Phaeophyta:* brown algae; *Sargassum* maintains a planktonic habitat in the Sargasso Sea.

IV. **Kingdom Animalia:** animals; multicellular heterotrophs with specialized cells, tissues, and organ systems; zooplankton. For temporary members of the zooplankton (or meroplankton), see the meroplankton listed in V.
   A. *Phylum Cnidaria:* coelenterates; radially symmetrical with tentacles and stinging cells.
      1. *Class Hydrozoa:* jellyfish as one stage in the life cycle, including such colonial forms as the Portuguese man-of-war.
      2. *Class Scyphozoa:* jellyfish.
   B. *Phylum Ctenophore:* comb jellies; translucent; move with cilia; often bioluminescent.
   C. *Phylum Chaetognatha:* arrowworms; free-swimming, carnivorous worms.
   D. *Phylum Mollusca:* mollusks; the snail-like pteropod is planktonic.
   E. *Phylum Arthropoda:* animals with paired, jointed appendages and hard outer skeletons.
      1. *Class Crustacea:* copepods and euphausiids.
   F. *Phylum Chordata:* animals, including vertebrates, with dorsal nerve cord and gill slits at some stage in development.
      1. *Subphylum Urochordata:* saclike adults with "tadpole" larvae; salps.

V. **Meroplankton:** larval forms from the phyla *Annelida* (segmented worms), *Mollusca* (shellfish and snails), *Arthropoda* (crabs and barnacles), *Echinodermata* (starfish and sea urchins), and *Chordata* (fish).

Classification Summary of the Nekton

The nekton, all members of the kingdom Animalia, can be classified in the following phyla and classes.

I. **Kingdom Animalia:** animals; multicellular heterotrophs with specialized cells, tissues, and organ systems.
   A. *Phylum Mollusca:* mollusks.
      1. *Class Cephalopoda:* squid; the octopus is a member of the benthos.
   B. *Phylum Chordata:* animals with a dorsal nerve cord and gill slits at some stage in development.
      1. *Subphylum Vertebrata:* animals with a backbone of bone or cartilage.
         a. *Class Agnatha:* jawless fish; lampreys and hagfish.
         b. *Class Chondrichthyes:* jawed fish with cartilaginous skeletons; sharks and rays.
         c. *Class Osteichthyes:* bony fish; all other fish, including commercial species such as salmon, tuna, herring, and anchovy.

d. *Class Reptilia:* air breathers with dry skin and scales; young develop in self-contained eggs; turtles, sea snakes, iguanas, and crocodiles.

e. *Class Mammalia:* warm-blooded; body covering of hair; produce milk; young born live.

   (1) *Order Cetacea:* whales.

       (a) *Suborder Mysticeti:* baleen whales, including blue, gray, right, sei, finback, and humpback whales.

       (b) *Suborder Odonticeti:* toothed whales, including the killer and sperm whale, porpoise, and dolphin.

   (2) *Order Carnivora:* sea otters, seals, sea lions, and walruses.

   (3) *Order Sirenia:* dugongs and manatees.

Classification Summary of the Benthos

Benthic organisms include members of the following kingdoms, divisions, and phyla.

I. **Kingdom Monera:** includes the bacteria, an important food source in mud and sand environments; also cyanobacteria or blue-green algae, especially abundant on coral reefs.

II. **Kingdom Protista:** convenience grouping of microscopic, usually unicellular plants and animals.

   A. *Division Chrysophyta:* golden-brown algae; mainly planktonic, but benthic diatoms are abundant.

   B. *Division Protozoa:* infauna include many members among sand and mud particles, including amoeboid and foraminiferan species.

III. **Kingdom Plantae:** plants; primarily nonmotile, multicellular, photosynthetic autotrophs.

   A. *Division Chlorophyta:* green algae; seaweeds such as *Ulva* and *Codium.*

   B. *Division Phaeophyta:* brown algae; coastal seaweeds, including the large kelps *Nereocystis, Macrocystis,* and *Postelsia. Sargassum* maintains a planktonic habit in the Sargasso Sea.

   C. *Division Rhodophyta:* red algae; littoral and sublittoral seaweeds of varied form.

   D. *Division Tracheophyta:* plants with vascular tissue; true roots, stems, and leaves present.

      1. *Class Angiospermae:* flowering plants; seeds enclosed in fruit; eelgrass, surf grass, and mangrove trees.

IV. **Kingdom Animalia:** animals; multicellular heterotrophs with specialized cells, tissues, and organ systems.

   A. *Phylum Porifera:* sponges; simple, nonmotile filter feeders.

   B. *Phylum Cnidaria:* coelenterates; radially symmetric, with tentacles and stinging cells.

      1. *Class Anthozoa:* sea anemones, corals, and sea fans.

   C. *Phylum Platyhelminthes:* flatworms; free-living and parasitic forms.

   D. *Phylum Nemertina:* ribbon worms.

   E. *Phylum Nematoda:* round worms, or nematodes.

   F. *Phylum Brachiopoda:* lampshells.

   G. *Phylum Annelida:* segmented worms.

      1. *Class Polychaeta:* free-living marine carnivores, including tube worms, *Nereis,* and clam worms.

   H. *Phylum Pogonophora:* deep-sea tube-dwelling worms; giant members found around hot vents on the sea floor.

   I. *Phylum Mollusca:* mollusks.

      1. *Class Polyplacophora:* chitons.

      2. *Class Scaphopoda:* tusk shells.

      3. *Class Gastropoda:* single-shelled mollusks; snails, limpets, and sea slugs.

      4. *Class Pelecypoda* or *Bivalvia:* bivalved mollusks; clams, mussels, scallops, and oysters.

      5. *Class Cephalopoda:* octopus. The squid is a member of the nekton.

   J. *Phylum Arthropoda:* animals with paired, jointed appendages and outer skeletons.

      1. *Class Crustacea:* crabs, shrimp, and barnacles.

   K. *Phylum Echinodermata:* radially symmetric, spiny-skinned animals with water-vascular system, including starfish, sea urchins, sand dollars, and sea cucumbers.

   L. *Phylum Hemichordata:* acorn worms.

   M. *Phylum Chordata:* animals, including vertebrates, with dorsal nerve cord and gill slits at some stage in development.

      1. *Subphylum Urochordata:* filter-feeding, saclike adults with "tadpole" larvae; sea squirt.

APPENDIX C
Units and Notation

Scientific Notation

The writing of very large and very small numbers is simplified by using exponents, or powers of 10, to indicate the number of zeroes required to the left or to the right of the decimal point. The numbers that are equal to some of the powers of 10 are as follows:

$$1,000,000,000. = 10^9 = \text{one billion}$$
$$1,000,000. = 10^6 = \text{one million}$$
$$1000. = 10^3 = \text{one thousand}$$
$$100. = 10^2 = \text{one hundred}$$
$$10. = 10^1 = \text{ten}$$
$$1. = 10^0 = \text{one}$$
$$0.1 = 10^{-1} = \text{one tenth}$$
$$0.01 = 10^{-2} = \text{one hundredth}$$
$$0.001 = 10^{-3} = \text{one thousandth}$$
$$0.000001 = 10^{-6} = \text{one millionth}$$
$$0.000000001 = 10^{-9} = \text{one billionth}$$

The number 149,000,000 is rewritten by moving the decimal point eight places to the left and multiplying by the exponential number 10^8, to form 1.49×10^8. In the same way, 605,000 becomes 6.05×10^5.

A very small number such as 0.000032 becomes 3.2×10^{-5} by moving the decimal point five places to the right and multiplying by the exponential number 10^{-5}. Similarly, 0.00000372 becomes 3.72×10^{-6}.

To add or subtract numbers written in exponential notation, convert the numbers to the same power of 10. For example,

$$\begin{array}{ccc} 1.49 \times 10^3 & 1.490 \times 10^3 & 14.90 \times 10^2 \\ +6.05 \times 10^2 = & 0.605 \times 10^3 = & 6.05 \times 10^2 \\ \hline & 2.095 \times 10^3 & 20.95 \times 10^2 \end{array}$$

$$\begin{array}{ccc} 2.36 \times 10^3 & 23.60 \times 10^2 & 2.360 \times 10^3 \\ -1.05 \times 10^2 = & -1.05 \times 10^2 = & -0.105 \times 10^3 \\ \hline & 22.55 \times 10^2 & 2.255 \times 10^3 \end{array}$$

To multiply, the exponents are added and the numbers are multiplied:

$$\begin{array}{c} 4.6 \ \times 10^3 \\ \times 2.2 \ \times 10^2 \\ \hline 10.12 \times 10^5 \ = \ 1.012 \times 10^6 \end{array}$$

To divide, subtract the exponents and divide the numbers:

$$\frac{6.0 \times 10^8}{2.5 \times 10^3} = 2.4 \times 10^5$$

The following prefixes correspond to the powers of 10 and are used in combination with metric units:

Exponential Value	Prefix	Symbol
10^{12}	Tera	T
10^9	Giga	G
10^6	Mega	M
10^3	Kilo	k
10^2	Hecto	h
10^1	Deka	da
10^{-1}	Deci	d
10^{-2}	Centi	c
10^{-3}	Milli	m
10^{-6}	Micro	μ
10^{-9}	Nano	n
10^{-12}	Pico	p
10^{-15}	Femto	f
10^{-18}	Atto	a

SI Units

The *Système International d'Unités*, or International System of Units, is a simplified system of metric units (known as SI units) adopted by international convention for scientific use.

Basic SI Units

Quantity	Unit	Symbol
Length	Meter	m
Mass	Kilogram	kg
Time	Second	s
Temperature	Kelvin	K

Derived SI Units

Quantity	Unit	Symbol	Expression in SI Base Units
Area	Meter squared	m^2	m^2
Volume	Meter cubed	m^3	m^3
Density	Kilogram per cubic meter	kg/m^3	kg/m^3
Speed	Meter per second	m/s	m/s
Acceleration	Meter per second per second	m/s^2	m/s^2
Force	Newton	N	$(kg)(m)/s^2$
Pressure	Pascal	Pa	N/m^2
Energy	Joule	J	$(N)(m)$
Power	Watt	W	$J/s; (N)(m)/s$

Length: The Basic SI Unit Is the Meter

Unit	Metric Equivalents	Traditional Equivalents	Other
Meter (m)	100 centimeters 1000 millimeters	39.37 inches 3.281 feet	0.546 fathom
Kilometer (km)	1000 meters	0.621 land mile	0.540 nautical mile
Centimeter (cm)	10 millimeters 0.01 meter	0.394 inch	
Millimeter (mm)	0.1 centimeter 0.001 meter	0.0394 inch	
Land mile (mi)	1609 meters	5280 feet	
Nautical mile (nm)	1852 meters	1.151 land mile 6076 feet	0.869 nautical mile 1 minute of latitude
Fathom (fm)	1.829 meters	6 feet	

Area: Derived from Length

Unit	Metric Equivalents	Traditional Equivalents	Other
Square meter (m^2)	10,000 square centimeters	10.76 square feet	
Square kilometer (km^2)	1,000,000 square meters	0.386 square land mile	0.292 square nautical mile
Square centimeter (cm^2)	100 square millimeters	0.151 square inch	

Volume: Derived from Length

Unit	Metric Equivalents	Traditional Equivalents	Other
Cubic meter (m^3)	1,000,000 cubic centimeters 1000 liters	35.32 cubic feet 264 gallons	
Cubic kilometer (km^3)	1,000,000,000 cubic meters	0.239 cubic land mile	0.157 cubic nautical mile
Liter (L)	1000 cubic centimeters	1.06 quarts, 0.264 U.S. gallon	
Milliliter (mL)	1.0 cubic centimeters		

Mass: The Basic SI Unit Is the Kilogram

Unit	Metric Equivalents	Traditional Equivalents
Kilogram (kg)	1000 grams	2.205 pounds
Gram (g)		0.035 ounce
Metric ton, or tonne (t)	1000 kilograms 1,000,000 grams	2205 pounds
U.S. ton	907 kilograms	2000 pounds

Time: The Basic SI Unit Is the Second

Unit	Metric and Traditional Equivalents
Minute	60 seconds
Hour	60 minutes; 3600 seconds
Day	86,400 seconds } mean solar day 24 hours
Year	31,556,880 seconds 8765.8 hours } mean solar year 365.25 solar days

Temperature: The Basic SI Unit Is the Kelvin

Reference Points	Kelvin (K)	Celsius (°C)	Fahrenheit (°F)
Absolute zero	0	–273.2	–459.7
Sea water freezes	271.2	–2.0	28.4
Fresh water freezes	273.2	0.0	32.0
Human body	310.2	37.0	98.6
Fresh water boils	373.2	100.0	212.0
Conversions	$°K = °C + 273.2$	$°C = \dfrac{(°F - 32)}{1.8}$	$°F = (1.8 \times °C) + 32$

Speed (Velocity): The Derived SI Unit Is the Meter per Second

Unit	Metric Equivalents	Traditional Equivalents	Other
Meter per second (m/s)	100 centimeters per second 3.60 kilometers per hour	3.281 feet per second 2.237 land miles per hour	1.944 knots
Kilometer per hour (km/hr)	0.277 meter per second	0.909 foot per second	0.55 knot
Knot (kt)	0.51 meter per second	1.151 land miles per hour	1 nautical mile per hour

Acceleration: The Derived SI Unit Is the Meter per Second per Second

Unit	Metric Equivalents	Traditional Equivalents
Meter per second per second (m/s^2)	12,960 kilometers per hour per hour 100 centimeters per second per second	8053 miles per hour per hour 3.281 feet per second per second

Force: The Derived SI Unit Is the Newton

Unit	Metric Equivalents	Traditional Equivalents
Newton (N)	100,000 dynes	0.2248 pound
Dyne (dyn)	0.00001 newton	0.000002248 pound

Pressure: The Derived SI Unit Is the Pascal

Unit	Metric Equivalents	Traditional Equivalents	Other
Pascal (Pa)	1 newton per square meter 10 dynes per square centimeter		
Bar	100,000 pascals 1000 millibars	14.5 pounds per square inch	0.927 atmosphere 29.54 inches of mercury
Standard atmosphere (atm)	1.013 bars 101,300 pascals	14.7 pounds per square inch	29.92 inches of mercury

Energy: The Derived SI Unit Is the Joule

Unit	Metric Equivalents	Traditional Equivalents
Joule (J)	1 newton-meter 0.2389 calorie	0.0009481 British thermal unit
Calorie (cal)	4.186 joules	0.003968 British thermal unit

Power: The Derived SI Unit Is the Watt

Unit	Metric Equivalents	Traditional Equivalents
Watt (W)	1 joule per second 0.2389 calorie per second 0.001 kilowatt	0.0569 British thermal unit per minute 0.001341 horsepower

Density: The Derived SI Unit Is the Kilogram per Cubic Meter

Unit	Metric Equivalents	Traditional Equivalents
Kilogram per cubic meter	0.001 gram per cubic centimeter (g/cm^3)	0.0624 pound per cubic foot

GLOSSARY

A

absorption   taking in of a substance by chemical or molecular means; change of sound or light energy into some other form, usually heat, in passing through a medium or striking a surface.

abyssal   pertaining to the great depths of the ocean below approximately 4000 m.

abyssal hill   low, rounded submarine hill less than 1000 m high.

abyssal plain   flat ocean-basin floor extending seaward from the base of the continental slope and continental rise.

active margin   *See* leading margin.

adsorption   attraction of ions to a solid surface.

advection   horizontal or vertical transport of seawater, as by a current.

advective fog   fog formed when warm air saturated with water vapor moves over cold water.

agar   substance produced by red algae; the gelatinlike product of these algae.

algae   marine and freshwater plants (including most seaweeds) that are single-celled, colonial, or multicelled, with chlorophyll but no true roots, stems, or leaves and with no flowers or seeds.

algin   complex organic substance found in or obtained from brown algae.

amplitude   for a wave, the vertical distance from sea level to crest or from sea level to trough, or one-half the wave height.

anion   negatively charged ion.

anoxic   deficient in oxygen.

Antarctic bottom water   densest oceanic water type, formed at the surface below sea ice, flows northward along the floor of the Atlantic Ocean.

Antarctic Circle   *See* Arctic and Antarctic Circles.

Antarctic intermediate water   water with a salinity of 34.4‰ and a temperature of 5°C, produced at the 40°S surface convergence in the Atlantic.

antinode   portion of a standing wave with maximum vertical motion.

aphotic zone   that part of the ocean in which light is insufficient to carry on photosynthesis.

aquaculture   cultivation of aquatic organisms under controlled conditions. *See also* mariculture.

Arctic and Antarctic Circles   latitudes 66½°N and 66½°S, respectively, marking the boundaries of light and darkness during the summer and winter solstices.

asthenosphere   upper, deformable portion of the earth's mantle, the layer below the lithosphere; probably partially molten; may be site of convection cells.

atmospheric pressure   pressure exerted by the atmosphere due to the weight of the column of air lying directly above any point on earth.

atoll   ring-shaped coral reef that encloses a lagoon in which there is no exposed preexisting land and which is surrounded by the open sea.

attenuation   decrease in the energy of a wave or beam of particles occurring as the distance from the source increases; caused by absorption, scattering, and divergence from a point source.

autumnal equinox   *See* equinoxes.

B

backshore   beach zone lying between the foreshore and the coast, acted upon by waves only during severe storms and exceptionally high water.

baleen   whalebone; horny material growing down from the upper jaw of plankton-feeding whales; forms a strainer, or filtering organ, consisting of numerous plates with fringed edges.

bar   offshore ridge or mound of sand, gravel, or other loose material, which is submerged, at least at high tide; located especially at the mouth of a river or estuary, or lying a short distance from and parallel to the beach.

barrier island   deposit of sand, parallel to shore and raised above sea level; may support vegetation and animal life.

barrier reef   coral reef that parallels land but is some distance offshore, with water between reef and land.

basalt   fine-grained, dark igneous rock, rich in iron, magnesium, and calcium; characteristic of oceanic crust.

basin   large depression of the sea floor having about equal dimensions of length and width.

bathymetric   pertaining to the study and mapping of seafloor elevations and the variations of water depth; pertaining to the topography of the sea floor.

beach   zone of unconsolidated material between the mean low water line and the line of permanent vegetation, which is also the effective limit of storm waves; sometimes includes the material moving in offshore, onshore, and longshore transport.

beach face   section of the foreshore normally exposed to the action of waves.

benthic   of the sea floor, or pertaining to organisms living on or in the sea floor.

benthos   organisms living on or in the ocean bottom.

berm   nearly horizontal portion of a beach (backshore) with an abrupt face; formed from the deposition of material by wave action at high tide.

berm crest   ridge marking the seaward limit of a berm.

biogenous sediment   sediment having more than 30 percent material derived from organisms.

bioluminescence   production of light by living organisms as a result of a chemical reaction either within certain cells or organs or outside the cells in some form of excretion.

biomass   amount of living matter, expressed in weight units, per unit of water surface area or volume.

bioturbation   reworking of sediments by organisms that burrow and ingest them.

blade   flat, photosynthetic, "leafy" portion of an alga or seaweed.

bloom   high concentration of phytoplankton in an area, caused by increased reproduction; often produces discoloration of the water. *See also* red tide.

bore   *See* tidal bore.

breaker   sea surface-water wave that has become too steep to be stable and collapses.

breakwater structure protecting a shore area, harbor, anchorage, or basin from waves; a type of jetty.

buffer substance able to neutralize acids and bases, therefore able to maintain a stable pH.

buoyancy ability of an object to float due to the support of the fluid the body is in or on.

C

calcareous containing or composed of calcium carbonate.

calorie amount of heat required to raise the temperature of 1 gram of water 1°C.

carbonate an ion composed of one carbon and three oxygen atoms, CO_3^{2-}.

carnivore flesh-eating organism.

carrageenan substance produced by certain algae, used as a thickening agent.

cation positively charged ion.

centrifugal force outward-directed force acting on a body moving along a curved path or rotating about an axis.

chemosynthesis formation of organic compounds with energy derived from inorganic substances such as ammonia, sulfur, and hydrogen.

chlorinity measure of the chloride content of seawater in grams per kilogram.

chlorophyll group of green pigments that are active in photosynthesis.

cilia microscopic, hairlike processes of living cells, which beat in coordinated fashion and produce movement.

Cnidaria phylum of radially symmetrical marine organisms with tentacles and stinging cells; includes jellyfish, sea anemones, and corals.

coast strip of land of indefinite width that extends from the shore inland to the first major change in terrain that is unaffected by marine processes.

coastal circulation cell longshore transport cell pattern of sediment moving from a source to a place of deposition.

coastal zone land and water areas including cliffs, dunes, beaches, bays, and estuaries.

coelenterate *See* Cnidaria.

colonial organism organism consisting of semi-independent parts that do not exist as separate units; groups of organisms with specialized functions that form a coordinated unit.

commensalism an intimate association between different organisms in which one is benefited and the other is neither harmed nor benefited.

conduction transfer of heat energy through matter by internal molecular motion; also heat transfer by turbulence in fluids.

continental drift motion of the continents due to plate tectonics.

continental margin zone separating the continents from the deep-sea bottom, usually subdivided into shelf, slope, and rise.

continental rise gentle slope formed by the deposition of sediments at the base of a continental slope.

continental shelf zone bordering a continent, extending from the line of permanent immersion to the depth at which there is a marked or rather steep descent to the great depths.

continental shelf break zone along which there is a marked increase of slope at the outer margin of a continental shelf.

continental slope relatively steep downward slope from the continental shelf break to depth.

contour line on a chart or graph connecting the points of equal value for elevation, temperature, salinity, and so on.

convection transmission of heat by the movement of a heated gas or liquid; vertical circulation resulting from changes in density of a fluid.

convection cell density-driven transfer of heat by circulation of liquid or gas in which warm, low-density material rises and cold, high-density material falls.

convergence situation in which fluids of different origins come together, usually resulting in the sinking, or downwelling, of surface water and the rising of air.

copepod small, shrimplike members of the zooplankton.

coral colonial animal that secretes a hard, outer, calcareous skeleton; the skeletons of coral animals form in part the framework for warm-water reefs.

core vertical, cylindrical sample of bottom sediments, from which the nature of the bottom can be determined; also the central zone of the earth, thought to be liquid or molten on the outside and solid on the inside.

corer device that plunges a hollow tube into bottom sediments to extract a vertical sample.

Coriolis effect apparent force acting on a body in motion, due to the rotation of the earth, causing deflection to the right in the Northern Hemisphere and to the left in the Southern Hemisphere; the force is proportional to the speed and latitude of the moving body.

cosmogenous sediment sediment particles with an origin in outer space; for example, meteor fragments and cosmic dust.

crest *See* berm crest, reef crest, wave crest.

crust outer shell of the solid earth; the lower limit is usually considered to be the Moho.

crustacean member of a class of primarily aquatic organisms with paired jointed appendages and a hard outer skeleton; includes lobsters, crabs, shrimp, and copepods.

ctenophores transparent, planktonic animals, spherical or cylindrical in shape with rows of cilia; comb jellies.

current horizontal movement of water.

current meter instrument for measuring the speed and direction of a current.

D

decomposer microorganisms (usually bacteria and fungi) that break down nonliving organic matter and release nutrients.

deep scattering layer (DSL) layer of organisms that move away from the surface during the day and toward the surface at night; the layer scatters or returns vertically directed sound pulses.

deep-sea reversing thermometer (DSRT) mercury-in-glass thermometer that records seawater temperature upon being inverted and retains its reading until returned to its upright position.

deep-water wave wave in water where depth is greater than one-half its wavelength.

delta area of unconsolidated sediment deposit, usually triangular in outline, formed at the mouth of a river.

density property of a substance defined as mass per unit volume and usually expressed in grams per cubic centimeter or kilograms per cubic meter.

depth recorder *See* echo sounder.

desalination process of obtaining fresh water from seawater.

detritus any loose material, especially decomposed, broken, and dead organic materials.

diatom microscopic unicellular alga with an external skeleton of silica.

diffraction process that transmits energy laterally along a wave crest.

dinoflagellate one of a class of planktonic organisms, often bioluminescent, may form red tides.

dispersion (sorting) sorting of waves as they move out from a storm center; occurs because long waves travel faster in deep water than short waves.

diurnal tide tide with one high water and one low water each tidal day.

divergence horizontal flow of fluids away from a common center, associated with upwelling in water and descending motions in air.

doldrums nautical term for the belt of light, variable winds near the equator.

downwelling sinking of water toward the bottom, usually the result of a surface convergence or an increase in density of water at the sea surface.

dredge cylindrical or boxlike sampling device made of metal, net, or both, which is dragged across the bottom to obtain biological or geological samples.

dugong *See* sea cow.

dune wind-formed hill or ridge of sand.

dynamic tide an actual tide as it occurs in the ocean basin.

E

earth sphere depth uniform depth of the earth below the present mean sea level, if the solid earth surface were smoothed off evenly (2440 m).

ebb tide falling tide; the period of the tide between high water and the next low water.

echo sounder (depth recorder) instrument used to measure the depth of water by measuring the time interval between the release of a sound pulse and the return of its echo from the bottom.

eddies circular movements of water.

Ekman spiral in a theoretical ocean of infinite depth, unlimited extent, and uniform viscosity, with a steady wind blowing over the surface, the surface water moves 45° to the right of the wind in the Northern Hemisphere. At greater depths the water moves farther to the right with decreased speed, until at some depth (approximately 100 m) the water moves opposite to the wind direction. Net water transport is 90° to the right of the wind in the Northern Hemisphere. Movement is to the left in the Southern Hemisphere.

Ekman transport net water transport in a theoretical ocean of infinite depth with a steady wind blowing over the surface, moves 90° to the right of the wind in the Northern Hemisphere and 90° to the left of the wind in the Southern Hemisphere.

electromagnetic radiation waves of energy formed by simultaneous electric and magnetic oscillations; the electromagnetic spectrum is the continuum of all electromagnetic radiation from low-energy radiowaves to high-energy gamma rays, including visible light.

El Niño wind-driven reversal of the Pacific equatorial currents resulting in the movement of warm water toward the coasts of the Americas, so called because it generally develops just after Christmas. *See also* La Niña.

epifauna animals living attached to the sea bottom or moving freely over it.

episodic wave abnormally high wave unrelated to local storm conditions.

equator 0° latitude, determined by a plane that is perpendicular to the earth's axis and is everywhere equidistant from the North and South Poles.

equilibrium tide theoretical tide formed by the tide-producing forces of the moon and sun on a water-covered earth.

equinoxes times of the year when the sun stands directly above the equator, so that day and night are of equal length around the world. The vernal equinox occurs about March 21, and the autumnal equinox occurs about September 22.

estuary semi-isolated portion of the ocean, which is diluted by freshwater drainage from land.

euphausiid planktonic, shrimplike crustacean. *See also* krill.

evaporation process by which a liquid becomes a vapor.

F

fathom a traditional unit of length equal to 1.83 m or 6 ft; used to measure water depth.

fault break or fracture in the earth's crust, in which one side has been displaced relative to the other.

fetch continuous area of water over which the wind blows in essentially a constant direction.

fjord narrow, deep, steep-walled inlet of the ocean formed by the submergence of a mountainous coast or by the entrance of the ocean into a deeply excavated glacial trough after the melting of the glacier.

flagellum (pl. flagella) long, whiplike extension from a living cell's surface that by its motion moves the cell.

flood tide rising tide; the period of the tide between low water and the next high water.

flushing time length of time required for an estuary to exchange its water with the open ocean.

fog visible assemblage of tiny droplets of water formed by condensation of water vapor in the air; a cloud with its base at the surface of the earth.

food chain sequence of organisms in which each is food for the next member of the sequence. *See also* food web.

food web complex of interacting food chains; all the feeding relations of a community taken together; includes production, consumption, decomposition, and the flow of energy.

foraminifera minute, one-celled animals that usually secrete calcarious shells.

foreshore portion of the shore that includes the low tide terrace and the beach face.

fringing reef reef attached directly to the shore of an island or a continent and not separated from it by a lagoon.

G

Gondwanaland the southern portion of Pangaea consisting of Africa, South America, India, Australia, and Antarctica.

grab sampler instrument used to remove a piece of the ocean floor for study.

granite crystalline, coarse-grained, igneous rock composed mainly of quartz and feldspar.

gravitational force mutual force of attraction between particles of matter (bodies).

gravity earth's gravity; acceleration due to the earth's mass is 981 cm/s²; g is used in equations.

greenhouse effect warming effect caused by the atmosphere's transparency to incoming sunlight and its opacity to outgoing radiation.

Greenwich mean time (GMT) *See* Universal Time.

group speed speed at which a group of waves travels (in deep water, group speed equals one-half the speed of an individual wave); the speed at which the wave energy is propagated.

guyot submerged, flat-topped seamount.

gyre circular movement of water, larger than an eddy; usually applied to oceanic systems.

H

half-life time required for half of an initial quantity of a radioactive isotope to decay.

heat budget accounting for the total amount of the sun's heat received on earth during one year as being exactly equal to the total amount lost from the earth due to radiation and reflection.

heat capacity quantity of heat needed to raise the temperature of a unit mass (1 g) of a substance by 1°C.

herbivore animal that feeds only on plants.

higher high water higher of the two high waters of any tidal day in a region of mixed tides.

higher low water higher of the two low waters of any tidal day in a region of mixed tides.

high-pressure zone region of air with greater than average density and atmospheric pressure.

high water maximum height reached by a rising tide.

holdfast organ of a benthic alga that attaches the alga to the sea floor.

holoplankton organisms living their entire life cycle in the floating (planktonic) state.

horse latitudes regions of calms and variable winds that coincide with latitudes at approximately 30° to 35°N and S.

hot spot surface expression of a persistent rising jet of molten mantle material.

hurricane severe, cyclonic, tropic storm at sea, with winds of 120 km/hr (73 mph) or more; generally applied to Atlantic Ocean storms. *See also* typhoon.

hydrogenous sediment sediment formed from substances dissolved in seawater.

hydrologic cycle movement of water among the land, oceans, and atmosphere due to changes of state, vertical and horizontal transport, evaporation, and precipitation.

hydrothermal vent seafloor outlet for high-temperature groundwater and associated minerals; a hot spring.

hypsographic curve graph of land elevation and ocean depth versus area.

I

iceberg mass of land ice that has broken away from a glacier and floats in the sea.

incidental catch fish caught incidentally while fishing for other species; also known as by-catch or trash fish, often wasted.

infauna animals that live buried in the sediment.

inner core inner solid portion of the earth's core.

international date line a theoretical line approximating the 180° meridian. Regions on the east side of the line are one day earlier in calendar date than regions on the west side.

intertidal *See* littoral.

intertidal zonation vertical distribution of plants and animals over tidal and subtidal areas.

invertebrate any animal without a backbone or spinal column.

ion positively or negatively charged atom or group of atoms.

island arc chain of volcanic islands formed when plates converge at a subduction zone.

isotope atoms of the same element having different numbers of neutrons.

J

jet stream a high-speed, generally westerly wind of the upper atmosphere between 30°N and 50°N.

jetty structure located to influence currents or protect the entrance to a harbor or river from waves (U.S. terminology). *See also* breakwater.

K

kelp any of several large, brown algae, including the largest known algae.

knot a traditional unit of speed equal to 0.51 m/s or 1 nautical mile per hour.

krill term used by whalers for the small, shrimplike crustaceans found in huge masses in polar waters and eaten by baleen whales.

L

lagoon shallow body of water, which usually has a shallow, restricted outlet to the sea.

La Niña condition of colder-than-normal surface water in the eastern tropical Pacific. *See also* El Niño.

larva (pl. larvae) immature juvenile form of an animal.

latitude line on the earth's surface paralleling the equator (0° latitude), value is expressed in degrees north and south of the equator.

Laurasia the northern portion of Pangaea composed of North America and Eurasia.

leading margin on a lithospheric plate moving toward a subduction zone, the margin of a continent closest to the subduction zone.

lithogenous sediment sediment composed of rock particles eroded mainly from the continents by water, wind, and waves.

lithosphere outer, rigid portion of the earth; includes the continental and oceanic crust and the upper part of the mantle.

littoral (intertidal) area of the shore between mean high water and low water; the intertidal zone.

longitude line on the earth's surface drawn between the poles whose value is expressed in degrees east and west of the prime meridian (0° longitude).

longshore current current produced in the surf zone by waves breaking at an angle with the shore; runs roughly parallel to the shoreline.

longshore transport movement of sediment by the longshore current.

loran navigational system in which position is determined by measuring the difference in the time of reception of synchronized radio signals; derived from the phrase "long-range navigation."

lower high water lower of the two high waters of any tidal day in a region of mixed tides.

lower low water lower of the two low waters of any tidal day in a region of mixed tides.

low-pressure zone region of air with less than average density and atmospheric pressure.

low tide terrace flat section of the foreshore seaward of the sloping beach face.

low water minimum height reached by a falling tide.

M

magma molten rock material that forms igneous rocks upon cooling; magma that reaches the earth's surface is referred to as lava.

magnetic pole either of the two points on the earth's surface where the magnetic field is vertical.

major constituents the most abundant salts in seawater, present in concentrations expressed in parts per thousand or grams per kilogram.

manatee *See* sea cow.

manganese nodules rounded, layered lumps found on the deep-ocean floor that contain as much as 20 percent manganese and smaller amounts of iron, nickel, and copper; a hydrogenous sediment.

mantle main volume of the earth between the crust and the core; increasing pressure and temperature with depth divide the mantle into layers.

mariculture cultivation of marine organisms under controlled conditions. *See also* aquaculture.

mean average, a value intermediate between other values.

mean sea level average height of the sea surface, based on observations of all stages of the tide over a nineteen-year period in the United States.

meridian circle of longitude passing through the poles and any given point on the earth's surface.

meroplankton floating developmental stages (eggs and larvae) of organisms that as adults belong to the nekton and benthos.

minus tide tide level that falls below the zero tide depth reference level.

mixed tide type of tide in which large inequalities between the two high waters and the two low waters occur in a tidal day.

Moho (Mohorovičić discontinuity) boundary between crust and mantle, marked by a rapid increase in seismic wave speed.

mollusks marine animals, usually with shells; includes mussels, oysters, clams, snails, and slugs.

monoculture cultivation of only one species of organism in an aquaculture system.

monsoon name for seasonal winds; first applied to the winds over the Arabian Sea, which blow for six months from the northeast and for six months from the southwest; now extended to similar winds in other parts of the world; in India, the term is popularly applied to the southwest monsoon and also to the rains that it brings.

motile moving or capable of moving.

mutualism an intimate association between different organisms in which both organisms benefit.

N

nautical mile unit of length equal to 1852 m or 1.15 land miles, or one minute of latitude.

neap tides tides occurring near the times of the first and last quarters of the moon, when the range of the tide is least.

nekton pelagic animals that are active swimmers; for example, adult squid, fish, and marine mammals.

neritic shallow water marine environment extending from low water to the edge of the continental shelf. *See also* pelagic.

node point of least or zero vertical motion in a standing wave.

North Atlantic deep water Atlantic Ocean water with a salinity of about 34.9‰ and a temperature of 2–4°C, formed at the surface at approximately 50–60°N.

nutrient in the ocean, any one of a number of inorganic or organic compounds or ions used primarily in the nutrition of primary producers; nitrogen and phosphorus compounds are examples.

O

oceanic pertaining to the ocean water seaward of the continental shelf; the "open ocean." *See also* pelagic.

ocean ranching raising of salmon to a juvenile stage, releasing of the juveniles to sea, and harvesting of adults on their return.

ocean thermal energy conversion (OTEC) method of extracting energy from the oceans that depends on the temperature difference between surface water and water at depth.

offshore direction seaward from the shore.

onshore direction toward the shore.

onshore current any current flowing toward the shore.

ooze fine-grained deep-ocean sediment composed of at least 30 percent sand or silt-sized calcareous or siliceous remains of small marine organisms, the remainder being clay-sized material.

orbit in water waves, the path followed by the water particles affected by the wave motion; also, the path of a body subjected to the gravitational force of another body, such as the earth's orbit around the sun.

osmosis tendency of water to diffuse through a semipermeable membrane to make the concentration of water on one side of the membrane equal to that on the other side.

outer core outer liquid portion of the earth's core, located between the inner core and the mantle.

overturn sinking of more dense water and its replacement by less dense water from below.

oxygen minimum zone in which respiration and decay reduce dissolved oxygen to a minimum, usually between 800 and 1000 m.

ozone a form of oxygen that absorbs ultraviolet radiation from the sun.

P

Pangaea single supercontinent of 200 million years ago.

parallel circle on the surface of the earth parallel to the plane of the equator and connecting all points of equal latitude; a line of latitude.

parasitism an intimate association between different organisms in which one is benefited and the other is harmed.

passive margin *See* trailing margin.

pelagic primary division of the sea, which includes the whole mass of water subdivided into neritic and oceanic zones; also pertaining to the open sea.

period *See* tidal period; wave period.

pH measure of the concentration of hydrogen ions in a solution; the concentration of hydrogen ions determines the acidity of the solution; $pH = -\log_{10}(H^+)$, where H^+ is the concentration of hydrogen ions in gram atoms per liter.

phosphorite form of phosphate precipitated from seawater, found in muds, sands, and nodules.

photic zone layer of a body of water that receives ample sunlight for photosynthesis; usually less than 100 m.

photosynthesis manufacture by plants of organic substances and release of oxygen from carbon dioxide and water in the presence of sunlight and the green pigment chlorophyll.

physiographic portrayal of the earth's features by perspective drawing.

phytoplankton microscopic plant forms of plankton.

pinniped member of the marine mammal group, characterized by four swimming flippers; for example, seals and sea lions.

plankton passively drifting or weakly swimming organisms.

plate tectonics theory and study of the earth's lithospheric plates, their formation, movement, interaction, and destruction; the attempts to explain the earth's crustal changes in terms of plate movements.

polar easterlies winds blowing from the poles toward approximately 60°N and S; winds are northeasterly in the Northern Hemisphere and southeasterly in the Southern Hemisphere.

polyculture cultivation of more than one species of organism in an aquaculture system.

polyp sessile stage in the life history of certain coelenterates (Cnidaria); sea anemones and corals.

precipitation falling products of condensation in the atmosphere, such as rain, snow, or hail; also the falling out of a substance from solution.

primary production (primary productivity) rate at which biomass is produced by photosynthetic and chemosynthetic organisms in the form of organic substances.

prime meridian meridian of 0° longitude, used as the origin for measurements of longitude; internationally accepted as the meridian of the Royal Naval Observatory, Greenwich, England.

productivity amount of organic material synthesized by organisms in unit time in a unit volume of water. *See also* primary production (primary productivity).

pteropod pelagic snail whose foot is modified for swimming.

P-waves primary, or faster, waves moving away from a seismic event; can penetrate solid rock; consist of energy transmitted by alternate compressions and dilations of the material. *See also* S-waves.

R

radar system of determining and displaying the distance of an object by measuring the time interval between transmission of a radio signal and reception of the echo return; derived from the phrase "Radio detecting and ranging."

radiation energy transmitted as rays or waves without the need of a substance to conduct the energy.

radiative fog fog formed as a result of warm moist air cooling at night when winds are calm, often in bays, inlets, and land valleys.

radiolarians minute, single-celled animals with siliceous skeletons.

radiometric dating determining ages of geological samples by measuring the relative abundance of radioactive isotopes and comparing isotope systems.

rain shadow low-precipitation zone on the sheltered side of islands and mountains.

red clay red to brown fine-grained lithogenous deposit, of predominantly clay size, which is derived from land, transported by winds and currents, and deposited far from land and at great depth; also known as brown mud and brown clay.

red tide red coloration, usually of coastal waters, caused by large quantities of microscopic organisms (generally dinoflagellates); some red tides result in mass fish kills, others contaminate shellfish, and still others produce no toxic effects.

reef crest highest portion of a coral reef on the exposed seaward edge of the reef.

reef flat portion of a coral reef landward of the reef crest and seaward of the lagoon.

reflection rebounding of light, heat, sound, waves, and so on, after striking a surface.

refraction change in direction or bending of a wave due to change in speed.

reservoir natural or artificial collection or accumulation of water.

residence time mean time that a substance remains in a given area before replacement, calculated by dividing the amount of a substance by its rate of addition or subtraction.

respiration metabolic process by which food or food-storage molecules yield the energy on which all living cells depend.

reverse osmosis process of obtaining fresh water from seawater by applying pressure to the salt water and forcing the water molecules through a membrane that does not pass salt molecules.

ridge long, narrow elevation of the sea floor, with steep sides and irregular topography.

rift valley trough formed by faulting along a zone in which plates move apart and new crust is created, such as along the crest of a ridge system.

rift zone *See* spreading center.

rip current strong surface current flowing seaward from shore; the return movement of water piled up on the shore by incoming waves and wind.

rise long, broad elevation that rises gently and generally smoothly from the sea floor.

S

salinity measure of the quantity of dissolved salts in seawater. It is formally defined as the total amount of dissolved solids in seawater in parts per thousand (‰) by weight when all the carbonate has been converted to oxide, all the bromide and iodide have been converted to chloride, and all organic matter is completely oxidized.

salinometer instrument for determining the salinity of water by measuring the electrical conductivity of a water sample of a known temperature.

salt wedge intrusion of salt water along the bottom; in an estuary, the wedge moves upstream on high tide and seaward on low tide.

satellite body that revolves around a planet; a moon; a device launched from earth into orbit.

satellite navigation system series of signal-emitting, earth-orbiting satellites that allow ships to determine their position relative to the satellites.

saturation value amount of material that can be held in solution without gain or loss.

scarp elongated and comparatively steep slope change separating flat or gently sloping areas on the sea floor or on a beach.

scattering random redirection of light or sound energy by reflection from an uneven sea bottom or sea surface, from water molecules, or from particles suspended in the water.

sea same as the ocean; subdivision of the ocean; surface waves generated or sustained by the wind within their fetch, as opposed to swell.

sea cow (dugong, manatee) large herbivorous marine mammal of tropic waters; includes the manatee and dugong.

seafloor spreading movement of crustal plates away from the mid-ocean ridges; process that creates new crustal material at the mid-ocean ridges.

sea ice ice formed from freezing of seawater.

sea level height of the sea surface above or below some reference level. *See also* mean sea level.

seamount isolated volcanic peak that rises at least 1000 m from the sea floor.

sea smoke type of fog caused by dry, cold air moving over warm water.

sea stack isolated mass of rock rising from the sea near a headland from which it has been separated by erosion.

sea state numerical or written description of the roughness of the ocean surface relative to wave height and wind speed.

Secchi disk white disk used to measure the transparency of the water by observing the depth at which the disk disappears from view.

sediment particulate organic and inorganic matter that accumulates in loose, unconsolidated form.

seiche standing wave oscillations of an enclosed or semienclosed body of water that continue, pendulum fashion, after the generating force ceases.

seismic pertaining to or caused by earthquakes or earth movements.

seismic sea wave *See* tsunami.

seismic wave shock wave or vibration produced by an earthquake or earth movement.

semidiurnal tide tide with two high waters and two low waters each tidal day.

sessile permanently fixed or sedentary; not free-moving.

set direction in which the current flows.

shallow-water wave wave in water where depth is less than one-twentieth the average wavelength.

siliceous containing silica.

slack water state of a tidal current when its velocity is near zero; occurs when the tidal current changes direction.

sofar channel natural sound channel in the oceans, in which sound can be transmitted for very long distances; the depth of minimum sound velocity; derived from the phrase "sound fixing and ranging."

solstices times of the year when the sun stands directly above 23½°N or S latitude. The winter solstice occurs about December 22, and the summer solstice occurs about June 22.

sonar method or equipment for determining, by underwater sound, the presence, location, or nature of objects in the sea; derived from the phrase "sound navigation and ranging."

sorting of waves *See* dispersion.

sound shadow zone area of the ocean into which sound does not penetrate because the density structure of the water refracts the sound waves.

South Atlantic surface water lens of surface water in the center of the South Atlantic current gyre.

spit (sand spit) low tongue of land, or a relatively long, narrow shoal extending from the shore.

spreading center region along which new crustal material is produced.

spring tides tides occurring near the times of the new and full moon, when the range of the tide is greatest.

standing crop biomass of a population present at any given time.

standing wave type of wave in which the surface of the water oscillates vertically between fixed points called nodes, without progression; the points of maximum vertical rise and fall are called antinodes.

steepness, of wave *See* wave steepness.

stipe portion of an alga between the holdfast and the blade.

storm center area of origin for surface waves generated by the wind; an intense atmospheric low-pressure system.

storm tide (storm surge) along a coast, the exceptionally high water accompanying a storm, owing to wind stress and low atmospheric pressure, made even higher when associated with a high tide and shallow depths.

stratosphere the layer of the atmosphere above the troposphere where temperature is constant or increases with altitude.

subduction zone plane descending away from a trench and defined by its seismic activity, interpreted as the convergence zone between a sinking plate and an overriding plate.

sublittoral (subtidal) benthic zone from the low tide line to the seaward edge of the continental shelf.

submarine canyon relatively narrow, V-shaped, deep depression with steep slopes, the bottom of which grades continuously downward across the continental slope.

submersible a research submarine, designed for manned or remote operation at great depths.

subsidence sinking of a broad area of the crust without appreciable deformation.

substrate material making up the base on which an organism lives or to which it is attached.

subtidal *See* sublittoral.

subtropical convergence downwelling zone produced by converging currents at approximately 30°–40°N and S (centers of oceanic gyres).

summer solstice *See* solstices.

supersaturated solution holding dissolved material in excess of the saturation value.

supralittoral benthic zone above the high tide level that is moistened by waves, spray, and extremely high tides.

surface tension tendency of a liquid surface to contract owing to bonding forces between molecules.

surf zone area of wave activity between the shoreline and the outermost limit of the breakers.

surimi refined fish protein used to form artificial crab, shrimp, and scallop meat.

S-waves secondary, or slower, transverse seismic waves; cannot penetrate a liquid; consist of elastic vibrations perpendicular to the direction of travel. *See also* P-waves.

swell long and relatively uniform wind-generated ocean waves that have traveled out of their generating area.

symbiosis living together in intimate association of two dissimilar organisms.

T

tectonic pertaining to processes that cause large-scale deformation and movement of the earth's crust.

temperate pertaining to intermediate latitudes, usually considered to be between 23½°N and S and 66½°N and S.

terranes fragments of the earth's crust bounded by faults, each fragment with a history distinct from each other fragment.

terrigenous of the land; sediments composed predominantly of material derived from the land.

thermohaline circulation vertical circulation caused by changes in density; driven by variations in temperature and salinity.

tidal bore high tide crest that advances rapidly up an estuary or river as a breaking wave.

tidal current alternating horizontal movement of water associated with the rise and fall of the tide.

tidal day time interval between two successive passes of the moon over a meridian, approximately 24 hours and 50 minutes.

tidal period elapsed time between successive high waters or successive low waters.

tidal range difference in height between consecutive high and low waters.

tide periodic rising and falling of the sea surface that results from the gravitational attractions of the moon and sun acting on the rotating earth.

tombolo deposit of unconsolidated material that connects an island to another island or to the mainland.

trace elements elements dissolved in seawater in concentrations of less than one part per million.

trade winds wind systems occupying most of the tropics, which blow from approximately 30°N and S toward the equator; winds are northeasterly in the Northern Hemisphere and southeasterly in the Southern Hemisphere.

trailing margin on a lithospheric plate moving away from a spreading center, the margin of a continent closest to the spreading center.

transform fault fault with horizontal displacement connecting the ends of an offset in a mid-ocean ridge. Some plates slide past each other along a transform fault.

trench long, deep, narrow depression of the sea floor with relatively steep sides, associated with a subduction zone.

triploid condition in which cells have three sets of chromosomes.

trophic relating to nutrition; a trophic level is the position of an organism in a food chain or food (trophic) pyramid.

tropical divergence   upwelling zone produced by diverging surface currents on either side of the equator.

Tropics of Cancer and Capricorn   latitudes 23½°N and 23½°S, respectively, marking the maximum angular distance of the sun from the equator during the summer and winter solstices.

troposphere   the lowest layer of the atmosphere, where the temperature decreases with altitude.

trough   long depression of the sea floor, having relatively gentle sides; normally wider and shallower than a trench. *See also* wave trough.

tsunami (seismic sea wave)   long-period sea wave produced by a submarine earthquake or volcanic eruption. It may travel across the ocean unnoticed from its point of origin and build up to great heights over shallow water at the shore.

tube worm   any worm or wormlike organism that builds a tube or sheath attached to a submerged substrate.

turbidite   sediment deposited by a turbidity current, showing a pattern of coarse particles at the bottom, grading gradually upward to fine silt.

turbidity   loss of water clarity or transparency owing to the presence of suspended material.

turbidity current   dense, sediment-laden current flowing downward along an underwater slope.

typhoon   severe, cyclonic, tropic storm originating in the western Pacific Ocean, particularly in the vicinity of the South China Sea. *See also* hurricane.

U

Universal Time   clock time set to 12 noon when the sun is directly above the prime meridian.

upwelling   rising of water rich in nutrients toward the surface, usually the result of diverging surface currents.

V

vernal equinox   *See* equinoxes.

vertebrate   animal with a skull surrounding a well-developed brain and a skeleton of cartilage or bone including a vertebral column or backbone extending through the main body axis: fish, amphibians, reptiles, birds, and mammals.

W

water budget   balance between the rates of water added and lost in an area.

wave   periodic disturbance that moves through or over the surface of a medium with a speed determined by the properties of the medium.

wave crest   highest part of a wave.

wave height   vertical distance between a wave crest and the adjacent trough.

wavelength   horizontal distance between two successive wave crests or two successive wave troughs.

wave period   time required for two successive wave crests or troughs to pass a fixed point.

wave steepness   ratio of wave height to wavelength.

wave train   series of similar waves from the same direction.

wave trough   lowest part of a wave.

westerlies   wind systems blowing from the west between latitudes of approximately 30° and 60°N and 30° and 60°S; they are southwesterly in the Northern Hemisphere and northwesterly in the Southern Hemisphere.

western intensification   tendency of currents flowing from low to high latitudes on the western sides of the Pacific and Atlantic Oceans to be strong, swift, and narrow.

wetlands   land on which water significantly dominates development of soil and types of organisms present.

wind wave   wave created by the action of the wind on the sea surface.

winter solstice   *See* solstices.

Z

zonation   parallel bands of distinctive plant and animal associations found within the littoral zones and distributed to take advantage of optimal conditions for survival.

zooplankton   animal forms of plankton.

zooxanthellae   symbiotic microscopic organisms (dinoflagellates) found in corals and other marine organisms.

CREDITS

PHOTOGRAPHS

Chapter 1

Opener: The Bettmann Archive; **1.1:** Courtesy of Tom Rice of Sea and Shore Museum, Port Gamble, WA; **1.5:** National Maritime Museum, London; **1.10a–d:** The Challenger Expedition: Dec. 21, 1872–May 24, 1876. Engravings from the Challenger Reports, Vol. 1, 1885; **1.11, 1.12:** Courtesy Norsk Sjofartsmuseum; **1.13:** Courtesy of the Mariner's Museum, Newport News, VA; **1.14:** National Oceanic Survey/NOAA; **1.15:** Courtesy Joe Creager, University of Washington; **1.16:** Woods Hole Oceanographic Institution; **1.17a:** NOAA; **1.17b:** United States Navy; **p. 20 (left):** Mary Rose Trust; **p. 20 (right):** Hans Hammarskiold for the Vasa Museum, Stockholm, Sweden. **p. 21 (a–b):** Woods Hole Oceanographic Institute.

Chapter 2

Opener: © Mark Newman/Tom Stack & Associates; **2.1:** NASA; **2.5:** Alyn and Alison Duxbury; **2.11:** Courtesy Stephen P. Miller. UC/Santa Barbara.

Chapter 3

Opener: © Kevin Schafer/Tom Stack & Associates; **3.6:** B. Heezen and M. Tharp: WORLD OCEAN FLOOR, © Marie Tharp, 1977, South Nyack, NY 10960. Reproduced by permission; **3.17b:** U.S. Geological Survey; **3.20:** P. W. Lipman, U.S. Geological Survey; **3.24:** Courtesy Ocean Drilling Program/Texas A & M University, College Station, TX; **3.25:** Courtesy Paul Johnson/School of Oceanography, University of Washington; **3.26, 3.27:** Courtesy John Delaney/School of Oceanography, University of Washington; **p. 64:** Peter J. Auster; **p. 65 (left):** © Woods Hole Oceanographic Institute; **p. 65 (right):** Peter J. Auster; **p. 67:** Courtesy James Morrison, Polar Science Center, HN-10, Applied Physics Laboratory, Seattle, WA.

Chapter 4

Opener: © Manfred Gottschalk/Tom Stack & Associates; **4.5:** Alyn and Alison Duxbury; **4.10:** Courtesy John Delaney/School of Oceanography, University of Washington; **4.11:** W. Haxby; Lamont-Doherty Geological Observatory; **4.12a–b:** R. W. Cook/INCO, Inc., NY; **4.13a–b:** Courtesy Ken Adkins, School of Fisheries, University of Washington; **4.16:** Courtesy Joe Creager, University of Washington; **4.17:** Alyn and Alison Duxbury; **4.18:** Courtesy Dean McManus, University of Washington; **4.19:** Courtesy Kathy Newell/School of Oceanography, University of Washington; **4.20b:** Courtesy Sara Barnes/School of Oceanography, University of Washington; **4.20c:** Alyn and Alison Duxbury; **4.20d:** Courtesy Dr. Richard Sternberg/School of Oceanography University of Washington; **4.21:** © Paul Wilbert/Wilbert-Peck Photography.

Chapter 5

Opener: © David Muench Photography; **5.6:** Alyn and Alison Duxbury; **5.8:** Courtesy EPC Laboratories, Inc., Danvers, MA; **5.17:** John Conomos/U. S. Geological Survey; **p. 109 (top left, right, bottom):** Alyn and Alison Duxbury.

Chapter 6

Opener: © Jack Swenson/Tom Stack & Associates; **6.6a–b:** M. Chahine/Jet Propulsion Laboratory, J. Sussking/Goddard Space Flight Center; **6.7a–b:** J. Zwally, Goddard Space Flight Center; **6.7c–d:** D. J. Cavan eri and P. Gloersen/Goddard Space Flight Center; **6.8:** Edward Joshberger/U. S. Geological Survey, Dept. of Interior; **6.19:** Jet Propulsion Laboratory, University of California, Los Angeles Atmospheric Environment Service; **6.23:** © William Rosenthal/Jeroboam, Inc.; **6.26:** National Hurricane Service, National Oceanic and Atmospheric Administration, U.S. Department of Commerce; **6.27:** National Weather Service.

Chapter 7

Opener: © Brian Parker/Tom Stack & Associates; **7.8:** Alyn and Alison Duxbury; **7.18:** Courtesy Otis Brown; **7.19:** Courtesy: Will Peterson/School of Oceanography, University of Washington; **7.20:** Courtesy Kathy Newell/School of Oceanography, University of Washington; **7.22c:** Courtesy Lockheed Missiles and Space Company; **p. 156 (left):** Bauer, Moynihan & Johnson, Seattle, WA; **p. 156 (right):** Courtesy S. Crosson, Victoria, B. C., Canada.

Chapter 8

Opener: © David Muench Photography; **8.1:** © Eda Rogers; **8.8:** Courtesy Exxon Corporation, photo by Leonard Wolhuter; **8.16a–b:** Alyn and Alison Duxbury; **8.17:** © Bob Barbour; **8.18a–c:** National Geophysical Data Center, NOAA; **8.27a–b:** Tourism New Brunswick Photo; **8.29:** Courtesy Nova Scotia Power Corporation.

Chapter 9

Opener: © David Muench Photography; **9.1a:** Courtesy Chesapeake Bay Foundation; **9.1b:** Courtesy Joe Creager/School of Oceanography, University of Washington; **9.1c:** NASA; **9.1d:** Alyn and Alison Duxbury; **9.1e:** Courtesy Dr. Sherwood Maynard/University of Hawaii; **9.1f:** © Terra-Mar Resource Information Services, Inc.; **9.2a:** Courtesy Dr. Richard Sternberg, School of Oceanography, University of Washington; **9.2b:** Alyn and Alison Duxbury; **9.2c:** Courtesy Landslides; **9.2d–e:** Alyn and Alison Duxbury; **9.2f:** Dr. Richard Sternberg, School of Oceanography, University of Washington; **9.2g, 9.3, 9.5, 9.6, 9.7:** Alyn and Alison Duxbury; **9.8:** Courtesy Dr. Richard Sternberg/School of Oceanography; University of Washington; **9.9:** U. S. Army Corps of Engineers; **9.17:** Courtesy Port of Seattle; **9.18:** Courtesy Landslides; **9.19:** Alyn and Alison Duxbury; **9.20:** © Bob Evans/La Mer Bleu Productions; **9.21a–c:** Courtesy Joe Lucas/Marine Entanglement

Research Program/National Marine Fisheries Service, NOAA; **9.21d:** Courtesy R. Herron/Marine Entanglement Research Program/National Marine Fisheries Service, NOAA; **9.22a–c:** Courtesy Hazardous Materials Response Branch/Jeremy Gault/NOAA; **9.22d:** NOAA; **9.22e:** Courtesy Hazardous Materials Response Branch/Jeremy Gault/NOAA; **9.22f:** Alyn and Alison Duxbury; **9.23:** Courtesy Clean Sound Cooperative/Seattle, WA; **p. 208:** Dr. Fred Nichols/USGS/Menlo Park, CA; **p. 209 (left):** © Caroline Kopp; **p. 209 (right):** Alyn and Alison Duxbury.

Chapter 10

Opener: © Biophoto Associates/Photo Researchers, Inc.; **10.7a–d:** Courtesy David English, School of Oceanography, University of Washington. These images were formed from data of the NASA Global Ocean Color Data Set which was implemented by Gene Feldman of NASA's Goddard Space Flight Center. The CZCS data was processed using the University of Miami DSP system developed by Otis Brown and Robert Evans with maintenance support of NASA.; **10.11:** Courtesy John Delaney; School of Oceanography, University of Washington; **10.12:** Courtesy Dr. James M. Brooks/Geochemical and Environmental Research Group/College Station, TX.

Chapter 11

Opener: © Mike Severns/Tom Stack & Associates; **11.5:** Courtesy Field Museum of Natural History, Chicago, Neg. # Z82860; **11.6:** Rita Horner, School of Oceanography, University of Washington, Seattle; **11.8:** Courtesy Enge's

Equipment Company/Morton Grove, IL; **11.10:** Courtesy Kathy Newell, School of Oceanography, University of Washington; **11.17:** Photo by Mark Holmes, USGS, School of Oceanography, University of Washington; **11.23a:** Courtesy Dr. Albert Erickson/Fisheries Research Institute/University of Washington; **11.23b:** Courtesy David Withrow/National Marine Fisheries Service, NOAA; **11.28:** H. Wes Pratt, Apex Predator Investigation NOAA- NMFS Narragansett Laboratory, RI; **11.30:** Bruce W. Buls, Seattle; **11.31:** Courtesy Sea Farm Washington, Inc.; **11.32:** Courtesy Anadromos, Inc., Corvallis, OR; **p. 259:** P. Dee Boersma, University of Washington, Dept. of Zoology; **p. 260 (top left):** Alyn and Alison Duxbury; **p. 260 (bottom left):** Alyn and Alison Duxbury; **p. 260 (top right):** Alyn and Alison Duxbury; **p. 261:** Alyn and Alison Duxbury.

Chapter 12

Opener: © Neil McDaniel; **12.10a–c:** Courtesy John Delaney, School of Oceanography, University of Washington; **12.11a–b:** Alyn and Alison Duxbury; **12.14:** Courtesy Brad Matsen, NATIONAL FISHERMAN Magazine, Seattle, WA; **12.15:** Courtesy Ken Chew, School of Fisheries, University of Washington.

LINE ART

3.8: From Barazangi and Dorman, *Bulletin of Seismological Society of America*, 1969. Reprinted by permission of the Seismological Society of America, 201 Plaza Professional Building, El

Cerrito, CA 94530. **5.4:** From Sverdrup/Johnson/Fleming, *The Oceans: Their Physics, Chemistry and General Biology*, © 1942, p. 105. Prentice-Hall, Inc., Englewood Cliffs, N.J. Reprinted by permission. **6.11:** From C. D. Keeling et al., "A Three Dimensional Model of Atmospheric CO_2 Transport Based on Observed Winds: Observational Data and Preliminary Analysis," in *Aspects of Climate Variability in the Pacific and the Western Americas*, Geophysical Monograph, Volume 55, November 1989, Appendix A, copyright by the American Geophysical Union. **7.16:** From Sverdrup/Johnson/Fleming, *The Oceans: Their Physics, Chemistry and General Biology*, © 1942, p. 501. Prentice-Hall, Inc., Englewood Cliffs, N.J. Reprinted by permission. **p. 157, 158:** From Curtis E. Ebbesmeyer and W. James Ingraham, Jr., "Shoe Spill in the North Pacific" in *American Geophysical Union EOS, Transactions*, vol. 73, no. 34, August 25, 1992, pages 361, 365, copyright by the American Geophysical Union.

ILLUSTRATIONS

Parrot Graphics

1.4, 1.6, 1.10e, 2.14, 2.15, 2.16, 2.17, 3.1, 3.3, 3.4, BX 3.4, 3.5, 3.8 a–b, 3.9, 3.11, 3.12, 3.14, 3.15, 3.17 a, 3.19 a–b, 3.22, 3.23 a–e, 4.2, 4.4, 4.9, BX 5.1–2, 5.12, 5.13, 5.14, 5.15, 5.16, 6.14, 7.12, 7.13, 8.5, 8.10, 9.10, 9.14, 11.24, 12.3, 12.13.

Alice Thiede

3.21, 7.15

INDEX

305